THE SHELL BITUMEN HANDBOOK

Shell Bitumen

Published by Shell Bitumen U.K.
Riversdell House
Guildford Street
Chertsey
Surrey KT16 9AU

Designed and Produced by
Design and Print Partnership Limited
East Molesey
Surrey

ISBN-0-9516625-0-3

First edition printed July 1990
Reprinted September 1991.

Foreword

Bitumen is a very widely used civil engineering material, the properties of which still remain a mystery to many of the large number of people involved in its various applications, notably in the construction and maintenance of roads. This situation arises because educational establishments generally do not provide adequate coverage in their crowded degree and other courses. Consequently, in practice, the use of bitumen has remained essentially a craft rather than a technology.

Engineers responsible for the design of works incorporating bitumen have a professional responsibility to acquire an adequate understanding of its physical properties. Given the poor background from formal education, the need to read and attend post experience or post graduate courses becomes essential.

The Shell Bitumen Handbook will be of great assistance in this context both to the individual and to those responsible for courses of instruction. The coverage of the book is very comprehensive and provides everything from bitumen chemistry to pavement design. Sections dealing with practical matters relating to U.K. road and airfield construction and maintenance are very detailed. They provide an up-to-date guide to practitioners at a professional level but with practical detail of use to the technician too. While concentrating on the U.K., a window on practices in other countries is also provided where appropriate. The book's illustrations are particularly helpful in bringing several complex issues to *life*. Where coverage of some topics is necessarily brief, extensive references are provided so that further reading can be pursued.

It is particularly appropriate that the Shell organisation should be responsible for producing a bitumen reference book. For several decades, it has been in the forefront of research on the civil engineering applications of bitumen against a background of pioneering work by people such as Nijboer, Van der Poel and Heukelom in their Amsterdam laboratories. Furthermore, the Shell Pavement Design Manual has, for many years, presented a comprehensive method for dealing with pavements worldwide using the mechanistic engineering approach, now also a feature of some U.K. practice following the publication of TRRL's report LR1132 in 1984.

An essential feature of civil engineering materials technology is that it should relate to the structure in which the material is to be used. This interaction between basic properties and applications in the road is strongly featured in the book.

Developments in asphalt technology are moving fast. To take full advantage of these, engineers will need a proper understanding of bitumen and its uses. The use of modified bitumens in appropriate situations is likely to be an increasing feature of pavement engineering, together with a move towards end-product specifications. Civil engineers can no longer rely on the philosophy of bitumen being a *black and sticky* material, about which they need to know little, if they want to take full advantage of these new developments.

As someone who has been involved in bitumen research, teaching and consulting for over 20 years, I have come to appreciate the contribution which Shell has made to our understanding of bitumen and its uses. I commend this latest contribution to all who wish to properly understand this important civil engineering material and use it economically and effectively.

Professor Stephen Brown
University of Nottingham
July 1990

Acknowledgements

I am indebted to friends both inside and outside Shell for all their valuable comments on the contents of the book. In particular I would like to thank both Terry Hoban (Technical Manager Shell Bitumen U.K.) and Ian Menzies (IMprimatur Ltd) for their advice and help in the drafting and production of the book.

In addition to Terry and Ian my thanks to the following for their helpful comments:

(In alphabetical order)

Derek Fordyce	–	Heriot-Watt University
Tony Harrison	–	Shell International Petroleum Company Limited
Geoffrey Lees	–	Leascon Surfacing Technology Limited
Jon Morten	–	Chemcrete Company Limited
Alan Mulder	–	Shell International Petroleum Company Limited
Ronald Regan	–	Printar Industries Limited
Chris Rendell	–	Wells (Trinidad Lake Asphalt) Limited
Roger Thomas	–	Shell International Petroleum Company Limited
Colin Underwood	–	Road Surface Dressing Association
Maurice White	–	British Aggregate Construction Materials Industries

and for supplying photographs I wish to thank:

Associated Asphalt Limited
Alfred McAlpine Quarry Products Limited
Bardon Quarries Limited and Bardon London Limited
Colas Roads Limited
ECC Quarries Limited
Redland Aggregates Limited
Tilcon Limited
Transport and Road Research Laboratory
Wimpey Asphalt Limited

The extracts from British Standard specifications are reproduced by permission of the British Standards Institution, Linford Wood, Milton Keynes, MK14 6LE from whom copies of the complete standards may be purchased.

In writing this book I have obtained information from a wide range of sources which I have endeavoured to acknowledge in the references. If there are any sources not listed, I hope the individuals concerned will accept my sincere apologies.

For more detailed technical information on bituminous materials I would like to direct the reader to the following:

The University of Nottingham residential course on Bituminous pavements: materials, designs and evaluation.

The joint BACMI/University of Newcastle upon Tyne residential course on Flexible pavements and bituminous materials.

An introduction to bituminous mixes and flexible pavement construction. Available from BACMI.

Contents

page

Chapter 1 – Introduction 13

1.1 The earliest uses of bituminous binders 13
1.2 The growth of bitumen consumption in the U.K. 15
1.3 Different types of binder 15
1.3.1 Lake asphalt 16
1.3.2 Rock asphalt 17
1.3.3 Tar 17
1.3.4 Bitumen 18

Chapter 2 – The manufacture, storage and handling of bitumen 21

2.1 The manufacture of bitumen 21
2.1.1 Fractional distillation of crude oil 21
2.1.2 The air blowing of short residues 22
2.1.2.1 Description of the continuous blowing process 22
2.1.2.2 Air rectified or semi-blown bitumens 24
2.1.2.3 Fully blown bitumens 24
2.1.2.4 The chemistry of the blowing process 24
2.2 Storage and handling temperatures of bitumen 25
2.2.1 Bitumen tanks 25
2.2.2 Storage and pumping temperatures 26
2.3 Health, safety and environmental aspects of bitumen 28
2.3.1 Potential hazards during the handling of bitumen 28
2.3.1.1 Toxicity of bitumen 28
2.3.1.2 Potential hazards via skin contact 30
2.3.1.3 First aid for skin burns 30
2.3.1.4 Potential hazards via inhalation 30
2.3.1.5 First aid for inhalation of bitumen fumes 31
2.3.2 Recommended precautions for personnel 31
2.3.2.1 Protective clothing 31
2.3.2.2 Personal hygiene 31
2.3.3 Fire prevention and fire fighting 31
2.3.4 Sampling 33
2.3.4.1 Dip sampling 33
2.3.4.2 Sample valves 33

Chapter 3 – Bitumen specifications and quality 35

3.1 Penetration grade bitumens 35
3.2 Oxidised bitumens 35
3.3 Hard bitumens 36
3.4 Cutback bitumens 36
3.5 Bitumen quality 36
3.5.1 Cohesion 39
3.5.2 Adhesion 39
3.5.3 Durability 40
3.5.3.1 Oxidative and evaporative hardening 40
3.5.3.2 Exudative hardening 42
3.5.4 Verification of the QUALAGON tests and test criteria 42

page

Chapter 4 – Bitumen emulsions 45

4.1 Emulsifiers 45
4.2 The manufacture of bitumen emulsions 46
4.3 Bitumen emulsion formulation 47
4.3.1 Emulsion stability 47
4.3.1.1 Storage stability 47
4.3.1.2 The breaking process 50
4.3.2 Emulsion adhesivity 50
4.3.3 Emulsion viscosity 51
4.4 Classification and specification of bitumen emulsions 51
4.5 Modification of bitumen emulsion properties 51
4.5.1 How to increase the viscosity of an emulsion 51
4.5.2 How to decrease the viscosity of an emulsion 54
4.5.3 How to change the breaking rate of an emulsion 54
4.5.4 How to influence the storage stability of an emulsion 54
4.5.5 How to change the particle size distribution of an emulsion 55
4.6 Uses of bitumen emulsions 57
4.6.1 Bitumen emulsions in road mixes 57
4.6.2 Miscellaneous uses of bitumen emulsions 57
4.6.2.1 Soil stabilisation 57
4.6.2.2 Jiffy pots 58
4.6.2.3 Slip coats 58
4.6.2.4 Protective coats 58
4.6.2.5 Crack filling and grouting 58

Chapter 5 – Mechanical testing and properties of bitumen 61

5.1 Standard specification tests for bitumens 61
5.1.1 The penetration test (IP49, ASTM D5, BS 2000: Part 49) 61
5.1.2 The softening point test (IP58, ASTM D36, BS 2000: Part 58) 62
5.2 The Fraass breaking point test 64
5.3 Viscosity 65
5.3.1 The sliding plate viscometer 67
5.3.2 Capillary viscometers 68
5.3.3 Cup viscometers 68
5.4 The bitumen test data chart 68
5.4.1 Class S bitumens 71
5.4.2 Class B bitumens 71
5.4.3 Class W bitumens 74
5.5 Temperature susceptibility – penetration index (PI) 74
5.6 Engineering properties of bitumen 75
5.6.1 The stiffness concept 75
5.6.2 Determination of the stiffness modulus of bitumen 80
5.6.2.1 Measurement of stiffness modulus 82
5.6.2.2 Prediction of stiffness modulus 82
5.6.3 Tensile strength 87
5.6.4 Fatigue 'strength' 87

page

Chapter 6 – Bitumen constitution, structure and rheology ... 89

6.1 Bitumen constitution ... 89
6.1.1 Asphaltenes ... 90
6.1.2 Resins ... 90
6.1.3 Aromatics ... 90
6.1.4 Saturates ... 90
6.2 Bitumen structure ... 91
6.3 The relationship between constitution and rheology ... 92
6.4 The relationship between broad chemical composition and physical properties ... 94

Chapter 7 – The influence of bitumen properties on performance in practice ... 99

7.1 The influence of bitumen properties during construction ... 100
7.1.1 Mixing and transport ... 100
7.1.2 Laying and compaction ... 104
7.2 The influence of bitumen properties on the performance of bituminous mixes in service ... 104
7.2.1 Performance at high service temperatures ... 104
7.2.1.1 Permanent deformation ... 104
7.2.1.2 Fatting-up ... 107
7.2.2 Performance at low service temperatures ... 109
7.2.2.1 Cracking ... 109
7.2.2.2 Thermal cracking ... 109
7.2.2.3 Load associated cracking ... 109
7.2.2.4 Reflective cracking ... 109
7.2.2.5 Fretting ... 112
7.2.2.5.1 Fretting of macadam wearing course ... 112
7.2.2.5.2 Fretting of hot rolled asphalt wearing course ... 116
7.2.3 The influence of the bitumen on the development of surface texture in hot rolled asphalt wearing course ... 118
7.2.3.1 Specification of permittivity ... 119

Chapter 8 – Durability of bitumen ... 123

8.1 Bitumen hardening ... 123
8.1.1 Oxidation ... 123
8.1.2 Loss of volatiles ... 123
8.1.3 Physical hardening ... 123
8.1.4 Exudative hardening ... 124
8.2 Hardening of bitumen during storage, mixing and in service ... 124
8.2.1 Hardening of bitumen in bulk storage ... 124
8.2.2 Hardening of bitumen during mixing with aggregate ... 124
8.2.3 Hardening of bitumen in a mix during hot storage, transport and laying ... 125
8.2.4 Hardening of bitumen on the road ... 127
8.3 Ageing tests ... 132

Chapter 9 – Adhesion of bitumen ... 135

9.1 Adhesion in the absence of water ... 135
9.2 Adhesion in the presence of water ... 135
9.2.1 Adhesion failures caused by the presence of water ... 136

		page
9.3	Fundamental properties of the bitumen/aggregate/water system	136
9.4	The factors affecting bitumen/aggregate adhesion	139
9.5	Disbonding mechanisms	140
9.5.1	Displacement	140
9.5.2	Detachment	140
9.5.3	Film rupture	141
9.5.4	Blistering and pitting	141
9.5.5	Hydraulic scouring	141
9.5.6	Pore pressure	142
9.5.7	Chemical disbonding	142
9.6	Adhesion testing	143
9.6.1	Static immersion tests	143
9.6.2	Dynamic immersion tests	143
9.6.3	Chemical immersion tests	144
9.6.4	Immersion mechanical tests	144
9.6.5	Immersion trafficking tests	144
9.6.6	Coating tests	145
9.7	Improving bitumen/aggregate adhesion	145
Chapter 10 –	**Modified bitumens**	149
10.1	The role of bitumen modifiers in bituminous mixes	149
10.2	The modification of bitumen	151
10.2.1	The modification of bitumen by the addition of sulphur	151
10.2.2	The modification of bitumen by the addition of rubbers	152
10.2.3	The modification of bitumen by the addition of organo-manganese compounds	152
10.2.4	The modification of bitumen by the addition of thermoplastic polymers	154
10.2.5	The modification of bitumen by the addition of thermoplastic rubbers	158
10.2.5.1	The influence of bitumen constitution on thermoplastic rubber/bitumen blends	159
10.2.5.2	The manufacture of thermoplastic rubber/bitumen blends	159
10.2.5.3	Properties of thermoplastic rubber/bitumen blends	160
10.2.6	Thermosetting binders	165
10.3	Cost-performance relationships for modified binders	167
Chapter 11 –	**Types and application of bituminous mixes**	169
11.1	Asphalt	170
11.1.1	Mastic asphalt	171
11.1.2	Hot rolled asphalt	171
11.2	Coated macadam	175
11.2.1	Open and medium-graded macadams	176
11.2.2	Dense and close-graded macadams	176
11.2.3	Heavy duty macadam	177
11.2.4	Fine-graded macadams	178
11.3	Asphaltic concrete	178
11.4	Pervious macadam/friction course	178
11.4.1	The development of pervious macadam	179
11.4.2	The benefits of pervious macadam	179
11.4.2.1	Reduction in spray	179

page

11.4.2.2 Reduction in noise 179
11.4.3 Disadvantages of pervious macadam 181
11.5 Comparison of the performance of various surfacings 184

Chapter 12 – The manufacture and testing of bituminous mixes 189

12.1 Types of mixing plants 189
12.1.1 Indirectly heated batch mixing plant 189
12.1.2 Directly heated batch mixing plant 193
12.1.3 Directly heated continuous mixing plants 194
12.2 Production control testing of bituminous mixes 196
12.3 Determination of recovered bitumen properties 198
12.4 Recycling bituminous materials 201
12.4.1 *In-situ* recycling 201
12.4.2 Off-site or plant recycling 202

Chapter 13 – Transport, laying and compaction of bituminous mixes 205

13.1 Transport 205
13.2 Laying bituminous materials 205
13.3 The compaction of bituminous materials 207
13.3.1 Methods of improving compaction 208
13.3.1.1 Increasing the layer thickness 208
13.3.1.2 Increasing the compactive effort 209
13.3.1.3 Modified rolling techniques 212
13.3.2 Compaction specifications for bituminous materials 213
13.3.2.1 Air voids specification 213
13.3.2.2 Percentage refusal density 213
13.4 The workability of bituminous materials 215
13.5 Surface texture 217

Chapter 14 – Mechanical testing of bituminous mixes 223

14.1 Fundamental tests 223
14.1.1 Repeated load triaxial test 223
14.1.2 Unconfined static uniaxial creep compression test 225
14.1.3 Repeated load indirect tensile test 225
14.1.4 Dynamic stiffness and fatigue tests 226
14.2 Simulative tests 229
14.2.1 Wheel-tracking tests 229
14.3 Empirical tests 229
14.3.1 The Marshall Test 230

Chapter 15 – The mechanical properties of bituminous mixes 235

15.1 Dynamic stiffness of bituminous materials 235
15.1.1 The prediction of mix stiffness 235
15.2 Permanent deformation of bituminous materials 237
15.2.1 The prediction of permanent deformation 239
15.3 Fatigue characteristics of bituminous mixes 241
15.3.1 The prediction of fatigue performance 243

page
Chapter 16 – The design and specification of bituminous mixes 245
16.1 Traditional recipe specification 246
16.2 The principles of mix design 246
16.3 The design of continuously graded mixes 247
16.3.1 The Asphalt Institute design method 247
16.3.2 The DoE (Airfields Branch) method 249
16.4 The design of gap-graded mixes 252
16.4.1 The BS 594 : 1973 mortar design method 252
16.4.2 The BS 598 : 1985 design method 252
16.5 An alternative mortar design method 254
16.6 An alternative approach to specifying Marshall properties 260

Chapter 17 – The design of bituminous pavements 265
17.1 The structural elements of a bituminous pavement 265
17.1.1 The pavement foundation 266
17.1.2 The roadbase 266
17.1.3 The surfacing 266
17.2 Factors involved in pavement design 267
17.2.1 Pavement loading 267
17.2.2 Environmental effects 268
17.3 Pavement design 269
17.3.1 Empirical and semi-empirical design methods 269
17.3.2 Analytical design methods 274
17.4 Improved roadbase and basecourse materials 277
17.5 Bituminous substitution 278

Chapter 18 – Surface dressing 281
18.1 Factors which influence the performance of surface dressings 282
18.2 Design of surface dressings 283
18.2.1 The Road Note 39 design approach 283
18.2.2 An engineering approach to the design of surface dressings 287
18.2.2.1 Embedment of chippings 290
18.2.2.2 Existing road surface texture 290
18.2.2.3 Surface texture and skid resistance 292
18.3 Types of binders used in surface dressing 292
18.3.1 Traditional cutback and emulsified bitumens 292
18.3.1.1 Bitumen emulsions 292
18.3.1.2 Cutback bitumen 293
18.3.1.3 Selection of binder 296
18.3.2 Proprietary polymer modified binders 297
18.3.3 Epoxy-resin thermoset binders 299
18.4 Aggregates and chippings 300
18.4.1 Chipping selection 300
18.4.2 Skid resistance 300
18.4.3 Binder/chipping adhesion 303
18.5 Surface dressing equipment 304
18.5.1 Sprayers 304

		page
18.5.2	Chipping spreaders	306
18.5.3	Rollers	306
18.5.4	Sweepers	306
18.6	Surface dressing techniques	306
18.6.1	Pad-coats	306
18.6.2	Double surface dressing techniques	308
18.6.2.1	The single spray double chip or racked-in method	308
18.6.2.2	The double spray double chip method	310
18.7	Traffic control and aftercare	311
18.8	Types of surface dressing failures	311
18.9	Slurry sealing	313
18.10	Safety	314
18.10.1	Handling temperatures	314
18.10.2	Operation of surface dressing sprayers using cutback bitumens	314
18.10.3	Operation of surface dressing sprayers using bitumen emulsions	315
18.10.4	Precautions, protective clothing and hygiene for personnel	315
18.11	The future for surface dressing	316
Chapter 19 –	**Miscellaneous uses of bituminous mixes**	319
19.1	Hydraulic applications	319
19.2	Railway applications	319
19.3	Coloured surfacings	320
19.3.1	Incorporating coloured pigments	320
19.3.2	Application of coloured chippings	321
19.3.3	Application of a coloured surface treatment	322
19.3.4	Coloured aggregate bound by a conventional bitumen	322
19.3.5	Coloured aggregate bound by a translucent binder	322
19.4	Bituminous materials for recreational areas	322
19.5	Motor racing tracks	323
19.6	Vehicle testing circuits	323
19.7	Asphalt kerbs	323
Chapter 20 –	**Miscellaneous information**	325
20.1	Physical constants of bitumen	325
20.1.1	Specific gravity	325
20.1.2	Coefficient of cubical expansion	325
20.1.3	Electrical properties	327
20.1.3.1	Electrical conductivity	327
20.1.3.2	Dielectric strength	327
20.1.3.3	Dielectric constant (permittivity)	327
20.1.4	Thermal properties	327
20.1.4.1	Specific heat	327
20.1.4.2	Thermal conductivity	327
20.2	Conversion factors for viscosities	327
20.3	Blending charts and formulae	328
20.4	Calculation of bitumen film thickness in a bituminous mix	332
20.5	The future	332
Index		333

Chapter 1 – Introduction

It is widely believed that the term 'bitumen' originated in Sanskrit, where the words 'jatu' meaning pitch and 'jatu-krit' meaning pitch creating, referred to the pitch produced by some resinous trees. The Latin equivalent is claimed by some to be originally 'gwitu-men' (pertaining to pitch) and by others, 'pixtu-men' (bubbling pitch), which was subsequently shortened to 'bitumen' then passing via French into English.

There are several references to bitumen made in the bible, although the terminology used can be rather confusing. In Genesis it refers to Noah's waterproofing of the ark, which was "pitched within and without with pitch", and of Moses' juvenile adventure in "an ark of bulrushes, daubed with slime and with pitch". Even more confusing are the descriptions of the building of the Tower of Babel. The Authorised Version of the bible says "they had brick for stone, and slime had they for mortar", the New International Version states that "they used bricks instead of stone and tar instead of mortar", Moffat's 1935 translation says "they had bricks for stone and asphalt for mortar", but the New English Bible states that "they used bricks for stone and bitumen for mortar". Even now it is not unknown for the terms bitumen, tar, asphalt and pitch to be used interchangeably.

1.1 The earliest uses of bituminous binders

In the vicinity of subterranean crude oil deposits surface seepages may occur at geological faults. The amount and nature of this naturally occurring material depends on a number of natural processes which will modify the properties of the material. This product may be considered to be a 'natural' bitumen, often being accompanied by mineral matter, the amount and nature of which will depend upon the circumstances which caused such an admixture to occur.

There are, of course, extensive crude oil deposits in the Middle East and for thousands of years there have been corresponding surface seepages of 'natural' bitumen. The ancient inhabitants of these parts were not slow to appreciate the excellent waterproofing, adhesive and preservative properties of the material placed so readily available at their disposal. For over 5,000 years(1) bitumen in one form or another has been used as a waterproofing and/or bonding agent; the earliest recorded use was by the Sumerians around 3,800 B.C. At Mohenjo Daro in the Indus valley there is a particularly well preserved water tank which dates back to around 3,000 B.C. In the walls of this tank not only are the stone blocks bonded with a 'natural' bitumen but there is also a vertical bituminous core in the centre of the wall. This same principle is used today in modern dam design. It is believed that Nebuchadnezzar was an able exponent of the use of bitumen because there is evidence that he used the product for waterproofing the masonry of his palace and as a grout for stone roads. The process of mummification used by the ancient Egyptians also testifies to the preservative qualities of bitumen, although it is a matter of dispute whether bitumens or resins were used.

The above ancient uses of 'natural' bitumens no doubt persisted in those inhabited parts of the world where these deposits were readily available. However, there seems to have been little development of the art elsewhere and in the U.K. it was not until the end of the 19th century that any of the present major uses of bitumen were introduced. However, there would appear to have been some knowledge of alternative binders in the interim period as it is on record that Sir Walter Raleigh, in 1595, proclaimed the lake asphalt he found in Trinidad to be "most excellent good" for ships caulking. In the middle of the 19th century attempts were made to utilise rock asphalt from European deposits for road surfacing and from this there was a slow development of the use of natural products for this purpose, followed by the advent of coal tar and later of bitumen manufactured from crude oil.

Agriculture

Damp-proofing and waterproofing buildings, structures
Disinfectants
Fence post coating
Mulches
Mulching paper
Paved barn floors, barnyards, feed platforms, etc.
Protecting tanks, vats, etc.
Protection for concrete structures
Tree paints
Water and moisture barriers (above and below ground)
Wind and water erosion control
Weather modification areas

Buildings

Floors
Damp-proofing and waterproofing
Floor compositions, tiles, coverings
Insulating fabrics, papers
Step treads

Roofing
Building papers
Built-up roof adhesives, felts, primers
Caulking compounds
Cement waterproofing compounds
Cleats for roofing
Glass wool compositions
Insulating fabrics, felts, papers
Joint filler compounds
Laminated roofing shingles
Liquid-roof coatings
Plastic cements
Shingles

Walls, siding, ceilings
Acoustical blocks, compositions, felts
Architectural decoration
Bricks
Brick siding
Building blocks, papers
Damp-proofing coatings, compositions
Insulating board, fabrics, felts, paper
Joint filler compounds
Masonry coatings
Plaster boards
Putty
Siding compositions
Soundproofing
Stucco base
Wallboard

Miscellaneous
Air drying paints, varnishes
Artificial timber
Ebonized timber
Insulating paints
Plumbing, pipes
Treated awnings

Hydraulics and erosion control

Canal linings, sealants
Catchment areas, basins
Dam groutings
Dam linings, protection
Dike protection
Ditch linings
Drainage gutters, structures
Embankment protection
Groynes
Jetties
Levee protection
Mattresses for levee and bank protection
Membrane linings, waterproofing
Reservoir linings
Revetments
Sand dune stabilisation
Sewage lagoons, oxidation ponds
Swimming pools
Waste ponds
Water barriers

Industrial

Aluminium foil compositions using bitumen
Backed felts
Conduit insulation, lamination
Insulating boards
Paint compositions
Papers
Pipe wrapping
Roofing, shingles

Automotive
Acoustical compositions, felts
Brake linings
Clutch facings
Floor sound deadeners
Friction elements
Insulating felts
Panel boards
Shim strips
Tacking strips
Underseal

Electrical
Armature carbons, windings
Battery boxes, carbons
Electrical insulating compounds, papers, tapes, wire coatings
Junction box compound
Moulded conduits

Compositions
Black grease
Buffing compounds
Cable splicing compound
Coffin linings
Embalming
Etching compositions
Extenders
Explosives
Fire extinguisher compounds
Joint fillers
Lap cement
Lubricating grease
Pipe coatings, dips, joint seals
Plastic cements
Plasticisers
Preservatives
Printing inks
Well drilling fluid
Wooden cask liners

Impregnated, treated materials
Armoured bituminised fabrics
Burlap impregnation
Canvas treating
Carpeting medium
Deck cloth impregnation
Fabrics, felts
Mildew prevention
Packing papers
Pipes and pipe wrapping
Planks
Rugs, asphalt base
Sawdust, cork, asphalt composition
Treated leather
Wrapping papers

Paints, varnishes, etc.
Acid-proof enamels, mastics, varnishes
Acid-resistant coatings
Air-drying paints, varnishes
Anti-corrosive and anti-fouling paints
Anti-oxidants and solvents
Base for solvent compositions
Baking and heat resistant enamels
Boat deck sealing compound
Lacquers, japans
Marine enamels

Miscellaneous
Belting
Blasting fuses
Briquette binders
Burial vaults
Casting moulds
Clay articles
Clay pigeons
Depilatory
Expansion joints
Flower pots
Foundry cores
Friction tape
Fuel
Gaskets
Gramophone records
Imitation leather
Mirror backing
Rubber, moulded compositions
Shoe fillers, soles
Table tops

Paving

(See also Hydraulics, Agriculture, Railways, Recreation)

Airport runways, taxiways, aprons, etc.
Asphalt blocks
Brick fillers
Bridge deck surfacing
Crack fillers
Curbs, gutters, drainage ditches
Floors for buildings, warehouses, garages, etc.
Highways, roads, streets, shoulders
Parking lots, driveways
Portland cement concrete underseal
Roof-deck parking
Sidewalks, footpaths
Soil stabilisation

Railways

Ballast treatment
Curve lubricant
Dust laying
Paved ballast, sub-ballast
Paved crossings, freight yards, station platforms
Rail fillers
Railway sleepers
Sleeper impregnating, stabilisation

Recreation

Paved surfaces for:
Dance pavilions
Drive-in movies
Gymnasiums, sports arenas
Playgrounds, school yards
Race tracks
Running tracks
Skating rinks
Swimming and wading pools
Tennis courts, handball courts

Bases for:
Synthetic playing field and running track surfaces.

Table 1.1 – More than 250 known uses of bitumen[6, 7]

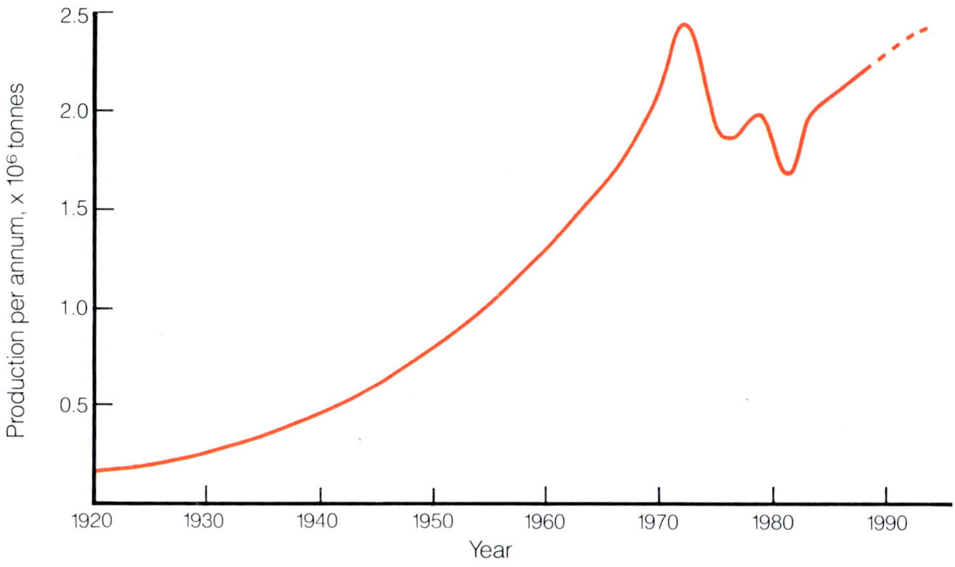

Figure 1.1 – U.K. bitumen production 1920-1990

1.2 The growth of bitumen consumption in the U.K.

Bitumen was imported into the U.K. on a small scale early in this century, but it was not until 1920, when Shell Haven refinery opened, that major bitumen production began in the U.K. Development was then rapid and with the facilities for bulk distribution, the basis was laid for the adoption of bitumen for its use in a wide range of applications, (see table 1.1).

In the early twenties bitumen consumption in the U.K. was about 200,000 tonnes per annum. The demand increased rapidly, reaching 1 million tonnes per annum by 1960 and with the start of the motorway building programme in the early sixties the demand for bitumen soared to 2 million tonnes per annum by 1970, reaching a peak of 2.4 million tonnes in 1973. Since 1973 the rather fickle governmental attitude to spending on roads both for new construction and maintenance has resulted in the capricious demand for bitumen shown in figure 1.1.

1.3 Different types of binders

As stated earlier much confusion still exists in differentiating between binding agents produced from coal (coal tars, pitches, etc) and those produced from crude oil. A detailed review of the differences between bitumens and coal tar products is given by Puzinauskas and Corbett[(2)].

Another semantic confusion arises from the difference in meaning of the word 'asphalt' in Europe and North America. In Europe it always means a mixture of bitumen and mineral aggregate, eg rolled asphalt, mastic asphalt, gussasphalt, etc. However, in North America 'asphalt' means bitumen.

The following brief description of the origin and manufacture of various binders should clarify the situation:

1.3.1 Lake asphalt

This is the most extensively used and best known form of 'natural' asphalt. It is found in well defined surface deposits, the most important of which is located in Trinidad. It is generally believed that this deposit was discovered in 1595 by Sir Walter Raleigh. However, it is now known that the Portuguese and possibly the Spaniards knew of the existence of the deposit before this date. The first definite reference in the literature refers to Sir Walter Raleigh's cousin Robert Dudley, who visited Trinidad and 'discovered' the asphalt deposit just three months before Raleigh's arrival. It is doubtful if either Raleigh or Dudley saw the 'Pitch Lake' on their first visits; they probably only found the overflow material on the sea-shore approximately one kilometer north of the lake itself, as the intervening country was dense and virtually impenetrable jungle.

There is a number of small deposits of asphalt on the island of Trinidad, but it is the 'Pitch Lake' which is located in the southern part of the island one kilometer from the sea, which constitutes one of the largest deposits in the world. The lake is approximately 100 acres (35 ha) in area and is estimated to be about 90 m deep, containing between 10 to 15 million tonnes of material. The surface of the lake is sufficiently hard to take the weight of crawler tractors and dumper trucks which transport excavated material from the surface of the lake to railway trucks which run along the edge of the lake.

Several theories have been suggested to explain the origin of the pitch lake; however, it is generally considered[(3)] to have originated as a surface seepage of a viscous bitumen in late Miocene times. Lowering of the earth's surface led to an incursion of the sea and as a result deposited silt and clay over the bitumen. Part of the silt and clay penetrated the bitumen, forming a plastic mixture of silt, clay, water and bitumen. Subsequently the land was elevated above sea level, lateral pressure deforming the material into its present shape of a pseudo-laccolite. Erosion removed the silt and clay covering, exposing the surface of the lake.

The enormous mass of the lake appears to be in constant, very slow, circulatory movement, from the centre outwards at the surface and it is assumed to return to the centre in the body of the lake. The level of the lake falls much less than expected from the quantity of material removed. Holes dug in the surface slowly fill up and usually disappear within 24 hours.

The as-dug material is refined by heating to 160°C, vaporising the water. The molten material is passed through fine screens to remove the coarse foreign and vegetable matter. This 'refined product' is termed Trinidad Epuré and typically has the following composition:

	% w
Binder	54
Mineral matter	36
Organic matter	10

The Epuré is too hard to be used in asphalt mixes as produced (it has a penetration of about 2 and a softening point of about 95°C) and is generally used in a 50/50 blend with 200 pen bitumen, the mixture having a penetration of about 50. This mixture is then used in normal hot rolled asphalt formulations, with adjustment to the mix composition to allow for the mineral matter contained in the lake asphalt.

In the fifties and sixties Trinidad Epuré/bitumen blends were used extensively in hot rolled asphalt wearing course because the addition of the Trinidad Epuré improved the 'weatherability' of the bitumen (see chapter 7, section 7.2.3). This prevented unacceptably smooth wearing surfaces developing during trafficking. However, changes in bitumen manufacturing techniques have improved the weathering characteristics of refined bitumen. This coupled with the fact that Trinidad Epuré/bitumen blends required several hours heated blending before use led to the rapid decline of Trinidad Epuré usage in the mid seventies.

1.3.2 Rock asphalt

Rock asphalt is formed by the impregnation of calcareous rocks such as limestone or sandstone with seepages of natural bitumen. This product can be mined or quarried and contains up to 12% by mass of bitumen. The principal sources of this deposit are from the Gard district in France, Neuchâtel in the Val de Travers region in Switzerland and at Ragusa in Italy. Such products were the earliest materials used for road paving. After crushing they were heated and then compacted into layers on site.

Today there is only minor usage of such natural asphalts in road paving or industrial applications. In nearly all cases it is combined with a flux oil or soft bitumen.

1.3.3 Tar

Bitumen derived from crude oil and tar derived from the manufacture of coke or smokeless solid fuel, are often considered synonymously. The reasons for this are two-fold. Firstly, to the naked eye, both materials appear to be very similar: black thermoplastics having relatively high viscosities at ambient temperatures. Secondly, these two materials are used for similar applications – road construction, roofing and as protective coatings for a number of industrial applications. However, there the similarity ends, since they are dissimilar, not only in their origin, but also in chemical constitution. Physical and chemical differences result in differences in behaviour and gaseous/vapour emissions during application and subsequent performance in service.

'Tar' is a generic word for the liquid obtained when natural organic materials such as coal or wood are carbonised or destructively distilled in the absence of air[1]. It is customary to preface the word tar with the name of the material from which it is derived: the products of this initial carbonisation process are referrred to as crude coal tar, crude wood tar, etc.

Two types of crude coal tar are produced as a by-product of the carbonisation of coal:

- *Coke oven tar*
  This material is produced at high temperatures (about 1200°C) in coke ovens during the manufacture of metallurgical or domestic coke. These coal tars have a high aromatic hydrocarbon content and a pitch content of around 50%.
- *Low temperature tar*
  This material is produced at comparatively low temperatures (600°C to 700°C) during the manufacture of smokeless solid fuel. This type of coal tar is less viscous than the high temperature coke oven tar, is paraffinic in nature and has a pitch content of about 35%.

Liquid by-products are produced by coal carbonisation. By cooling the hot gases and vapours from the oven or retort, aqueous liquor and crude tar are condensed and, since they are substantially immiscible, are separated by decantation. By scrubbing the cooled gas with a suitable solvent, a light hydrocarbon mixture is extracted.

The crude tar is refined by fractional distillation which separates it into a number of distillate oil fractions and a residue. The residue is termed 'pitch'. 'Refined tar' is produced by blending the pitch with distillate oil fractions. British Standard 76[4] includes eight grades of road tar from 30°C EVT to 58°C EVT in 4°C increments. The EVT (equi-viscous temperature) is the temperature at which 50 ml of the tar has a flow time of fifty seconds through the standard 10 mm orifice in the standard tar viscometer (see chapter 5, photograph 5.5). Thus the higher the EVT, the more viscous the tar. 54°C EVT tar would be suitable for a dense roadbase macadam for heavily trafficked roads, whereas a 38°C EVT tar would be used in an open-graded macadam on a lightly trafficked road.

In the last twenty years dramatic changes have occurred in the supply of crude coal tar in the U.K. In the mid sixties over 2 million tonnes of crude coal tar were produced per annum, of which 50% was manufactured using carbonisation ovens used to produce town gas. However, the introduction of North Sea gas in the late sixties rapidly diminished the output from carbonisation ovens and by 1975 tar from this source had totally disappeared.

Now the greatest proportion of crude coal tar available for distillation is produced by high temperature carbonisation ovens in the iron, steel and coal industries. Although the demand for these products is difficult to predict, it seems likely that the availability of crude coal tar from this source will remain stable between 0.3 to 0.4 million tonnes per annum into the next decade.

At present approximately 90% of the refined tar and pitch produced is used in four applications: road tars, briquetting, electrode binders and pipe enamels. The use of refined tars in road construction is basically limited to:

- Tar/bitumen mixtures[(5)] for surface dressing which are usually manufactured using more miscible low temperature tar. It is generally believed that the addition of tar to bitumen improves wetting of the aggregate and thereby improves adhesion. Thus tar/bitumen blends remain popular with some specifying authorities. However, health and safety considerations and the development of chemical adhesion agents are likely to restrict the use of this type of binder in the future. Tar/bitumen blends for surface dressings are discussed further in chapter 18.
- Refined tar in dense tar surfacing. Refined tar is less susceptible to the effects of diesel and oil spillage than bitumen. As a result dense macadams manufactured using refined tar are often specified for vehicle parking areas, etc.

Refined tar is more temperature susceptible than bitumen; as a result at high service temperatures, say 60°C, it is softer than the equivalent grade of bitumen. Thus bitumen-bound surfacings are more resistant to plastic deformation than tar-bound surfacings. At low service temperatures tar is stiffer than bitumen and is therefore more susceptible to brittle fracture.

1.3.4 Bitumen

Bitumen is manufactured from crude oil. It is generally agreed that crude oil originates from the remains of marine organisms and vegetable matter deposited with mud and fragments of rock on the ocean bed. Over millions of years, organic material and mud accumulated into layers hundreds of metres thick, the immense weight of the upper layers compressing the lower layers into sedimentary rock. Conversion of the organisms and vegetable matter into the hydrocarbons of crude oil is thought to be the result of the application of heat from within the earth's crust, pressure applied by the upper layers of sediments, possibly aided by the effects of bacterial action and radio-active bombardment. As further layers of sediment were deposited on the sedimentary rock where the oil had formed, the additional pressure squeezed the oil sideways and upwards through porous rock. Where porous rock extended to the earth's surface, oil seeped through, resulting in the surface seepages described earlier. Fortunately, the majority of oil and gas was trapped in porous rock which was overlain by impermeable rock forming gas and oil reservoirs. Here the oil remained until its presence was detected by seismic surveys and recovered by drilling through the impermeable rock.

The four main oil producing areas in the world are the U.S.A., the Middle East, the countries around the Caribbean and the U.S.S.R. Crude oils differ both as to their physical and chemical properties. Physically they vary from viscous black liquids to free-flowing straw coloured liquids. Chemically they may be predominantly paraffinic, naphthenic or aromatic, with combinations of the first two being common.

There are nearly 1500 different crudes produced worldwide. Based on the yield and the quality of the product produced, only a few of these are considered suitable for the manufacture of bitumen. In the U.K. bitumens are principally manufactured using either Middle East or South American crudes.

BS 3690 : Part 1 : 1989[5] defines bitumen as "A viscous liquid, or a solid, consisting essentially of hydrocarbons and their derivatives, which is soluble in trichloroethylene and is substantially non-volatile and softens gradually when heated. It is black or brown in colour and possesses waterproofing and adhesive properties. It is obtained by refinery processes from petroleum, and is also found as a natural deposit or as a component of naturally occurring asphalt, in which it is associated with mineral matter."

1.4 References

(1) Abraham, H: *Asphalts and allied substances, their occurrence, modes of production, uses in the arts and methods of testing.* Fifth edition, vol 1. D van Nostrand Company. January 1945.

(2) Puzinauskas, V P and Corbett, L W: *Differences between petroleum asphalt, coal tar pitch and road tar.* The Asphalt Institute, Research Report 78-1, January 1978.

(3) Attwooll, A W and Broome, D C: *Trinidad Lake Asphalt.* Third edition. The Baynard Press, London, 1962.

(4) British Standard 76 : 1974 : *Specification for tars for road purposes.*

(5) British Standard 3690 : Part 1 : 1989 : *Bitumens for building and civil engineering. Part 1. Specification for bitumens for road purposes.*

(6) Shell Briefing Service: *Shell in the bitumen industry.* Shell International Petroleum Company Limited, June 1973.

(7) The Asphalt Institute: *The asphalt handbook.* Manual series No 4, 1989.

Photograph 1.1 – Shell Haven refinery

Chapter 2 – The manufacture, storage and handling of bitumen

2.1 The manufacture of bitumen

Crude oil is a complex mixture of hydrocarbons differing in molecular weight and consequently in boiling range. Before it can be used, crude oil has to be separated, purified, blended, and sometimes chemically or physically changed.

2.1.1 Fractional distillation of crude oil

The first process in the refining of crude oil is fractional distillation. This is carried out in tall steel towers known as fractionating or distillation columns. The inside of the column is divided at intervals by horizontal steel trays with holes to allow vapour to rise up the column. Over the holes are small domes called bubble caps which deflect the vapour downwards so that it bubbles through liquid condensed on the tray, thus improving the efficiency of distillation and reducing the necessary height of the column.

On entering the crude distillation plant the crude oil is heated in a furnace to temperatures between 300°C and 350°C before being passed into the lower part of a column operating at slightly above atmospheric pressure. The material entering the column is a mixture of liquid and vapour; the liquid comprises the higher boiling point fractions of the crude and the vapour the lower boiling point fractions. The vapours rise up the column through the holes in the trays losing heat as they rise. When each fraction reaches the tray where the temperature is just below its own boiling point it condenses and changes back into a liquid. As the fractions condense on the trays they are continuously drawn off via pipes.

The lightest fractions of the crude remain as a vapour and are taken from the top of the distillation column, heavier fractions are taken off the column as side-streams with the heaviest fractions remaining as a liquid and therefore left at the base of the column. The lightest fractions produced by the crude distillation process include propane and butane which are gases under atmospheric conditions. Moving down the column a slightly heavier product, naphtha, is produced which is a feedstock for gasoline production and the chemical industry. Then there is kerosine, which is used primarily for aviation fuel and to a lesser extent for domestic fuel. Heavier again is gas oil, which is used as a fuel for diesel engines and central heating. The heaviest fraction taken from the crude oil distillation process is long residue which is a complex mixture of high molecular weight hydrocarbons, which requires further processing before it can be used as a feedstock for the manufacture of bitumen.

The long residue is further distilled at reduced pressure in a vacuum distillation column. This is carried out under a vacuum of 10 mm to 100 mm Hg at 350°C to 400°C to produce gas oil, distillates and short residue. If this second distillation were carried out by increasing the temperature of the process, cracking or thermal decomposition of the long residue would occur.

The short residue is the feedstock used in the manufacturing of over 20 different grades of bitumen. The viscosity of the short residue is a function of both the origin of the crude oil and the temperature and pressure in the vacuum column during processing. Depending on the origin of the crude oil the conditions in the column are adjusted to produce a short residue with a penetration in the range 100 to 300 dmm.

Figure 2.1 shows a schematic representation of the distillation process and its relationship with other fundamental refining processes such as: reforming and cracking for the manufacture of motor gasoline, and lub oil plants for the manufacture of lubricating oils, waxes, etc.

2.1.2 The air blowing of short residues

The physical properties of the short residue may further be modified by 'air blowing'. This is an oxidation process which involves the blowing of air through the short residue, either on a batch or a continuous basis, with the short residue at a temperature of 240°C to 320°C.

2.1.2.1 Description of the continuous blowing process

After preheating, the short residue is introduced into the blowing column just below the normal liquid level. Air is blown through the bitumen by means of an air distributor located at the bottom of the column. The air is not only the reactant but also serves to agitate and mix the bitumen, thereby increasing the surface area and rate of reaction. Oxygen is consumed by the bitumen as the air ascends through the material. Steam and water are sprayed into the vapour space above the bitumen level, the former to suppress foaming and dilate the oxygen content of waste gases and the latter cools the vapours to prevent after-burning. The 'blown' product flows from the bottom of the blowing column into a surge drum via an external draw-off line. In this way the minimum level of product in the blowing column is controlled.

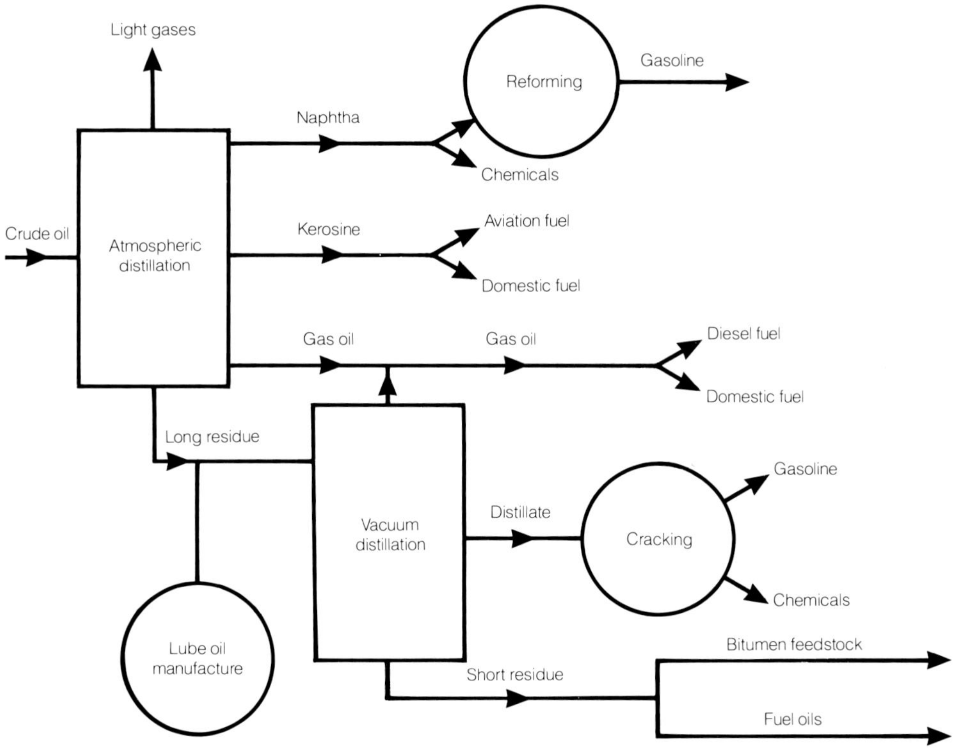

Figure 2.1 – Schematic diagram of the manufacture of bitumen

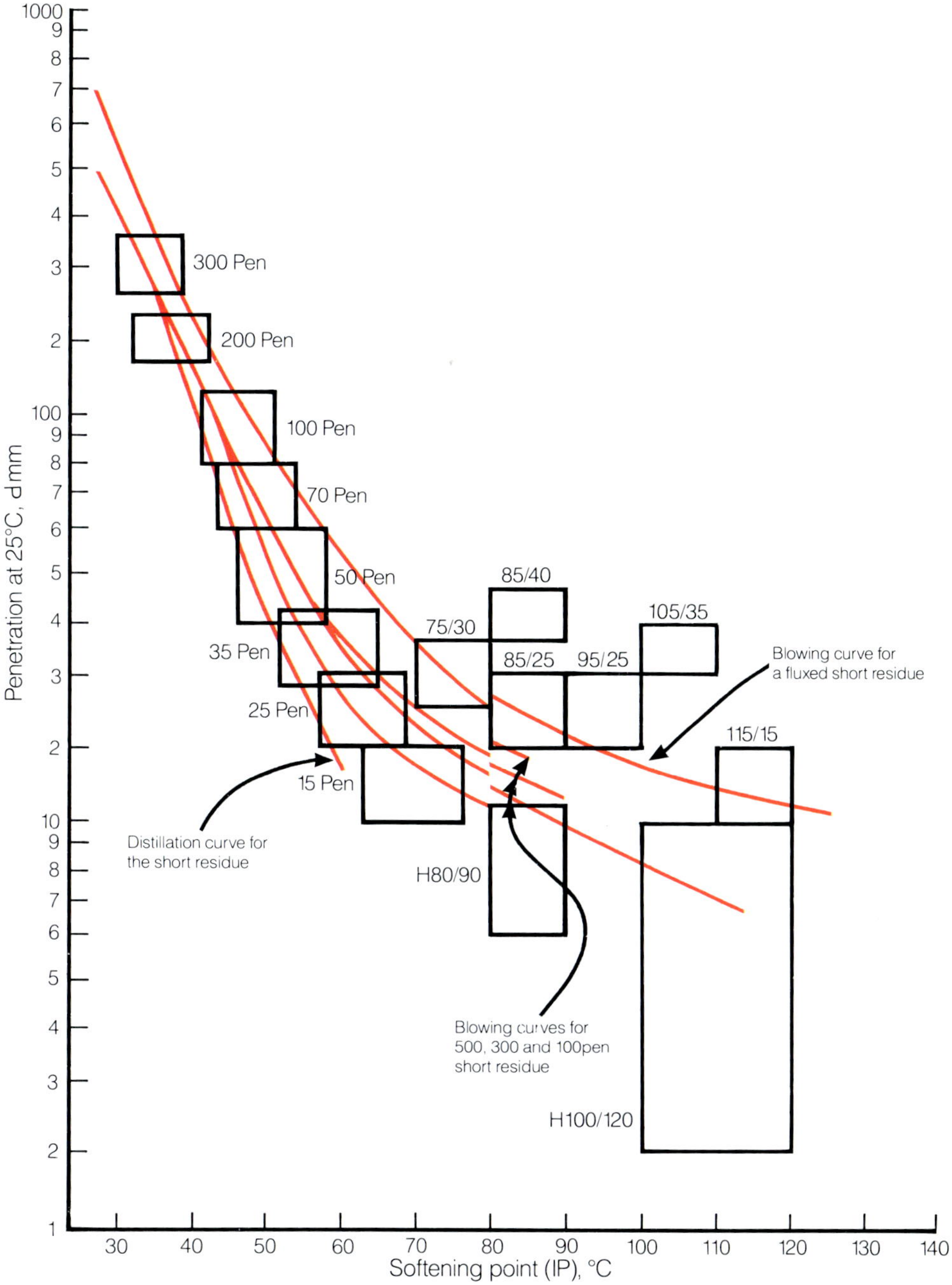

Figure 2.2 – Distillation and blowing curves for a Middle East short residue

From the surge drum the blown product is passed through heat exchangers to achieve the desired 'rundown' temperatures and to provide an economical means of preheating the short residue, before pumping the product to storage. The penetration and softening point of the blown bitumen are affected by:

the viscosity of the feedstock;

the temperature in the blowing column;

the residence time in the blowing column;

the origin of the crude oil used to manufacture the feedstock;

the air-to-feed ratio.

Figure 2.2 shows distillation and blowing 'curves' for a Middle East feedstock. In both the distillation and blowing processes the softening point increases and the penetration falls. However, in the distillation processs the temperature susceptibility (or penetration index) of the material is effectively unchanged. Thus the distillation line on the figure is virtually a straight line, whereas the curves for the blown bitumen flatten substantially as the softening point of the bitumen increases, showing that the temperature susceptibility of the material is substantially reduced, ie the penetration index is increased.

The discontinuity at 80°C is due to the fact that the softening point of bitumen must be determined in water if the softening point is below 80°C and in glycerol if it is greater than 80°C; as glycerol has a lower specific heat than water, it produces a higher softening point (see chapter 5, section 5.1.2).

The blowing process dehydrogenates the short residue, resulting in oxidation and polycondensation, increasing the overall molecular size of the asphaltenes already present in the feed and forming additional asphaltenes from the maltene phase (see chapter 6, section 6.4). The reaction is exothermic.Therefore close temperature control of this process is required, which is achieved by regulating the air-to-short residue ratio in the blowing column.

2.1.2.2 Air-rectified or semi-blown bitumens

To produce penetration grade bitumens suitable for road construction, bitumens manufactured from some crudes require a limited amount of air blowing. This process is termed semi-blowing or air-rectification. Used judiciously, semi-blowing can be applied to reduce the temperature susceptibility of the bitumen, ie increase its penetration index.

2.1.2.3 Fully blown bitumens

Fully blown or oxidised bitumens are produced by vigorous air-blowing of short residue or short residue blended with a heavy distillate. The position of the blowing curve on figure 2.2 is primarily dependent on the viscosity of the feed, ie the softer the feed the higher the curve. The severity of blowing depends on the temperature in the column and to a lesser extent on the residence time. Thus by controlling the viscosity of the feed and the conditions in the column all the blown grades of bitumen can be manufactured.

2.1.2.4 The chemistry of the blowing process

The aim of the blowing process is the formation of asphaltenes. Three phenomena can be identified:

- *reactions during which the size of the molecules increases*; formation of esters is particularly important; they not only account for about 60% of the oxygen in blown bitumen but also link up two different molecules and thus contribute to the formation of material of higher molecular weight; this mechanism results in an increase in the asphaltene content and a change in the colloid-chemical constitution and rheological properties of the bitumen;

- *reactions during which the size of the molecule is unchanged*; formation of cyclic hydrocarbons by means of dehydrogenation with H_2O as a side product;
- *reactions during which the size of the molecule decreases*; separation of side branches from the molecules with blown distillate produced as a side product.

2.2 Storage and handling temperatures of bitumens[1,2]

When handled properly, bitumens can be reheated or maintained at elevated temperatures for a considerable time without adversely affecting their properties. However, maltreatment of bitumen by overheating or by using conditions which promote oxidation, can adversely affect bitumen properties and could influence the long-term performance of materials made with it. The degree of hardening (or, under certain circumstances, softening) produced as a result of mishandling is a function of a number of parameters such as temperature, the presence of air, the surface-to-volume ratio of the bitumen, method of heating and the duration of exposure to these conditions:

2.2.1 Bitumen tanks[3]

All bitumens should be stored in tanks specifically designed for the purpose. To minimise the possible hardening of the bitumen during storage certain aspects of the design of the tank should be considered:

- To minimise the risk of overheating the bitumen, the tank should be fitted with accurate temperature sensors and readout devices. These should be positioned in the region of the heaters and preferably removable so that regular cleaning and maintenance can be easily carried out.
- Oxidation and loss of volatiles from bitumen are both related to the exposed surface-to-volume ratio of the storage tank, which for a cylindrical vessel equals the reciprocal height of the filled part of the tank:

  $$\frac{\text{Surface}}{\text{Volume}} = \frac{\pi r^2}{\pi r^2 h} = \frac{1}{h}$$

  where h = height of bitumen

  r = radius of the tank.

  Thus, the dimensions of the bulk storage tank should aim to minimise the surface-to-volume ratio and therefore vertical storage tanks with a high h to r ratio are preferable to horizontal tanks.
- It is common practice for bitumen in bulk storage tanks at mixing plants to be recirculated around a ring main to heat the pipework delivering the bitumen to the processing point. Return lines in a recirculation system should re-enter the storage tank below the bitumen surface to prevent hot bitumen cascading through the air. Often the bitumen is returned to the bulk storage tank through a pipe fitted into the upper part of the tank flush with the side, roof or protruding just into the airspace at the top of the tank. If the bitumen enters the tank above the bitumen all the factors which promote oxidation are present:

  high temperature;

  access to oxygen;

  high exposed surface-to-volume ratio.

Fortunately, the residence time of bitumen in the tank is usually sufficiently low for any hardening to be insignificant. However, if material is stored for a prolonged period, recirculation should be used only intermittently and the bitumen should be tested before use to ensure its continued suitability for the application.

In preference to manual dipping, tanks should be fitted with automatic level indicators with low and high level alarms, the former to prevent exposure of hot heater tubes and thereby creating a fire/ explosion risk and the latter to avoid overfilling the tank. Manual dipping should be carried out only to check the accuracy of automatic devices. Whether or not a high level alarm is fitted, a maximum safe working level for the tank should be predetermined. This should allow a margin for thermal expansion of the product in the tank and for the expansion which might result from a small accidental ingress of water.

Before intake of additional product a check should be made to ensure that there is sufficient ullage in the tank for the quantity to be taken in, without exceeding the maximum safe working level.

Every tank should be clearly labelled with the grade of bitumen it contains. When the grade of bitumen in a tank is changed it is important to ensure that the tank is empty and relabelled before the new grade is delivered.

2.2.2 Storage and pumping temperatures

Bitumen should always be stored and handled at the lowest temperature possible, consistent with efficient use. As a guide,working temperatures for specific operations are given in table 2.1[1,2,3]. These temperatures have been calculated on the basis of viscosity measurements and are supported by operational experience. For normal operations, ie the blending and transferring of liquid bitumen, temperatures of 10°C to 50°C above the minimum pumping temperature are recommended, but to prevent auto ignition of the bitumen 230°C must *never* be exceeded.

The time bitumen resides in a storage tank at elevated temperature and is recirculated should be minimised to prevent hardening of the bitumen. If bitumen must be stored for an extended period, ie say for more than 1 week without addition of fresh material, the temperature should be reduced to approximately 20°C to 25°C above the softening point of the bitumen and, where possible, recirculation stopped.

When bitumen is being reheated in bulk storage, care must be taken to heat the bitumen intermittently over an extended period to prevent localised overheating of the product around the heating pipes or coils. This is particularly important where direct flame tube heating is used because surface temperatures in excess of 300°C can be reached. In such instances initially only a limited amount of heat should be applied, sufficient to raise the temperature of the product to just above its softening point. This will allow the material to soften, then the heat can be reapplied to bring the temperature of the product to the required working temperature. This technique is beneficial because when the bitumen is a fluid, albeit a viscous fluid, convection currents are more able to dissipate the heat throughout the bulk and therefore localised overheating is less of a problem. Circulation of the tank contents should begin as soon as the product is sufficiently fluid, thereby further reducing the likelihood of local overheating. If prolonged direct flame heating is used, cracking of the bitumen may result. With hot oil, steam or properly designed electric heaters, reheating from cold should not cause these problems.

Grade	(a) Minimum pumping temperature, °C	Typical application temperature (b) Mixing/coating, °C	(c) Spraying, °C	Maximum safe handling temperature, °C
Penetration grades				
450	90	130	160	190
300	95	135	165	190
200	100	140	175	190
100	105	155	190	200
70	110	160	–	200
50	115	165	–	200
40 HD	125	175	–	200
35	125	175	–	220
25	135	185	–	220
15	140	190	–	220
Oxidised grades				
75/30	150	195	–	230
85/25	165	210	–	230
85/40	165	210	–	230
95/25	175	220	–	230
105/35	195	220	–	230
115/15	205	230	–	230
Hard grades				
H 80/90	160	200	–	230
H 110/120	190	230	–	230
Cutback grades				
50 secs	65	105	150	160
100 secs	70	110	160	170
200 secs	80	120	170	180

(a) Maximum pumping viscosity – 2 Pa.s
(b) Mixing/coating viscosity – 0.2 Pa.s
(c) Spraying viscosity
– penetration grades – 0.06 Pa.s
– cutback grades – 0.03 Pa.s

Note: Bitumen should always be stored and handled at the lowest temperature possible consistent with its efficient use.

Table 2.1 – Recommended handling temperatures[1,2,3]

Sources	Benzo(a)pyrene content, µg/1000 m³ of air
Asphalt mixing plant	13
Power station, gas	100
Power station, coal	300
Diesel engine	5,000
Refuse burning	11,000
Coke oven volatiles	35,000
Domestic boiler, coal	100,000

Table 2.2 – Benzo(a)pyrene emissions from various sources[5]

2.3 Health, safety and environmental aspects of bitumen[1,2,3]

Throughout industry in recent years there has been general increasing awareness of the need to improve health and safety procedures in the workplace. This applies also in the construction industry where the majority of bitumen is used.

Based on practical experience over a number of years and from specific field studies there is no indication that bitumen presents an occupational health risk to workers who handle it regularly, nor does it present an environmental health problem to those who come into contact with bitumen during its manufacture and application. Thus bitumen presents a low order of potential hazard provided that good handling practices are observed. These are described in detail in the Institute of Petroleum Model Code of Safe Practice, Part 11, Bitumen Safety Code[3] and in Shell Bitumen U.K.'s product safety data sheets.

The following highlights some of the potential hazards associated with the handling of bitumen. However, it is not intended to be used as a substitute for health and safety information given by individual suppliers or the Institute of Petroleum Model Code of Safe Practice.

2.3.1 Potential hazards during the handling of bitumen[3,4]

As bitumen is normally handled at temperatures well in excess of 100°C the major hazard is heat burns on skin in direct contact with the hot bitumen. However, low temperature skin contact of some grades of bitumen and inhalation of fumes from bitumen also create potential health hazards, particularly as substances which may be toxic can be present from materials associated with bitumen, eg hydrogen sulphide, or added to bitumen, eg cutback diluents, emulsifying agents, adhesion agents, coal tar products, etc.

2.3.1.1 Toxicity of bitumen

Bitumens are complex hydrocarbon materials containing components of many chemical forms, the majority of which are of high molecular weight including polycyclic aromatic hydrocarbons (PCAs). The toxicity of these components needs to be considered in view of their known carcinogenicity. In experimental animal studies polycyclic aromatics with 3 to 7 (usually 4 to 6) fused rings, with molecular weights in the range 200 to 450, have been shown to be biologically active carcinogens. In particular benzo(a)pyrene and benz(a)anthracene are considered to be powerful carcinogens. However, the concentrations of these carcinogens in bitumen are extremely low, as indicated in table 2.2[5].

Bitumen and coal tar are often erroneously associated with each other because of their similar superficial appearance and parallel applications. Consequently there is some confusion regarding the toxicity of these two products. Numerous animal studies have been carried out to evaluate the cancerous effects of prolonged contact with bitumen and coal tar. A study carried out by Wallcave [6] showed that contact with coal tar pitch induced a high incidence (over 90%) of epidermal carcinomas and papillomas on the skin of mice, whereas less than 0.5% of the mice in contact with bitumen developed epidermal carcinomas and only 2% developed papillomas. These results reflect the level of carcinogens in the two materials. The benzo(a)pyrene content of the coal tar pitch was between 8,400 ppm and 12,500 ppm, whereas bitumen was between 0.1 ppm and 27 ppm (see table 2.3[6]).

The presence of carcinogens in bitumen does not necessarily constitute a risk to health in practice but the information is vital so that appropriate precautions can be taken by personnel handling the product.

		Bitumens from various sources								Two examples of coal tar pitch	
PCAs*, ppm	Formula	A	B	C	D	E	F	G	H	A	B
Anthracene	$C_{14}H_{10}$	–	–	–	–	–	–	–	–	8 600	10 000
Phenanthrene	$C_{14}H_{10}$	2.3	0.4	3.5	1.3	0.6	35	1.1	2.3	31 000	29 000
Pyrene	$C_{16}H_{10}$	0.6	1.8	4.0	8.3	0.9	38	0.3	0.08	20 000	29 000
Fluoranthene	$C_{16}H_{10}$	+	+	2.0	+	+	5	–	–	40 000	43 000
Benzofluorenes	$C_{17}H_{12}$	+	+	+	+	+	+	+	–	7 300	5 100
Benz (a) anthracene	$C_{18}H_{12}$	0.15	2.1	1.1	0.7	0.9	35	0.2	–	8 900	12 500
Triphenylene	$C_{18}H_{12}$	0.25	6.1	3.1	3.4	3.8	7.6	1.0	0.3	1 500	1 100
Chrysene	$C_{18}H_{12}$	0.2	8.9	2.3	3.9	3.2	34	0.7	0.04	7 400	10 000
Benzo (a) pyrene	$C_{20}H_{12}$	0.5	1.7	1.3	2.5	1.6	27	0.1	–	8 400	12 500
Benzo (e) pyrene	$C_{20}H_{12}$	3.8	13	2.9	3.2	6.5	52	1.6	0.03	5 400	7 000
Benzo (k) fluoranthene	$C_{20}H_{12}$	+	–	+	+	+	–	–	–	7 100	9 000
Perylene	$C_{20}H_{12}$	–	39	2.2	6.1	2.9	3.0	0.1	–	2 000	3 300
Anthanthrene	–	–	Tr	Tr	Tr	+	1.8	–	–	1 300	2 100
Benzo (ghi) perylene	$C_{22}H_{12}$	2.1	4.6	1.0	1.7	2.7	15	0.6	Tr	3 200	3 300
Indeno (1,2,3-cd) pyrene	$C_{22}H_{12}$	Tr	–	Tr	Tr	Tr	1.0	–	–	7 300	9 300
Picene	$C_{22}H_{14}$	+	+	+	+	+	1.0	+	–	NE	2 000
Coronene	$C_{24}H_{12}$	1.9	0.8	0.5	0.2	0.9	2.8	0.9	–	700	700

* PCA is equivalent to PNA.

Tr	Trace.
NE	Not estimated but present in substantial amounts.
+	Not estimated but present in small amounts.
–	Not detected.

Table 2.3 – Polycyclic aromatics (PCAs) in different bitumens and coal tar pitches[6]

2.3.1.2 *Potential hazards via skin contact*

Other than heat burns, the hazards associated with skin contact of most bitumens are negligible. Studies by the International Agency for Research on Cancer (IARC)[7], part of the World Health Organisation, concluded that bitumens have been widely used for many years and there was no direct evidence to associate them with long-term skin disorders in man. However, they do contain polycyclic aromatic compounds some of which have been shown in animal studies to have carcinogenic properties. Therefore it is prudent to avoid intimate and prolonged skin contact with bitumen.

Cutback bitumens and bitumen emulsions are handled at lower temperatures; therefore the likelihood of skin contact is increased (see chapter 18, section 18.10). If personal hygiene is poor, regular skin contact may occur; if this is prolonged over many years there is a slight risk of skin cancer. However, good personal hygiene would eliminate this risk, (see section 2.3.2.2.) Bitumen emulsions cause irritation to the skin and eyes and allergic responses in some individuals.

Materials are sometimes added to bitumen which decrease its viscosity and therefore increase bioavailability. Some of these materials will be more hazardous than bitumen and suitable extra precautions must be taken.

2.3.1.3 *First aid for skin burns*

Skin burns from the accidental contact with hot bitumen should be immediately plunged under cold running water and held there for at least 10 minutes, or until thoroughly cold. No attempt should be made to remove firmly adhering bitumen from the skin. Once the bitumen has cooled, it will do no further harm and in fact provides a sterile covering over a burnt area. As healing takes place, the bitumen plaque will detach itself, usually after a few days. When it is absolutely necessary to remove adhering bitumen from the skin, because of the site and size of contact or the nature of the material, eg cutback bitumen, liberal amounts of warm medicinal paraffin can be used. If solvent treatment is required, it should be followed by washing with soap and water, and then the application of a proprietary refatting agent or skin cleansing cream.

2.3.1.4 *Potential hazards via inhalation*

Generally, when bitumens are heated in bulk or mixed with hot aggregate, fumes are emitted. The fumes contain particulate matter, hydrocarbon vapour and very small amounts of hydrogen sulphide.

The Asphalt Institute have measured fumes emitted from hot asphalt immediately after discharge from the pug mill[8,9]. The concentration of bitumen fumes ranged from 0.2 to 5.4 mg/m^3 with an average of 1.6 mg/m^3. The level of known carcinogenic polycyclic aromatic hydrocarbons in all cases was very low. A similar study on fume emission during paving[10] showed that for paver operators particulate emission levels varied from 0.15 to 5.6 mg/m^3 and for other operators the levels varied from 0.25 to 3.5 mg/m^3 with mean levels of 0.9 mg/m^3.

The U.K. occupational exposure limits for bitumen fumes in the working atmosphere are[11]:

Long-term exposure limit (8 h time weighted average) = 5 mg/m^3

Short-term exposure limit (10 min time weighted average) = 10 mg/m^3

Thus the measured concentrations of total particulate matter only occasionally exceeded the long-term U.K. exposure limit and never exceeded the short term limit.

When working with hot bitumen in the open air, hydrogen sulphide does not present a hazard as the concentration is too low to be a significant risk to health. However, it is possible for hydrogen sulphide to accumulate to lethal concentrations above hot bitumen in the vapour space of bulk storage tanks.

The familiar odour of 'bad eggs' cannot be relied upon to warn operatives of the presence of dangerous concentrations of hydrogen sulphide because the gas deadens the sense of smell at concentrations greater than 200 ppm.

The U.K. occupational exposure limits for hydrogen sulphide in the working atmosphere are [11]:

Long-term exposure limit (8 h time weighted average) = 10 ppm (14 mg/m^3)

Short-term exposure limit (10 min time weighted average) = 15 ppm (21 mg/m^3)

In all cases unnecessary exposure to bitumen fumes should be avoided and where there is *any* doubt, tests to determine the concentration of bitumen fumes or hydrogen sulphide in the working atmosphere should be carried out, eg Draegar tube analysis.

2.3.1.5 First aid for inhalation of bitumen fumes

Persons affected by inhalation of bitumen fumes should be removed to fresh air as soon as possible. If the symptoms are severe or if symptoms persist, medical help should be sought without delay.

2.3.2 Recommended precautions for personnel

2.3.2.1 Protective clothing

The main hazard from handling hot bitumen is the heat burn which results from prolonged contact with the product. Thus it is essential to wear clothing which provides adequate protection:

heat resistant gloves with close-fitting cuffs;

eye and face protection, (see photograph 2.1);

an oversuit with close-fitting cuffs and overlapping at the boots and gloves;

heat-resistant, non-sparking boots with reinforced toes, close-fitting at the top;

head covering with neck flap over collar.

Garments soiled with bitumen should either be replaced or dry cleaned thereby avoiding permeation of the product to the underclothing.

Soiled rags or tools should not be placed in the pockets of overalls as contamination of the lining of the pocket will result.

2.3.2.2 Personal hygiene

Personnel handling bitumen and bituminous materials should be provided with *and use* barrier creams to protect exposed skin, particularly hands and fingers. Skin should be thoroughly washed after any contamination and always before going to the toilet, eating or drinking.

The application of barrier creams, prior to handling bitumen, assists in subsequent cleaning should accidental contact occur. However, barrier creams are not substitutes for gloves or other impermeable clothing and therefore they should not be used as the sole form of protection. Solvents such as petrol, diesel oil, white spirit, etc, should not be used for removing bitumen from the skin as these may spread the contamination. Use an approved skin cleanser and warm water.

2.3.3 Fire prevention and fire fighting

The adoption of safe handling procedures will substantially reduce the risk of fire. However, if a fire occurs, it is essential that personnel are properly trained and well equipped to extinguish the fire, thereby ensuring that risk of injury to personnel and damage to plant is minimised. Detailed advice on fire prevention and fire fighting is given in the Institute of Petroleum Bitumen Safety Code.[3]

Small bitumen fires can be put out using dry chemical powder, foam, vaporising liquid or inert gas extinguishers, fog nozzle spray hoses or steam lances. Direct water jets should be avoided because frothing may occur which tends to spread the hot bitumen and, therefore, the fire.

Internal tank fires where the roof of the tank is largely intact can be extinguished by injection of steam or a 'fog' of water into the vapour space. However, this method should be used only by trained operatives because the water vaporises instantly on contact with the hot bitumen. This initiates foaming which may result in the tank overflowing, creating an additional hazard. Alternatively foam extinguishers may be used. The foam ensures that the water is well dispersed, thereby reducing the risk of froth-over. The disadvantage with this type of extinguisher is that the foam breaks down rapidly when applied to hot bitumen.

Portable aqueous film-forming foam extinguishers and dry chemical powder extinguishers are suitable for 'first attack' of small bitumen fires. These should be placed at strategic locations which are permanent and conspicuous around bitumen handling areas. The type and location of 'main attack' equipment should be discussed with the local fire brigades before installation.

Photograph 2.1 – Protective clothing worn during a bitumen delivery

2.3.4 Sampling

Sampling of bitumen is particularly hazardous because of the risk of heat burns from spills and splashes of the material. It is therefore essential that protective clothing should always be worn when sampling bitumen.

2.3.4.1 Dip sampling

A sample of bitumen is obtained by dipping a weighted can or 'thief' on the end of a rope or rod through the manlid into bitumen stored in a bulk tank. The sample is then transferred to a suitable permanent container. The method is simple but is appropriate only for small samples. Dip sampling from cutback tanks should be avoided because of the presence of flammable atmospheres in tank vapour spaces. Safe access to and egress from the sample point, should be provided. Gantry access should be provided where dip samples are required from vehicle tanks.

2.3.4.2 Sample valves

Properly designed sample valves are very useful for sampling from pipelines or from tanks. Their design should ensure that they are kept hot by the product in the pipeline or tank to avoid blockage when in the closed position.

Sample valves should preferably be the screw driven, plunger type. When closed, the plunger of this type of valve extends into the fresh product, so that when the valve is opened a representative sample of product is obtained without 'fore-runnings'. With ball and plug type valves fore-runnings have to be collected and disposed of before a representative sample can be obtained.

2.4 References

(1) Refined Bitumen Association: *Safe handling of petroleum bitumens.* Technical Bulletin No 5, March 1981.

(2) Refined Bitumen Association: *Recommendations to users for the storage and handling of bitumen.* Technical Bulletin No 7, August 1988.

(3) Institute of Petroleum: *Model code of safe practice in the petroleum industry. Part 11. Bitumen Safety Code,* 1990.

(4) Chipperfield, E H: *Bitumen production, properties and uses in relation to occupational exposures.* IARC review on bitumen carcinogenicity. Institute of Petroleum, Report IP 84-006, 1984.

(5) Hangebrauck, R P, von Lehmden, J U and Meeker, J E: *Sources of polynuclear hydrocarbons in the atmosphere.* U.S. Department of Health, Education and Welfare. Publication No 999-AP-33, 1967.

(6) Wallcave, L, Garcia, H, Feldman, R, Lijinsky, W and Shubik, P: *Skin tumourigenesis in mice by petroleum asphalts and coal tar pitches of known polynuclear aromatic hydrocarbon content.* Toxicology and Applied Pharmacology, vol 18, pp 41-52, January 1971.

(7) International Agency for Research on Cancer: *Evaluation of the carcinogenic risk of chemicals to humans. (Polynuclear aromatic compounds, part 4, bitumens, coal tars and derived products, shale-oils and soots).* Vol 35, January 1985.

(8) The Asphalt Institute: *Asphalt hot-mix emission study.* Research Report RR-75-1, March 1975.

(9) Puzinauskas, V P and Corbett, L W: *Report on emissions from asphalt hot mixes.* The Asphalt Institute, Research Report RR-75-1A, May 1975.

(10) Puzinauskas, V P: *Exposure of paving workers to asphalt emissions.* The Asphalt Institute, Research Report RR-80-1, July 1980.

(11) Health and Safety Executive Guidance Note EH40: *Occupational exposure limits.*

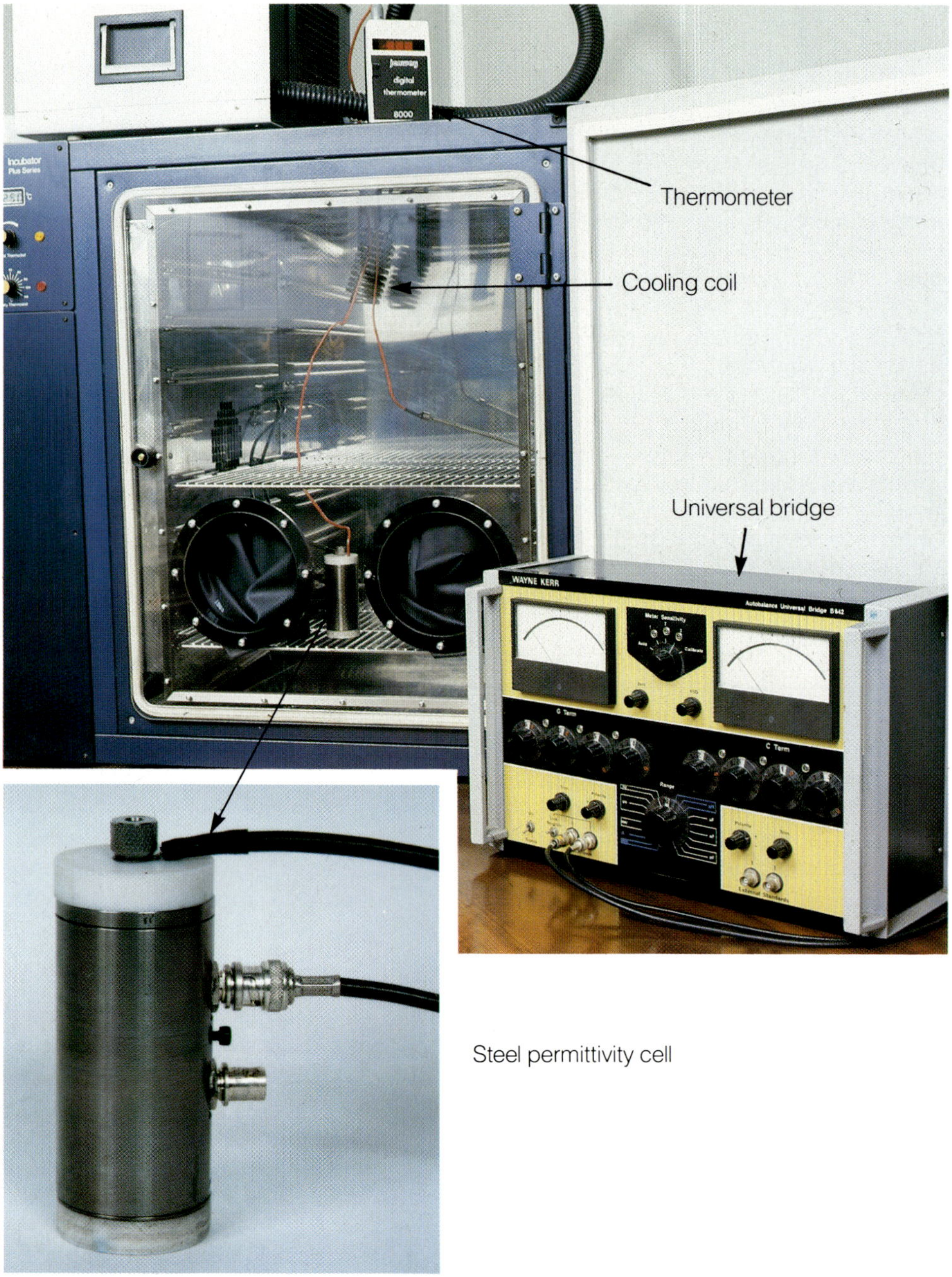

Photograph 3.1 – The permittivity test

Chapter 3 – Bitumen specifications and quality

In the U.K. bitumens are manufactured to British Standard (BS) 3690 [1,2,3]. Part 1 covers bitumens for road purposes, part 2 covers bitumens for industrial uses and part 3 covers blends of bitumen with coal tar, and with lake asphalt or pitch. Four types of bitumen are manufactured by Shell Bitumen U.K. which are characterised by a combination of penetration, softening point and viscosity, viz:

penetration grades	– penetration and softening point;
oxidised and hard grades	– softening point and penetration;
cutback grades	– viscosity.

In addition to the tests specified in BS 3690, Shell Bitumen U.K. carry out a further range of laboratory tests to ensure that the quality of the bitumen is maintained to a consistent standard and to ensure total suitability for purpose.

3.1 Penetration grade bitumens

Penetration grade bitumens are specified by the penetration[4] and softening point[5] tests (see chapter 5, photographs 5.1 and 5.2 respectively), but are designated by penetration only, eg 100 pen bitumen has a penetration of 100 ± 20 and a softening point of 46 ± 5°C. The grades range from 15 pen through to 450 pen, with corresponding softening point ranges for each grade. In addition to penetration and softening point, limits are given for solubility in trichloroethylene[6], loss on heating[7] and for grades used in the manufacture of hot rolled asphalt wearing course, ie 35 pen, 50 pen, 70 pen, 100 pen and HD 40, there is a minimum permittivity requirement[8]. The loss on heating test ensures that there are no volatile components in the bitumen that could result in excessive hardening during storage, application or in service. The solubility limit ensures that contaminants such as carbon (coke) and mineral matter are kept to negligible proportions. In the permittivity test (see chapter 3, photograph 3.1) the space between two plates in a parallel plate capacitor is filled with bitumen and a potential difference is applied across the two plates. The permittivity of bitumen is defined as the ratio of:

$$\frac{\text{The capacitance of a capacitor with the bitumen as the dielectric}}{\text{The capacitance of the same capacitor with a vacuum as the dielectric.}}$$

The Transport and Road Research Laboratory showed that the ability of bitumens to weather under the combined effects of ultra-violet light in sunlight, oxygen, rain and oil deposition from traffic was related to the permittivity or dielectric constant of the bitumen (see chapter 7, section 7.2.3).

Table 1 of BS 3690 : Part 1 details the specification requirements for penetration grade bitumens and this table has been reproduced in table 3.1.

The majority of penetration grade bitumens is used in road construction, the harder grades, 35 pen to 100 pen, being used in asphalt where bitumen stiffness is of primary importance and the softer grades, 100 pen to 450 pen, in macadams where the lubricating properties during application and bonding of the aggregate in service are more important.

3.2 Oxidised bitumens

Oxidised bitumens are used almost entirely for industrial applications, eg roofing, flooring, mastics, pipe coatings, paints, etc, and are specified and designated by both softening point and penetration tests, eg 85/40 is an oxidised grade bitumen with a softening point of 85 ± 5°C and a penetration of 40 ± 5. Oxidised bitumens also have to comply with solubility and loss on heating criteria. Table 3 of BS 3690: Part 2, details the specification requirements for the six U.K. oxidised grade bitumens and this table has been reproduced in table 3.2. The softening points of oxidised grades of bitumen are much higher than those of the corresponding penetration grade bitumen and therefore the temperature susceptibility, ie penetration index is much higher, ie +2 to +8.

3.3 Hard bitumens

Hard bitumens are also used entirely for industrial applications, eg coal briquetting, paints, etc. They are also specified by both softening point and penetration tests but are designated by a softening point range only, and a prefix H, eg H 80/90 is a hard grade bitumen with a softening point between 80°C and 90°C. The penetration index varies from 0 to +2. Table 4 of BS 3690: Part 2, details the specification requirements of hard grade bitumens; this has been reproduced in table 3.3.

3.4 Cutback bitumens

Cutback bitumens are manufactured by blending either 100 pen or 200 pen bitumen with kerosine to comply with a viscosity specification. In the U.K. cutback bitumens are specified and designated by the flow time through a standard tar viscometer (STV)[9] as shown in chapter 5, photograph 5.5. Three grades are available, 50 seconds, 100 seconds and 200 seconds. The majority of cutback bitumen is used in surface dressing (see chapter 18) but a significant amount is also used for the manufacture of both standard and deferred set macadams.

Table 2 of BS 3690: Part 1 details the specification requirements of cutback bitumens and this table has been reproduced in table 3.4. In addition to STV and solubility, cutback bitumens have to comply with a distillation specification[10] and a penetration requirement on the residual bitumen. This ensures that during application and in service the diluent will evaporate at a consistent and predictable rate and that the residual bitumen will have the appropriate properties in service.

The suffix X on the SHELPHALT range of cutback bitumens indicates that they have been doped with a specially formulated heat stable passive adhesion agent. This additive assists 'wetting' of the aggregate and resists stripping of the binder from the aggregate in the presence of water (see chapter 18, section 18.3.1.2).

3.5 Bitumen quality

For many years Shell has been investigating the relationship between laboratory measured properties of penetration grade bitumens and their performance in asphalt mixes on the road. With increasing traffic loadings and more demanding performance requirements the need to be able to predict long-term behaviour is essential. Performance on the road depends on many factors, including the design, application and the quality of the individual components. Although bitumen is, in terms of its volume, a relatively minor component of a bituminous mix, it has a crucial role acting as a durable binder and conferring visco-elastic properties to the mix.

Essentially, satisfactory performance of a bitumen on the road can be ensured if four properties are controlled:

rheology;
cohesion;
adhesion;
durability.

Property		Test method	Grade of bitumen									
			15 pen	25 pen	35 pen	40 pen HD	50 pen	70 pen	100 pen	200 pen	300 pen	450 pen
Penetration at 25°C dmm		BS 2000 : Part 49	15 ± 5	25 ± 5	35 ± 7	40 ± 10	50 ± 10	70 ± 10	100 ± 20	200 ± 30	300 ± 45	450 ± 65
Softening point, °C	min	BS 2000 : Part 58	63	57	52	58	47	44	41	33	30	25
	max		76	69	64	68	58	54	51	42	39	34
Loss on heating for 5 h at 163°C		BS 2000 : Part 45										
(a) loss by mass, %	max		0.1	0.2	0.2	0.2	0.2	0.2	0.5	0.5	1.0	1.0
(b) drop in penetration, %	max		20	20	20	20	20	20	20	20	25	25
Solubility in trichloroethylene, % by mass	min	BS 2000 : Part 47	99.5	99.5	99.5	99.5	99.5	99.5	99.5	99.5	99.5	99.5
Permittivity at 25°C and 1592 Hz	min	BS 2000 : Part 357	–	–	2.630	2.630	2.630	2.630	2.630	–	–	–

Table 3.1 – BS 3690 : Part 1 Specification for penetration grade bitumens(1)

Property		Test method	Grade of bitumen					
			75/30	85/25	85/40	95/25	105/35	115/15
Softening point, °C		BS 2000 : Part 58	75 ± 5	85 ± 5	85 ± 5	95 ± 5	105 ± 5	115 ± 5
Penetration at 25°C dmm		BS 2000 : Part 49	30 ± 5	25 ± 5	40 ± 5	25 ± 5	35 ± 5	15 ± 5
Loss on heating for 5 h at 163°C, loss by mass %	max	BS 2000 : Part 45	0.2	0.2	0.5	0.2	0.5	0.2
Solubility in trichloroethylene, % by mass	min	BS 2000 : Part 47	99.5	99.5	99.5	99.5	99.5	99.5

Table 3.2 – BS 3690 : Part 2 Specification for oxidised grade bitumens(2)

Property		Test method	Grade of bitumen	
			H 80/90	H 100/120
Softening point, °C		BS 2000: Part 58	85 ± 5	110 ± 10
Penetration at 25°C, dmm		BS 2000: Part 49	9 ± 3	6 ± 4
Loss on heating for 5 h at 163°C, loss by mass %	max	BS 2000: Part 45	0.05	0.05
Solubility in trichloroethylene, % by mass	min	BS 2000: Part 47	99.5	99.5

Table 3.3 – BS 3690: Part 2 Specification for hard grade bitumens[2]

Property		Test method	Grade of cutback		
			50 secs	100 secs	200 secs
Viscosity (STV) at 40°C, 10 mm cup, secs		BS 2000: Part 72	50 ± 10	100 ± 20	200 ± 40
Distillation		BS 2000: Part 27			
(a) distillate to 225°C, % by volume	max		1	1	1
360°C, % by volume			10 ± 3	9 ± 3	7 ± 3
(b) penetration at 25°C of residue from distillation to 360°C, dmm		BS 2000: Part 49	100 to 350	100 to 350	100 to 350
Solubility in trichloroethylene, % by mass	min	BS 2000: Part 47	99.5	99.5	99.5

Table 3.4 – BS 3690: Part 1 Specification for cutback bitumens[1]

The rheology of the bitumen at service temperatures is adequately characterised by penetration and penetration index[11]. By studying correlations between field performance and measured properties of experimental and commercial bitumens Shell Research Limited have developed a set of laboratory tests which can assess the quality of a bitumen. The set of tests include six on the bitumen itself and three on a bituminous mix. For ease of assessment these nine results are presented on a regular polygon, which is called the bitumen QUALAGON*[12], (see figure 3.1). Within the QUALAGON are tests which cover the three remaining key performance elements:

cohesion – low temperature ductility;
adhesion – retained Marshall;
durability – rolling thin film oven test (oxidation stability);
true boiling point – gas liquid chromatography (volatility);
exudation droplet test (homogeneity);
field trials (oxidation stability).

* QUALAGON is a Shell trademark

3.5.1 Cohesion

The cohesive strength of penetration grade bitumens is characterised by low temperature ductility (ASTM test method D 113, see photograph 3.2). In this test three dumb-bells of bitumen are immersed in a water bath and stretched at a constant speed of 50 mm per minute until fracture occurs. The distance the specimen is stretched before failure is reported as the ductility. The test temperature is adjusted depending on the penetration of bitumen, eg 10°C for 80 pen to 100 pen, 13°C for 60 pen to 70 pen and 17°C for 40 pen to 50 pen. Under these conditions the test has been found to discriminate between bitumens of different cohesive strengths.

3.5.2 Adhesion

In the QUALAGON the adhesion characteristics of a bitumen are assessed by a retained Marshall test. In the test eight Marshall specimens are manufactured using a prescribed aggregate, aggregate grading and bitumen content. The eight specimens are divided into two groups of four so that the average void content of the two groups is equal. One group of four is tested immediately using the standard Marshall method and the remaining four specimens are vacuum treated under

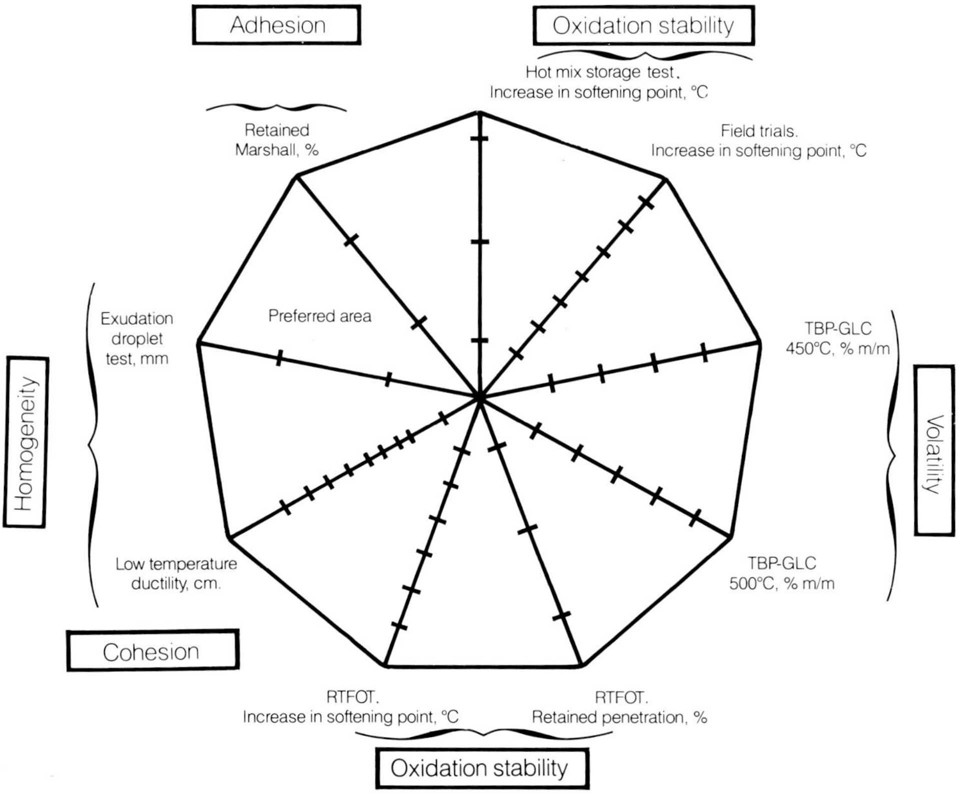

Figure 3.1 – The penetration grade QUALAGON[12]

water at a temperature of between 0°C to 1°C, to saturate as far as possible the pore volume of the mix with water. Subsequently, the specimens are stored in a water bath at 60°C for 48 hours. The Marshall stability of these four specimens is then determined. The ratio of the Marshall stability of the treated specimens divided by the initial Marshall stability is termed the retained Marshall stability.

Although a relatively wide scatter in absolute stability values and poor reproducibility are inherent in Marshall testing, it has been found that consideration of the percentage relative to the result from the standard procedure reduces the differences between results from different laboratories.

From both laboratory tests and observation of performance in practice, key links have been identified between functional properties and the constitution of the bitumen. This work has indicated that if the molecular weight distribution and chemical constitution of the bitumen is unbalanced it can exhibit inhomogeneity which may adversely affect both the cohesive and adhesive properties of the bitumen.

3.5.3 Durability

Durability can be defined as the ability to maintain satisfactory rheology, cohesion and adhesion in long-term service. As a part of the quality criteria for bitumen the following have been identified as the prime durability factors:

oxidative hardening;
evaporative hardening;
exudative hardening.

3.5.3.1 Oxidative and evaporative hardening

Hardening due to oxidation has long been held to be the main cause of bitumen ageing. In mixes where the finely structured voids are interconnected, hardening *in-situ* by oxidation is important. However, under such circumstances hardening due to evaporation of the more volatile components of the bitumen can also be significant.

- Tests on the bitumen

  The rolling thin film oven test (ASTM test method D 2872, see photograph 3.3) measures hardening by both oxidation and evaporation. In this test a thin film of bitumen is continuously rotated around the inner surface of a glass jar at 163°C for 75 minutes, with an injection of hot air into the jar every 3 to 4 seconds. The amount of bitumen hardening during the test, ie drop in penetration and increase in softening point, correlates well with that observed during the manufacture of a bituminous mix. The oxidation stability of bitumens is plotted on the QUALAGON as the percentage retained penetration and the increase in softening point.

  Evaporative hardening can be assessed by measuring the front-end volatility of the bitumen by true boiling point – gas liquid chromatography (TBP-GLC). In this test a small sample of bitumen (150 mg) is dissolved in carbon disulphide and is separated on two chromatographic columns. The first column separates the heavy-end components such as asphaltenes, heavy polar aromatics, etc from the bitumen. The hydrocarbons eluted from this column are then separated on a second column. The chromatograph is calibrated using components of known boiling points and the results are plotted as a ratio of the percent mass divided by mass on intake against boiling temperature.

  TBP-GLC is a rapid and accurate method of front-end volatility analysis and is plotted on the QUALAGON as the % m/m recovered at 450°C and 500°C thus taking account of the shape of the volatility curve. The loss of 0.2% m in the rolling thin film oven test correlates well with the TBP-GLC recovery limits at 450°C and 500°C[12].

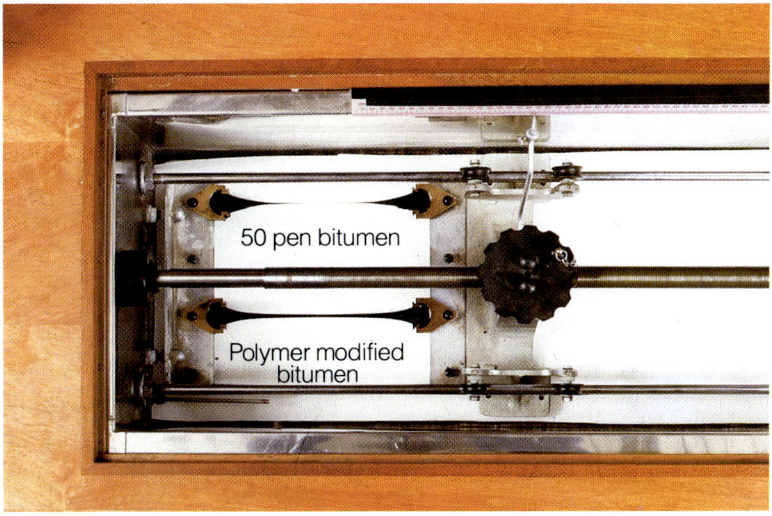

Photograph 3.2 – A plan view of the ductility test

Photograph 3.3 – The rolling thin film oven test

- Tests on the mix

  The hardening due to oxidation and evaporation of a thin film of bitumen in contact with aggregate is assessed by two mix tests: the hot mix storage test and by the change in the softening point of the bitumen during commercial mix manufacture. The hot mix storage test simulates ageing conditions during mixing and hot storage. A prescribed mix is manufactured in the laboratory and a specified quantity of this mix is stored for 16 hours at 160°C in a sealed tin. Thus the volume of air entrained in the sample is known and is constant from test to test. The bitumen is recovered from both the as-mixed and stored material and the penetration and softening points are determined on these two recovered bitumens. The ageing of the bitumen during mixing and storage is expressed as the difference between the softening points of the bitumen after storage and that of the original bitumen.

  At laboratory scale this is a very severe test and the change in softening point is very much larger than that found during actual bulk storage. Nevertheless, the test correlates with the hardening tendency of a mix at high temperature when in prolonged contact with air.

3.5.3.2 Exudative hardening

As discussed in paragraph 3 of section 3.5.2, if the constitution of a bitumen is unbalanced it may, when in contact with a porous aggregate, exude an oily component into the surface pores of the aggregate, resulting in a hardening of the bitumen film remaining on the aggregate surface. Exudation is primarily a function of the amount of low molecular weight paraffinic components present in the bitumen relative to the amount and type of asphaltenes.

To measure quantitatively the exudation tendency of a bitumen, Shell Research Limited developed the exudation droplet test[12]. In this test bitumen droplets are applied to the recesses in specially made white marble plates. The plates are stored at 60°C for 4 days under a nitrogen blanket. During this period oily rings develop around the bitumen droplet which can be measured under ultra-violet light using a microscope. Ring widths vary from a few tenths of a millimetre for a balanced bitumen to several millimetres for an unbalanced one.

Hardening in service as a result of exudation can be substantial and depends not only on the exudation tendency of the bitumen but also on the porosity of the aggregate. If the aggregate is of low porosity the quantity of exudate absorbed is negligible, irrespective of the exudation tendency of the bitumen. Similarly, if the exudation tendency of the bitumen is low, the exudate absorbed will be negligible, irrespective of the porosity of the aggregate.

3.5.4 Verification of the QUALAGON tests and test criteria

Several years' experience using the QUALAGON in conjunction with rheological data have confirmed that the set of criteria on which it is based provide a satisfactory quantitative measure of bitumen quality and road performance.

Good quality bitumens generally have properties within the 'preferred' area of the QUALAGON. However, this does *not* mean that a bitumen which is partially outside the preferred area is necessarily a 'bad' or 'poor quality' bitumen. The QUALAGON limits are not pass/fail criteria and the whole figure must be interpreted with care and judgement. For this reason the QUALAGON is *not* suitable for specification purposes but is an invaluable research tool providing an excellent guide to performance in service.

3.6 References

(1) British Standard 3690: Part 1: 1989: *Bitumens for building and civil engineering. Part 1. Specification for bitumens for road purposes.*

(2) British Standard 3690: Part 2: 1989: *Bitumens for building and civil engineering. Part 2. Specification for bitumens for industrial purposes.*

(3) British Standard 3690: Part 3: 1983: *Bitumens for building and civil engineering. Part 3. Specification for bitumen mixtures.*

(4) British Standard 2000: Part 49: 1983: *Penetration of bituminous materials.*

(5) British Standard 2000: Part 58: 1983: *Softening point of bitumen (ring and ball).*

(6) British Standard 2000: Part 47: 1983: *Solubility of bituminous binders.*

(7) British Standard 2000: Part 45: 1982: *Loss on heating of bitumen and flux oil.*

(8) British Standard 2000: Part 357: 1983: *Permittivity of bitumen.*

(9) British Standard 2000: Part 72: 1982: *Viscosity of cutback bitumen and road oil.*

(10) British Standard 2000: Part 27: 1982: *Distillation of cut-back asphaltic (bituminous) products.*

(11) de Bats, F Th and van Gooswilligen, G: *Practical rheological characterisation of paving grade bitumens.* Proceedings of the fourth Eurobitume symposium, pp 304-310, Madrid, October 1989.

(12) van Gooswilligen, G, de Bats, F Th and Harrison, T: *Quality of paving grade bitumen – A practical approach in terms of functional tests.* Proceedings of the fourth Eurobitume symposium, pp 290-297, Madrid, October 1989.

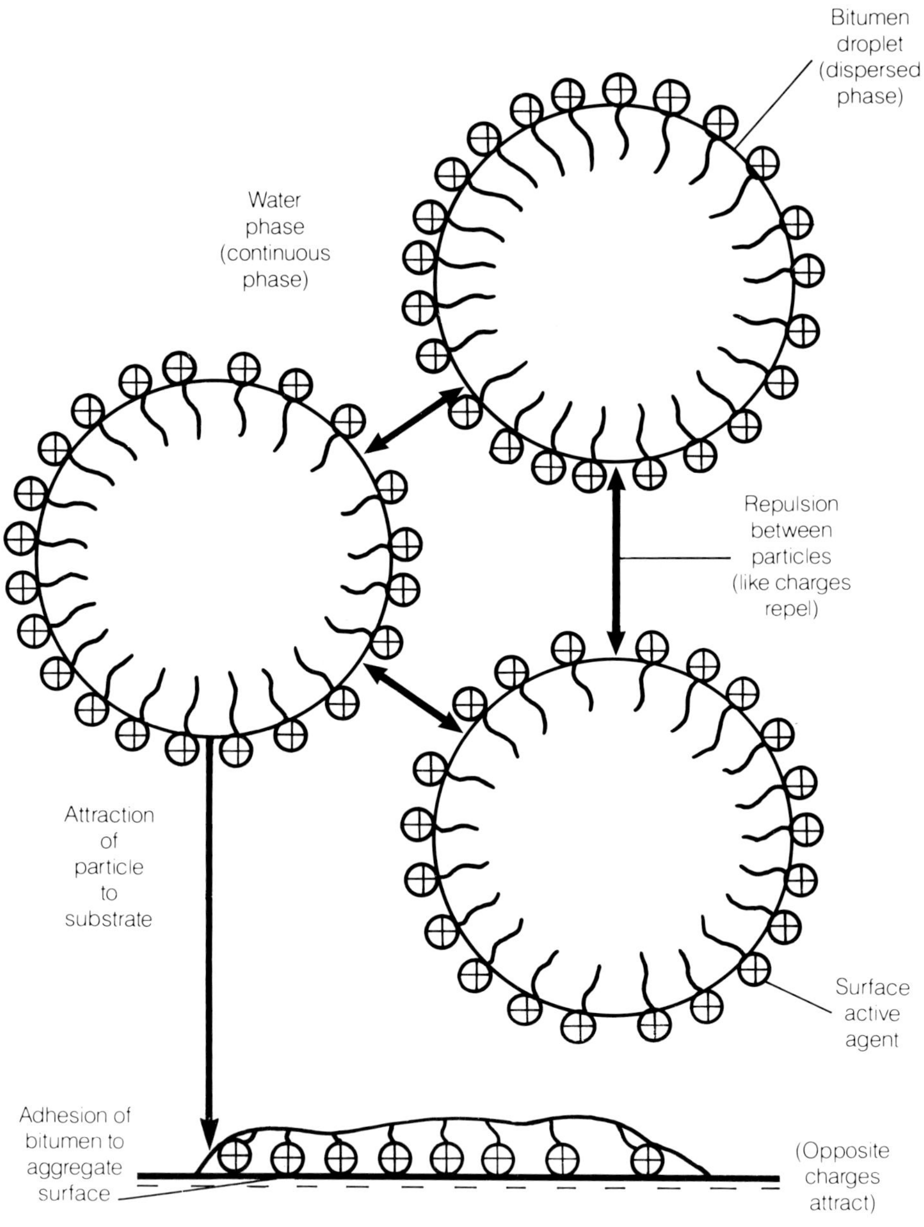

Figure 4.1 – Schematic diagram of charges on bitumen droplets

Chapter 4 – Bitumen emulsions

Bitumen emulsions are heterogeneous two-phase systems consisting of two immiscible liquids, bitumen and water. The bitumen is dispersed throughout the continuous water phase in the form of discrete globules, typically 0.1 micron to 5 microns in diameter, which are held in suspension by electrostatic charges stabilised by an emulsifier.

Bitumen emulsions can be divided into four classes of which the first two are by far the most important:

anionic emulsions;
cationic emulsions;
non-ionic emulsions;
clay-stabilised emulsions.

The terms anionic and cationic stem from the electrical charges surrounding the bitumen globules. This identification system originates from one of the basic laws of electricity – like charges repel, unlike charges attract. If an electrical current is passed through an emulsion containing negatively charged particles of bitumen they will migrate to the anode. Hence, the emulsion is referred to as anionic. Conversely, positively charged bitumen particles will move to the cathode and the emulsion is known as cationic. The bitumen in non-ionic emulsions is neutral and therefore they will not migrate to either pole. These are rarely used for road emulsions.

Clay-stabilised emulsions are used for industrial rather than for road applications. Here the emulsifiers are fine powders such as clays and bentonites.

It was in 1906 when Schade van Westrum(1) took out a patent covering the application of bituminous dispersions in water for road building. Initially efforts were made to form emulsions by purely mechanical means; however, it rapidly became apparent that mechanical action alone was insufficient, and from these pioneering days anionic or cationic emulsifiers have been used to manufacture bitumen emulsions(2). Initially the naturally occurring organic acids in the bitumen were utilised by the simple expedient of adding to the aqueous phase sodium or potassium hydroxide. The subsequent reaction between the acid and the alkali formed an anionic soap which was surface active and therefore capable of stabilising the dispersion(3). A great variety of chemicals has been used to promote the stability of bitumen emulsions. These include the residues from fatty acid distillation, rosin acids, hydroxystearic acid, lignin sulphonates, etc, which were generally blended with the bitumen prior to emulsification.

Since the early fifties cationic emulsifiers have become increasingly popular because of their ability to be adsorbed onto a large variety of solid surfaces. This affinity for solid surfaces is an important property of cationic emulsions in road construction because good adhesion of bitumen to different types of mineral aggregate is essential. The cationic emulsifiers most widely used are the linear stearyl mono-amines and di-amines, amidoamines and imidazolines(4).

4.1 Emulsifiers

An emulsifier consists of a long hydrocarbon chain which terminates with either a cationic or an anionic functional group. The paraffinic portion of the molecule has an affinity for bitumen and the ionic portion an affinity for water. The emulsifier ion orients itself on the surface of the bitumen droplet such that the hydrocarbon chain is firmly bound in the bitumen with the ionic portion located at the surface. Hence, the droplets are positively charged in the cationic emulsions and negatively charged in anionic emulsions.

The emulsifier is not only a stabilising agent but also an adhesion promoter. The positive cationic part of the emulsifier forms a strong electrostatic bond with a negatively charged aggregate surface. The bitumen particles are attracted to the aggregate surface and the positive charges of the bitumen are neutralised by the negative charges of the oxygen valencies on the aggregate surface. These particles coalesce and adhere to the aggregate (see figure 4.1).

Emulsifier solutions are prepared by dissolving the amines, diamines or amino alkoxylated amines in hydrochloric acid or acetic acid. This is carried out by carefully controlling the pH, producing the amine salts. For example:

$$\underset{\text{Amine}}{R-NH_2} + \underset{\text{Hydrochloric acid}}{HCl} \rightarrow \underset{\text{Amine hydrochloride}}{RNH_3^+ + Cl^-}$$

It has been shown that when an emulsion is formed with an amine salt, the positively charged cation orientates itself on the surface of the bitumen droplet. The negatively charged chloride ions are then attracted to the positively charged surface and together, with the water, create a so-called electrical double layer. This is illustrated in figure 4.2. The thickness of this layer has a great influence on the stability and viscosity of the emulsion.

Anionic emulsions are made with fatty acids which have been saponified with sodium hydroxide. The emulsifier solution creates an excess of sodium hydroxide which also neutralises the natural acids contained in the bitumen. For example:

$$\underset{\text{Fatty acid}}{R\,COOH} + \underset{\text{Sodium hydroxide}}{Na\,OH} \rightarrow \underset{\text{Fatty acid soap}}{R\ COO^- + Na^+ + H_2O}$$

4.2 The manufacture of bitumen emulsions

The majority of bitumen emulsions are manufactured using colloid mills. The mill consists of a high-speed rotor which revolves at 1000 rpm to 6000 rpm in a stator; the clearance between the rotor and the stator is typically 0.25 mm to 0.50 mm, and is usually adjustable. The hot bitumen and emulsifier solution are fed in separately but simultaneously into the colloid mill. The temperatures of the two components vary depending on the grade and percentage of bitumen in the emulsion, the type of emulsifier, etc. The viscosity of the bitumen entering the colloid mill should not exceed 0.2 Pa.s (2 poise); to achieve this viscosity, bitumen temperatures in the range 100°C to 140°C are used. The temperature of the water phase is adjusted so that the temperature of the emulsion produced is not greater than 90°C. As the bitumen and emulsifier solution enter the colloid mill they are subjected to intense shearing forces which cause the bitumen to break into small globules. The individual globules are then coated with the emulsifier which gives the surface of the droplets an electrical charge and the resulting electrostatic forces prevent the globules from coalescing.

Schematic diagrams of continuous and batch emulsion manufacturing plants are shown in figures 4.3a and 4.3b respectively. The bitumen and aqueous phases are separately metered into the colloid mill by metering pumps or under manual control in conjunction with flow meters.

The method of adding the emulsifier to the water varies; some emulsifiers, such as amines, must be mixed and reacted with an acid, eg hydrochloric acid, to attain water solubility, whereas others, such as fatty acids, must be mixed and reacted with an alkali, eg sodium hydroxide, to attain water solubility.

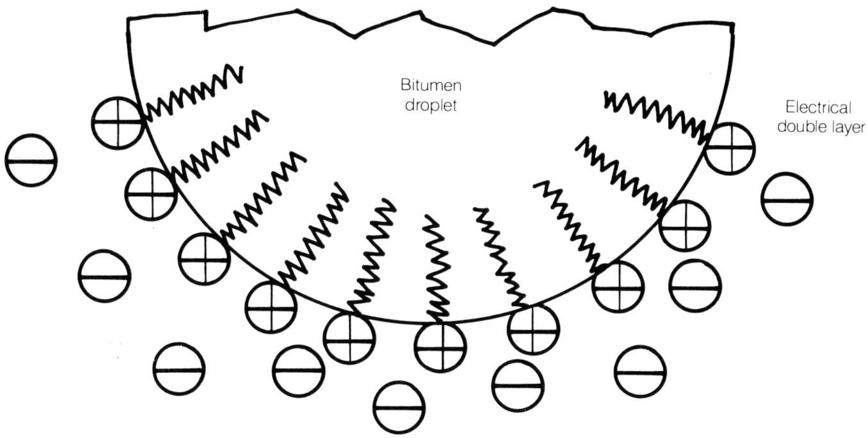

Figure 4.2 – Bitumen droplet with an electrical double layer

4.3 Bitumen emulsion formulation

Bitumen emulsion formulation involves a complex stability problem. A stable dispersion is required to facilitate storage and transportation without breaking. However, the emulsion should break rapidly after application to the road surface. From a technological viewpoint the most important properties of bitumen emulsions can be described as[5]:

- the stability of the emulsion;
- the adhesivity of the emulsion;
- the viscosity of the emulsion.

4.3.1 Emulsion stability

This encompasses the two conflicting stability requirements of bituminous emulsions; storage stability and breaking rate.

4.3.1.1 Storage stability

An inadequacy in the storage stability of an emulsion is initially indicated by settlement of the emulsion, resulting in a bitumen lean upper layer and a bitumen rich lower layer. Settlement is the result of gravity acting upon the denser discontinuous bitumen droplets. The velocity of the downward movement of these particles can be estimated by using Stokes' Law[6], which expresses the settlement velocity (V) as follows:

$$V = \frac{2/9\, g r^2 (d_1 - d_2)}{\eta}$$

where g = gravitational force

r = particle radius

d_1 = specific gravity of the bitumen

d_2 = specific gravity of the aqueous phase

η = viscosity of the aqueous phase.

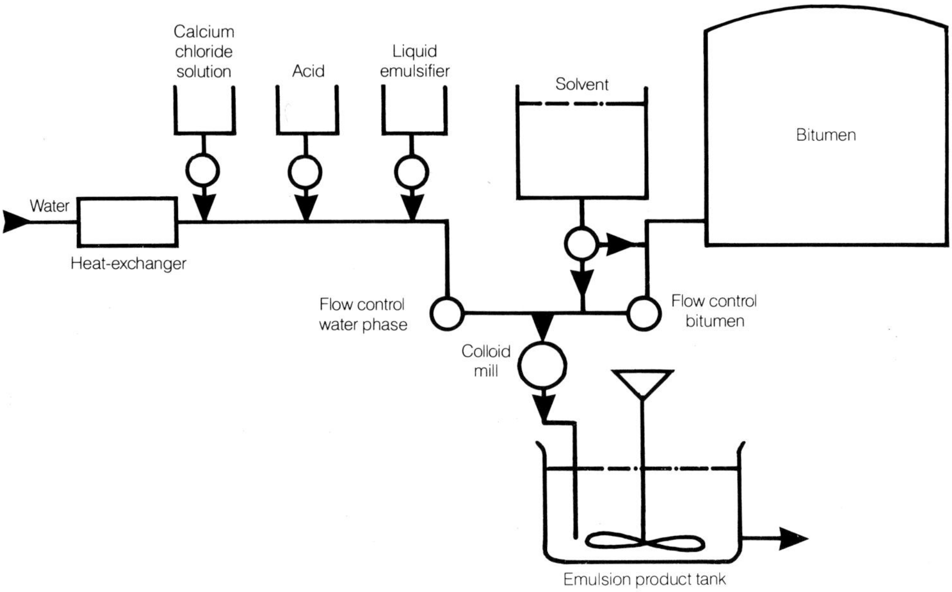

Figure 4.3a – Schematic diagram of a continuous emulsion plant

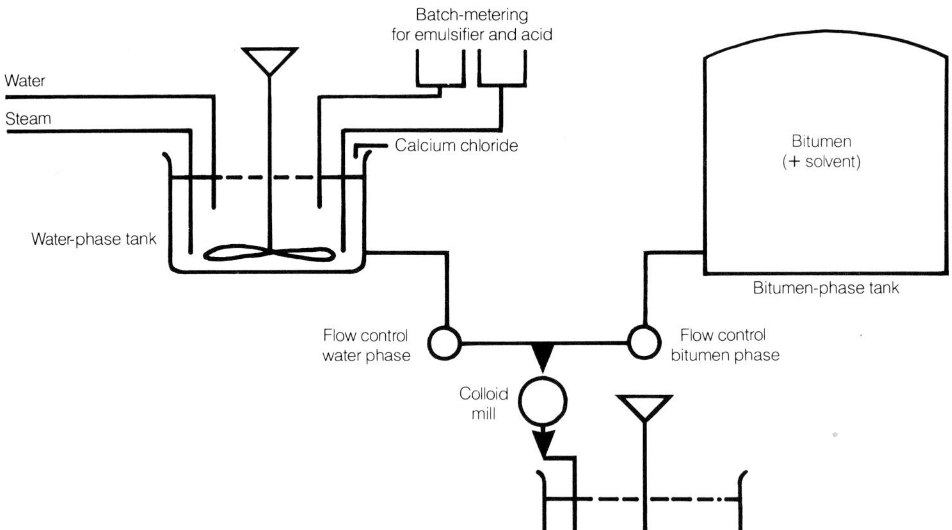

Figure 4.3b – Schematic diagram of a batch emulsion plant

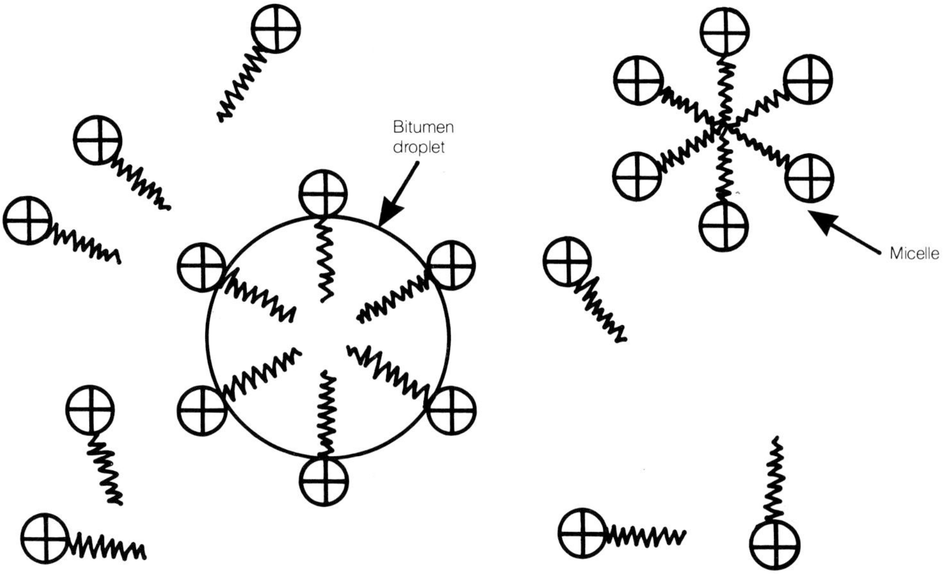

Figure 4.4a – Emulsifier ions forming micelles in a stable solution

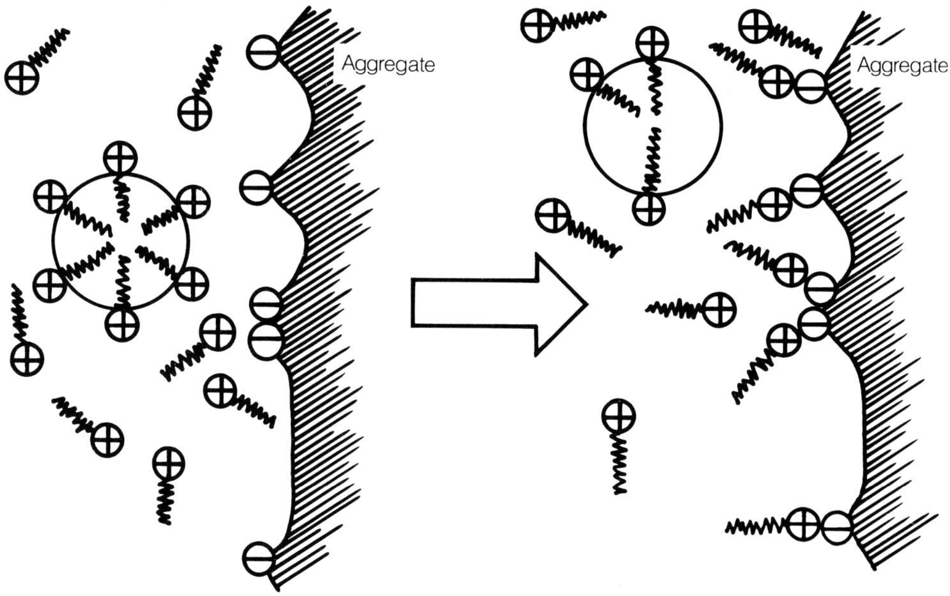

Figure 4.4b – Schematic diagram of the breaking process

Stokes' Law, however, applies to particles which are free to move. In a bitumen emulsion, the droplets are so tightly packed that Stokes' Law will inevitably overestimate the particle velocity.

In addition to gravity there are two other forces acting in an emulsion: repulsive and attractive forces. The repulsive force originates from the repulsion between the electrostatic double layers on the droplets created by the ionised emulsifier. It can be increased by raising the concentration of emulsifier and decreased by the addition of an excessive amount of negatively charged ions. The attractive force is associated with the mass of the droplets. If the droplets are large, or if the distribution of particle sizes is wide, the attractive force becomes a repulsive force.

Coalescence follows settlement in two stages. Firstly, bitumen droplets agglomerate into clumps (like bunches of grapes) creating a flock; this reversible phenomenon is called flocculation. This is followed by coalescence, which is irreversible, in which the flocks fuse together to form larger globules. This process can be spontaneous or it can be induced by mechanical action.

4.3.1.2 The breaking process

The emulsion contains emulsifier molecules in both the water phase and on the surface of the droplets. Some of the emulsifier ions have formed micelles and in a stable emulsion an equilibrium exists as shown in figure 4.4a. If some of the emulsifier ions are removed from the solution the balance is restored by ions from the micelles and the surface of the droplets. This occurs when an emulsion comes into contact with a mineral aggregate. The negatively charged aggregate surface rapidly absorbs some of the ions from the solution, weakening the charge on the surface of the droplets, which initiates the breaking process as shown in figure 4.4b. A point is reached where the charge on the surface of the droplets is so depleted that rapid coalescence takes place. The aggregate is now covered in hydrocarbon chains and therefore the liberated bitumen adheres strongly to its surface.

4.3.2 Emulsion adhesivity

It is a prime requirement in all applications where bitumen is used as an adhesive between solid surfaces, that the bitumen 'wets' the surface to create the maximum contact area. With dry substrates the 'critical surface tension of wetting' of the aggregate must be high enough to ensure that the bitumen spreads easily over the surface; the resultant adhesion generally exceeds the cohesion of the bitumen. However, when an aggregate surface is covered with water, the wetting of the aggregate becomes a three phase phenomenon which can occur only if the balance of the interfacial energies favours wetting by the bitumen. Cationic emulsifiers are particularly efficient at reducing free surface energy of a polar aggregate, forming a thermodynamically stable condition of minimum surface energy by the emulsifier being attracted to the aggregate surface[7,8].

Most cationic emulsifiers are antistripping agents; consequently, initial bonding is assured. However, the quality of the bond between the bitumen and the aggregate depends on a number of factors, viz:

- the type and amount of emulsifier;
- the bitumen grade and constitution;
- the pH of the emulsifier solution;
- the particle size distribution of the emulsion;
- the aggregate type.

4.3.3 Emulsion viscosity

The viscosity of a bitumen emulsion system is of prime importance because the majority of emulsions are applied in the form of a spray. The distribution of an emulsion from a spray bar is a function of the viscosity of the emulsion. As a consequence the viscosity of both anionic and cationic emulsions is specified in BS 434: Parts 1 and 2: 1984[9,10]. The viscosity of an emulsion is dependent on a number of factors which are discussed in sections 4.5.1 and 4.5.2.

4.4 Classification and specification of bitumen emulsions

Bitumen emulsions are classified in BS 434: Part 1: 1984 by a three part code. The first part of the code, either A or K, indicates anionic or cationic emulsion. The second, 1 to 4, indicates the breaking rate or stability, the higher the number the greater the stability. The third part of the code, 40 to 70, indicates the bitumen content of the emulsion. For example:

K1-70 is a cationic emulsion, rapid setting with a bitumen content of 70%.

A2-50 is an anionic emulsion, semi-stable with a bitumen content of 50%.

BS 434 principally specifies bitumen emulsions in terms of viscosity (in degrees Engler or Redwood II seconds) and bitumen content.

4.5 Modification of bitumen emulsion properties

There is a number of measures an emulsion manufacturer can take to modify the basic properties of a bitumen emulsion, ie viscosity, storage stability, breaking rate and particle size distribution. Contrary to popular belief, the bitumen is not the sole controlling agent, although its influence is, of course, very important.

There are three basic ways in which changes to emulsion properties can be induced:

by changing the grade, concentration or origin of the bitumen;

by changing the emulsion formulation;

by changing the type of emulsifier and/or its concentration.

4.5.1 How to increase the viscosity of an emulsion

- *Increasing the bitumen content*
  This is subject to some limitations; firstly increasing the bitumen content could be considered too expensive and, secondly, if the bitumen content is already high, a small increase in concentration can induce a dramatic viscosity increase, (see figure 4.5).
- *Modification of the aqueous phase*
  The viscosity of a bitumen emulsion is highly dependent on the aqueous phase composition. It has been shown that viscosity can be increased by decreasing the acid content, increasing the emulsifier content or by increasing the neutralisation ratio between the acid and amine contents.
- *Increasing the flow rate through the mill*
  By increasing the flow rate through the mill the particle size distribution of the emulsion will be changed. At bitumen contents less that 65% the viscosity of the emulsion is

virtually independent of flow rate. However, at bitumen contents greater than 65% where the globules of bitumen are packed relatively closely together, inducing a change in the particle size distribution by changing the flow rate has a marked effect on viscosity, as shown in figure 4.6.

- *Decreasing the viscosity of the bitumen*
  If the viscosity of the bitumen entering the colloid mill is lowered the particle size of the emulsion will be reduced which will tend to increase the viscosity of the emulsion.

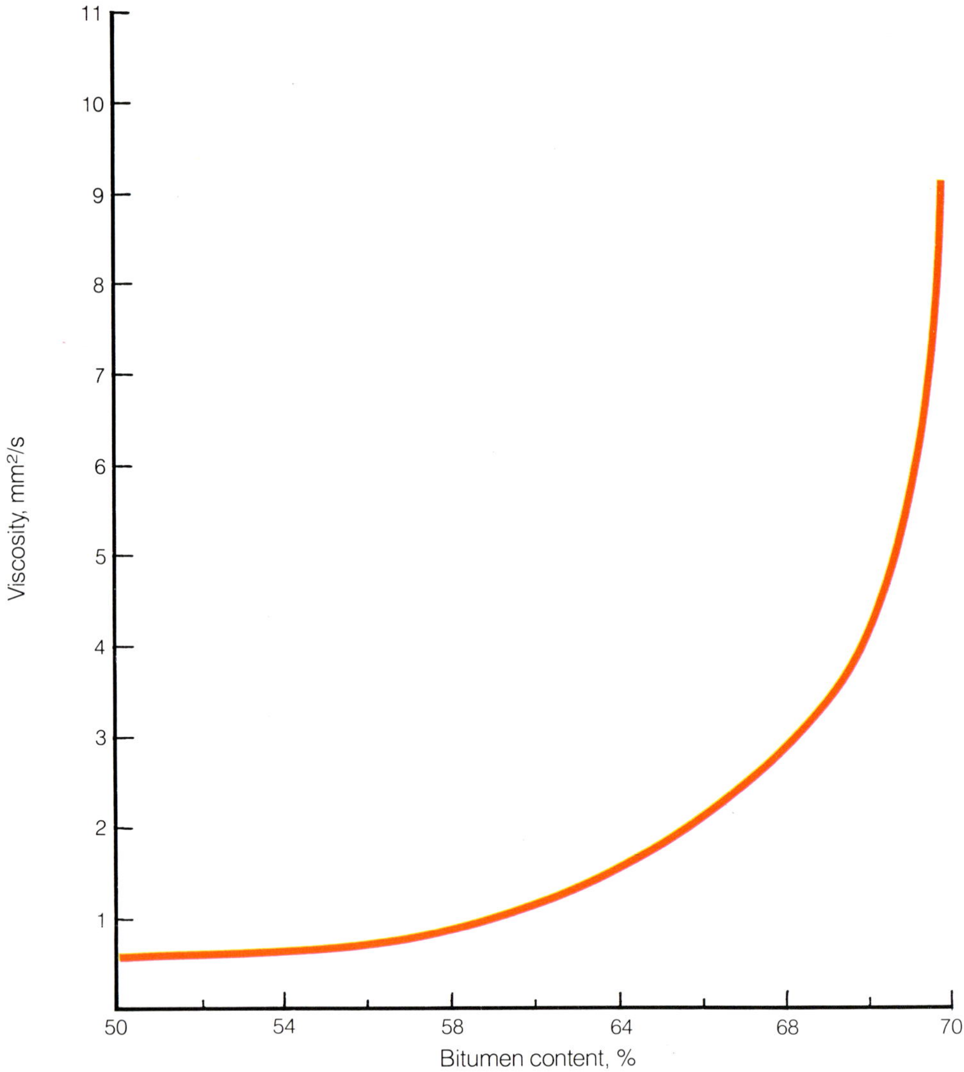

Figure 4.5 – Emulsion viscosity as a function of bitumen content

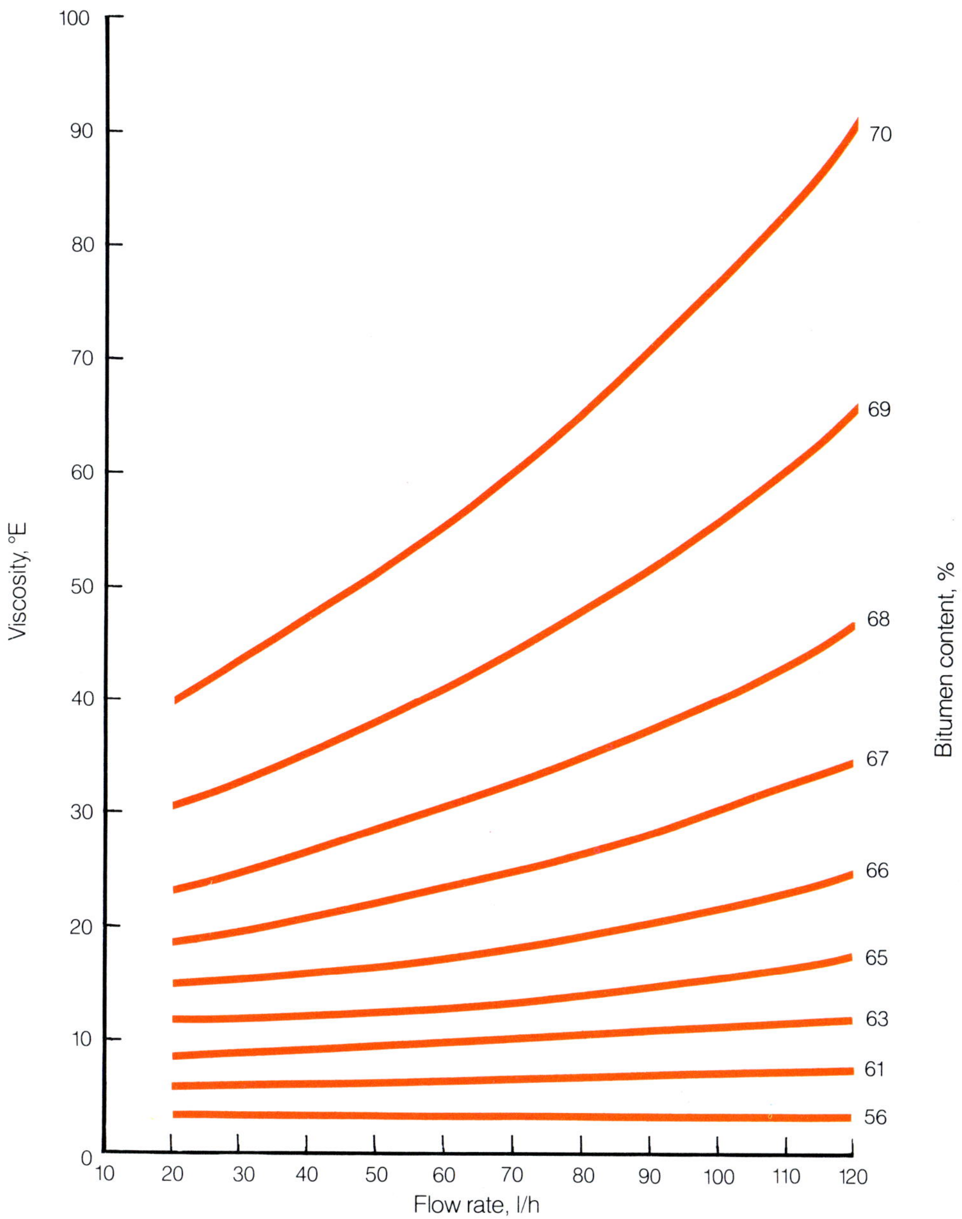

Figure 4.6 – Emulsion viscosity as a function of flow-rate for different bitumen contents

4.5.2 How to decrease the viscosity of an emulsion

- *Reducing the bitumen content*
  This technique is again governed by the specified bitumen content of the emulsion. Below 60% bitumen content the effect is small; the degree of latitude is, of course, governed by the minimum bitumen content.
- *Modification to the emulsification formula*
  To decrease emulsion viscosity it is necessary either to increase the acid content, or to decrease the amine content. However, it should be borne in mind that other emulsion properties are strongly dependent on the aqueous phase composition.
- *Decreasing the flow rate through the mill*
  This is the converse of the theory for increasing emulsion viscosity.

4.5.3 How to change the breaking rate of an emulsion

The breaking rate of an emulsion is strongly dependent on the type of chippings and their size distribution; however, the breaking rate can be modified by:

- *Modification of the aqueous phase composition*
  The breaking rate of a bitumen emulsion has been shown to be increased by reducing the acid content, increasing the emulsifier content or by decreasing the ratio between the acid and emulsifier contents.
- *Increasing the bitumen content*
  Increasing the bitumen content of an emulsion induces an increase in its breaking rate, the size of the effect depending on the aqueous phase composition.
- *Addition of breaking agents*
  The use of breaking agents can accelerate the breaking of an emulsion. For surface dressing emulsions it is possible to spray a chemical breaking agent just after application of the emulsion on the road. It is usually an amphoteric substance. For emulsions used in cold mix preparation, lime can be dispersed on the chippings before mixing. Breaking agents generally have a double role as they may also improve adhesion between the aggregate and the bitumen.
- *Other parameters*
  Several other parameters may influence the breaking rate of emulsion and can be summarised as follows:

  the type of emulsifier;

  the particle size and distribution; the finer the particle size and smaller the dispersion the slower the breaking rate;

  temperature; the higher the ambient temperature the faster the breaking rate of the emulsion.

4.5.4 How to influence the storage stability of an emulsion

It is usual for any deficiency in the storage stability of an emulsion to show itself in the form of settlement for which there are several causes:

- *Bitumen with a high specific gravity*

  Bitumen with a high specific gravity will tend to settle when emulsified; this can be remedied by:

  the specific gravity of the bitumen can be reduced by the addition of cutback kerosine before emulsification; however, this will result in an increased emulsion viscosity and a reduced viscosity of the binder on the road surface;

  the aqueous phase specific gravity can be increased with a salt such as calcium chloride.

- *Emulsion with a low viscosity*

  Low viscosity emulsions are more prone to settlement than high viscosity emulsions because the particles have more freedom to move. The storage stability can be improved by increasing the viscosity as discussed above. Increasing the emulsifier content will also decrease the settlement rate.

- *Electrolyte content of the bitumen*

  The presence of cations in the bitumen can reduce the storage stability of the emulsion. In cationic emulsions a high sodium concentration in the bitumen may induce breaking during storage. This can be counteracted by the addition of a salt to the aqueous phase.

- *Particle size distribution of the emulsion*

  Bitumen emulsions with a broad spectrum of particle sizes are more prone to settlement than those with a relatively narrow size distribution. This is because large particles settle more rapidly due to the repulsion forces between the particles. Therefore emulsions with a relatively narrow particle size distribution are more storage stable.

4.5.5 How to change the particle size distribution of an emulsion

The distribution of the emulsion droplet size is dependent on the interfacial tension between the bitumen and the aqueous phase (the lower the interfacial tension, the easier the bitumen disperses) and on the energy used in dispersing the bitumen. For a given mechanical energy imput, harder bitumens will produce coarser emulsions and high penetration or cutback bitumens will produce finer emulsions. It is possible to influence the particle size and distribution to achieve a finely divided emulsion with a narrow spread of particle sizes:

- *The addition of acid to the bitumen*

  The addition of naphthenic acids to a non-acidic bitumen is important for the production of anionic emulsions. The acids react with the alkaline aqueous phase to form soaps which are surface active and which stabilise the dispersion.

  The addition of naphthenic acids causes a decrease in the mean particle size of the emulsion without changing its size distribution. In addition, the specific surface area of the bitumen emulsion is increased, and consequently, the amount of emulsifier absorbed at the bitumen surface is increased. This results in a reduction in the emulsifier concentration in the aqueous phase and consequently the breaking rate is increased.

- *Manufacturing conditions*

  The manufacturing conditions substantially influence the particle size distribution of the emulsion, viz:

temperature; increasing the temperature of either the aqueous phase or the bitumen, ie reducing its viscosity, normally increases the mean particle size of the emulsion;

bitumen content; increasing the bitumen content increases the mean particle size and tends to reduce the range of particle sizes;

composition of the aqueous phase; for cationic emulsions manufactured with hydrochloric acid and an amine emulsifier the particle size can be decreased by increasing either the acid or emulsifier content; if the ratio of acid to amine is kept constant, the particle size can also be reduced by increasing the amine/acid content; the size distribution does not appear to be related to the concentration of these two components;

operating conditions of the colloid mill; the gap and rotational speed of the colloid mill strongly influence the particle size and distribution of the emulsion; a small gap will result in a small particle size with a relatively narrow range of sizes; high rotational speed will produce a small particle size.

The above is an overview of the factors which influence the properties of emulsions and a number of alternative approaches are available to the emulsion manufacturer to adjust emulsion properties. However, it is clear that it is virtually impossible to adjust one property of the emulsion without influencing others. This interdependence is shown in figure 4.7.

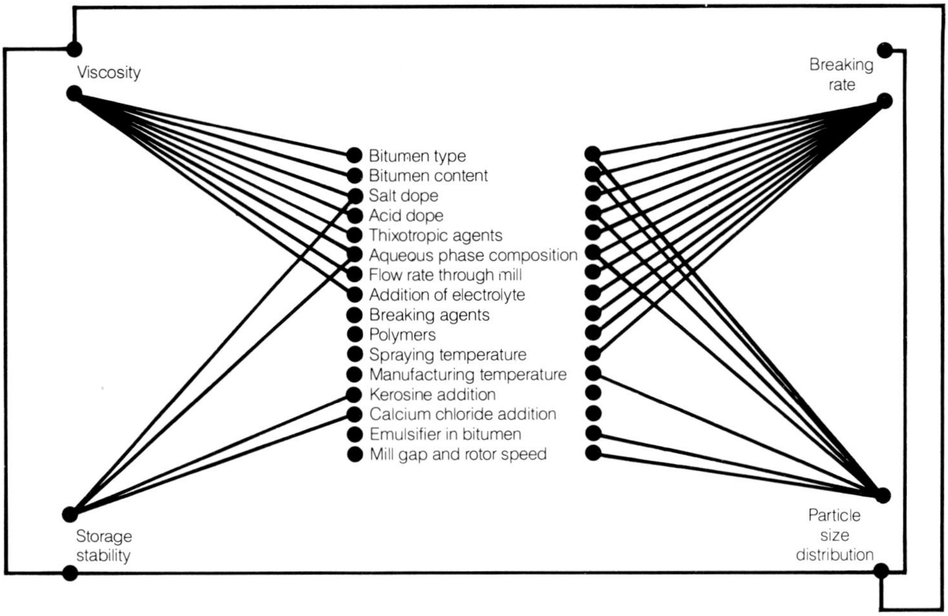

Figure 4.7 – Inter-relationships between manufacturing variables and emulsion properties

4.6 Uses of bitumen emulsions[11]

The vast majority of bitumen emulsions are used in surface dressing applications (chapter 18), but their versatility makes them suitable for a wide variety of different applications from road mixes to binding peat flower pots.

4.6.1 Bitumen emulsions in road mixes[12]

Road mixes manufactured using bitumen emulsions have been used in France since the fifties, and today over one million tonnes of this type of mix are laid annually. The predominant mix type is *grave-emulsion* which is used in the road base. However, some wearing course and basecourse mixes manufactured using emulsions have also been successfully laid.

The major problem with the use of bitumen emulsions in road mixes is that relatively high void contents are needed to allow the water to escape during compaction and service. Furthermore, strength is developed relatively slowly. For both of these reasons emulsion mixes are suitable only for sites carrying modest traffic loads. As a result the material has been used very little in the U.K.

Traditionally, deferred or delayed set macadams have been manufactured using a penetration grade bitumen 'cutback' with various flux oils. These materials are principally used for trench reinstatements and remedial patching. The development of coating grade emulsions[13] has resulted in some deferred set materials being manufactured using bitumen emulsions.

Another application of bitumen emulsions in road mixes is the Retread process[14]. Briefly this process consists of scarifying the existing road to a suitable depth, breaking down the scarified material to the required size and reshaping. The aggregate is then oversprayed 2 or 3 times with a bitumen emulsion. After each application the material is harrowed to distribute the emulsion. When harrowing is complete the material is compacted with an 8 to 10 tonne dead weight roller and finally the surface is sealed with a surface dressing.

Bitumen emulsions are also used for tack coating which is an established technique for ensuring that an adhesive and cohesive bond develops between the layers in a road pavement[15].

4.6.2 Miscellaneous uses of bitumen emulsions[16]

Bitumen emulsions are used for a wide variety of uses not only in road construction; they are also used in civil engineering, horticultural and agricultural applications. Examples of some of these applications are given below and whilst the list is by no means complete it illustrates the versatility of bitumen emulsions.

4.6.2.1 Soil stabilisation[17]

Fresh top-soil on embankments or ploughed agricultural land is susceptible to surface erosion (and slippage on embankments). It is essential therefore to either stabilise the surface with a binding agent or establish a vegetation root system.

Bitumen emulsion sprayed on the surface of the top-soil binds the surface together and assists seed germination by:

retaining moisture in the soil;
improving thermal insulation;
protecting seeds from the elements and birds.

4.6.2.2 Jiffy pots[18]

Peat flower pots, which are filled with a growing medium, are well known in horticulture. One of their attractions is that roots of plants grown in them penetrate through the peat walls so that they can be easily repotted or planted out. The pot used to be enclosed in a plastic mesh to enable it to be handled without disintegrating. This external support has now been replaced by binding the peat with a bitumen emulsion which binds the fibres together giving it strength but at the same time allowing the material to expand when water is absorbed.

4.6.2.3 Slip-coats

Bitumen emulsions are used to create an impermeable membrane between the base and upper layers of concrete, the object being to retain the strength of the upper layer by preventing water seepage into the lower layers. This avoids rigid adhesion between layers of different ages and strengths so that they can cure without setting up internal stresses.

4.6.2.4 Protective coats

Bitumen emulsions are used for protecting buried concrete, pipelines and iron work. To improve the adhesive and cohesive characteristics of the cured binder film, a rubber latex modified emulsion is normally used.

4.6.2.5 Crack filling and grouting[19]

Bitumen emulsions, preferably containing rubber latex, are a relatively inexpensive and effective method of sealing cracks in bituminous materials to stop ingress of water into the structural layers of a pavement. It is important that cracks should be treated as soon as practicable to limit the damage, especially in high rainfall areas or during winter when freeze/thaw cycles of water in cracks can result in further deterioration.

Grouting is a process involving the construction or stabilisation of road surfacings and footpaths whereby emulsion is applied to compacted dry or damp aggregate, the low viscosity of the emulsion permitting in-depth penetration through the interstices of the aggregate. The technique concerns the construction of a combined base/wearing course surfacing from a thickness of 50 mm to 75 mm in one course or up to 100 mm in two courses, as may be specified.

Suitable emulsion types for different applications are given below.

Surface dressing	–	K1-70, K1-60
Tack coats	–	K1-40, K2-40
Slurry seals	–	K3-60
Open-graded macadams	–	K2-60, K2-70
Retreading	–	K2-60
Mist spraying	–	K1-40, K2-40
Concrete curing	–	K1-40, K2-40

4.7 References

(1) van Westrum, S: German Patent 173.639, 1906.

(2) Albert, K and Berend, L: *Chem. Fabrik.* Austrian Patent 72451, 1916.

(3) Bradshaw, L C: *Paint Technology*, vol 24, pp 19-23, 1960.

(4) Schwitzer, M K: *Chemistry and Industry*, vol 21, pp 822-831, 1972.

(5) Mogel, R B: *Encycl. Chem. Process Des.*, vol 4, pp 1-23, 1977.

(6) Stokes, G G: *Trans. Cambridge Phil. Soc.*, vol 9, 1851.

(7) Highways Research Board, Special Report No 98, Washington, 1968.

(8) Heukelom, W and Wijga, P W O: *Bitumen testing*. Koninklijke/Shell Laboratorium Amsterdam.

(9) British Standard 434: Part 1: 1984: *Bitumen road emulsions (anionic and cationic). Part 1. Specification for bitumen road emulsions.*

(10) British Standard 434: Part 2: 1984: *Bitumen road emulsions (anionic and cationic). Part 2. Code of practice for use of bitumen and emulsions.*

(11) Road Emulsion Association: *Bitumen road emulsion*. Technical data sheet No 1.

(12) Bradshaw, L C: *Bitumen emulsions in road mixes*. Shell Bitumen Review No 45, pp 8-11, 1974.

(13) Road Emulsion Association: *Coating emulsions*. Technical data sheet No 11.

(14) Road Emulsion Association: *The Retread Process*. Technical data sheet No 10.

(15) Road Emulsion Association: *Tack coating*. Technical data sheet No 5.

(16) Road Emulsion Association: *Miscellaneous uses of bitumen road emulsions*. Technical data sheet No 12.

(17) Buchta, H: *Terrafix in agricultural applications*. Shell Bitumen Review No 61, pp 7-8, 1986.

(18) Melvold, O: *Bitumen in horticulture*. Shell Bitumen Review No 52, p 20, 1975.

(19) Road Emulsion Association: *Patching and grouting*. Technical data sheet No 8.

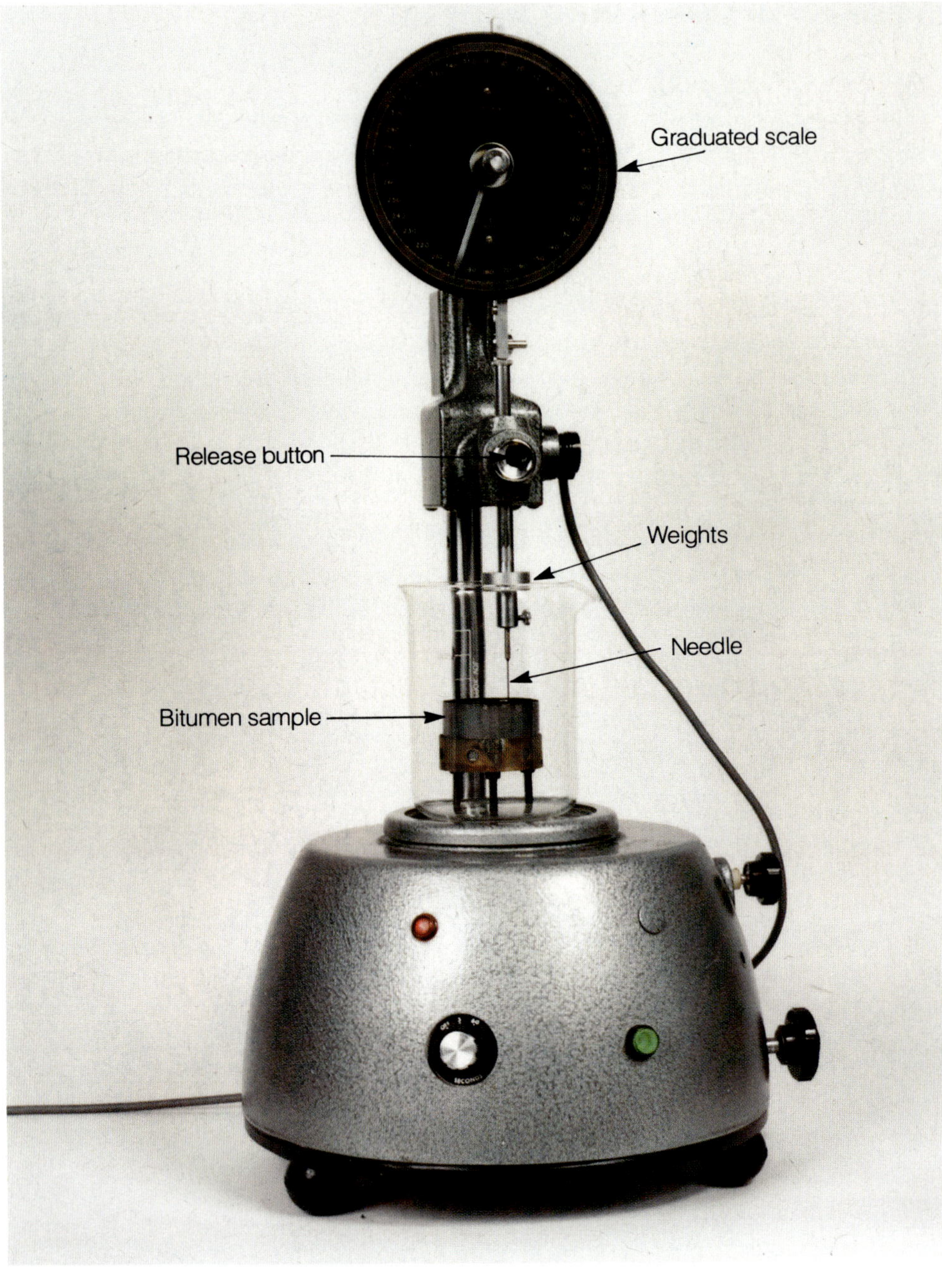

Photograph 5.1 – The penetration test

Chapter 5 – Mechanical testing and properties of bitumen

A wide range of tests is performed on bitumens, from specification tests to more fundamental rheological and mechanical tests. Bitumen is a complex material with a complex response to stress. The response of a bitumen to stress is dependent on both temperature and loading time. Thus, the nature of the test and what it indicates about the properties of a bitumen must be interpreted in relation to the nature of the material.

5.1 Standard specification tests for bitumens

As an almost infinite variety of bitumens could be manufactured, it is necessary to have tests which can characterise different grades. The two tests used in the U.K. to characterise bitumens are penetration and softening point. These two tests are used to specify different grades of bitumen. Although they are arbitrary empirical tests, it is possible to estimate important engineering properties, such as high temperature viscosity and low temperature stiffness of the bitumen from these two simple tests (see sections 5.4 and 5.6.2.2). The use of the penetration test for characterising the consistency of bitumen dates from the late 19th century(1).

As the penetration and softening point tests are empirically derived it is essential that they are always carried out under exactly the same conditions. The Institute of Petroleum (IP)(2), the American Society for Testing and Materials (ASTM)(3) and British Standards (BS)(4,5) publish standard methods of testing bitumen. In many cases the methods are identical and therefore methods are published jointly. However, some methods differ in detail, for example the IP and ASTM softening point method, and in these cases a correction factor is provided to relate test results obtained using these two test methods.

The majority of the methods quote limits for assessing the acceptability of test results. Limits of variability of results obtained by a single operator (repeatability) and by different operators in different laboratories (reproducibility) are specified. Thus tolerance is given to allow for differences between operators and equipment at different locations.

5.1.1 The penetration test *(IP 49, ASTM D 5, BS 2000 : Part 49)*

The consistency of a penetration or oxidised bitumen is measured by the penetration test. In this test a needle of specified dimensions is allowed to penetrate into a sample of bitumen, under a known load (100 g), at a fixed temperature (25°C), for a known time (5 seconds) (see photograph 5.1). The distance the needle penetrates, in units of decimillimetre, dmm (0.1 mm), is termed the penetration. Therefore the greater the penetration of the needle the softer the bitumen. Penetrations less than 2 and greater than 500 cannot be determined with accuracy and even within this range the specified procedure has to be followed closely to obtain reliable results. This test is the basis upon which penetration grade bitumens are classified into standard penetration ranges.

It is essential that the test methods are followed precisely, as even a slight variation can cause large differences in the result. The most common errors are: poor sampling and sample preparation; badly maintained apparatus and needles; and incorrect temperature and timing. Temperature control is *critical*, control to ±0.1°C is essential. Needles must be checked regularly for straightness, correct profile and cleanliness. Automatic timing devices are needed for accuracy but these must be checked regularly.

For each test three individual measurements of penetration are made. The average of the three values is recorded to the nearest whole unit. The recorded penetration is reported if the difference between the individual three measurements does not exceed a specified limit.

The acceptable repeatability and reproducibility quoted in IP49 for the penetration test are as follows:

- *Repeatability*
  If the penetration is less than 50 – 1 dmm
  If the penetration is greater than 50 – 3% of the mean of the two results

- *Reproducibility*
  If the penetration is less than 50 – 4 dmm
  If the penetration is greater than 50 – 8% of the mean of the two results.

5.1.2 The softening point test *(IP 58, ASTM D 36, BS 2000: Part 58)*

The consistency of a penetration or oxidised bitumen can also be measured by determining its softening point. In this test a steel ball (3.5 g) is placed on a sample of bitumen contained in a brass ring; this is suspended in a water or glycerol bath (see photograph 5.2). Water is used for bitumens with a softening point of 80°C or below, and glycerol is used for softening points greater than 80°C. The bath temperature is raised at 5°C per minute, the bitumen softens and eventually deforms slowly with the ball through the ring. At the moment the bitumen and steel ball touch a base plate 25 mm below the ring, the temperature of the water is recorded. The test is performed in duplicate and the mean of the two measured temperatures is reported, to the nearest 0.2°C for a penetration grade bitumen and 0.5°C for an oxidised bitumen. If the difference between the two results exceeds 1.0°C the test must be repeated. The reported temperature is designated the softening point of the bitumen, and represents an equi-viscous temperature. In the ASTM version of the softening point test the bath is not stirred, whereas in the IP version the water or glycerol is stirred; consequently the softening points determined by using these two methods differ. The ASTM results are generally 1.5°C higher than for the IP or BS method[6].

As with the penetration test, the procedure for carrying out the softening point test must be followed precisely to obtain accurate results. Sample preparation, rate of heating and accuracy of temperature measurement are critical. For example it has been shown[7] that varying the heating rate from 4.5°C to 5.5°C, the tolerance permitted by the IP 58 test method can result in a difference of 1.6°C in the measured softening point. Automatic softening point machines are available which ensure close temperature control and which automatically record the result at the end of the test.

The acceptable repeatability and reproducibility quoted in IP 58 for the softening point test is as follows:

- *Repeatability*

  For penetration grade bitumens: 1.0°C

  For oxidised bitumens:

Softening point, °C	*Repeatability, °C*
Under 80	1.5
Over 80-100	2.0
Over 100-120	2.5
Over 120-140	3.0

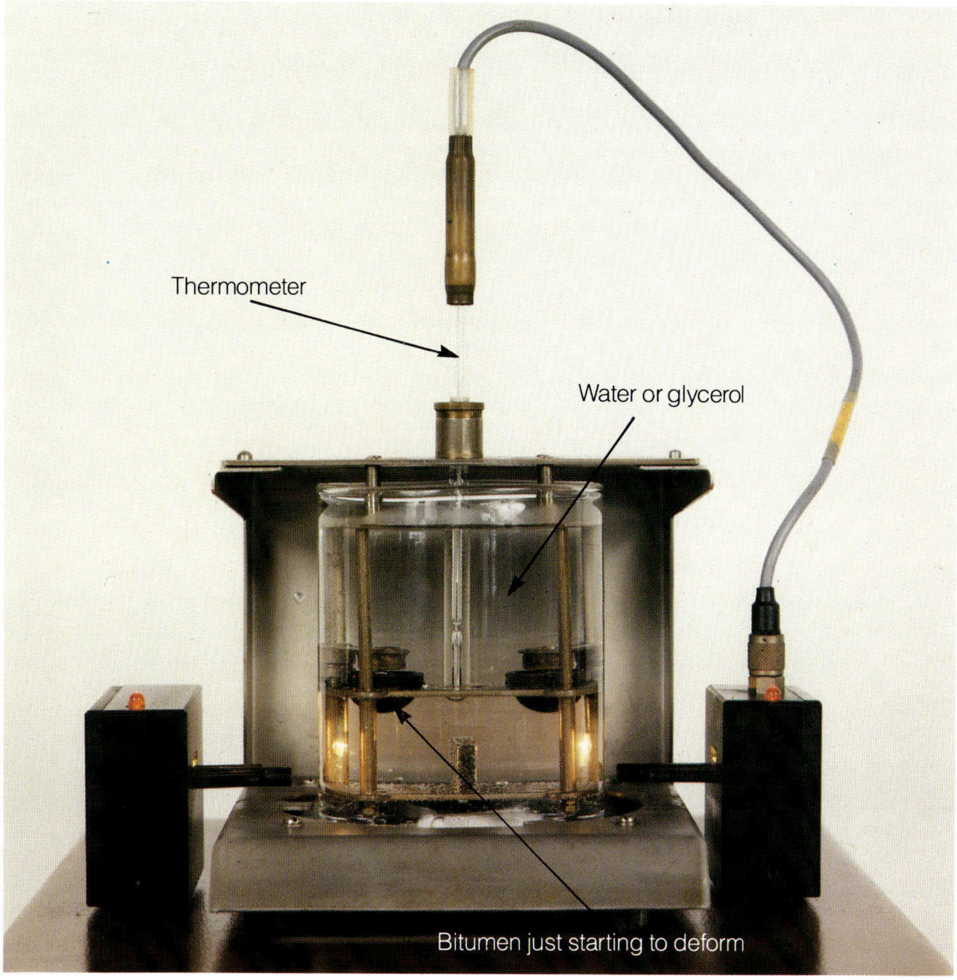

Photograph 5.2 – The softening point test

- *Reproducibility*

For penetration grade bitumens:	2.5°C
For oxidised bitumens:	5.5°C

The consistency of bitumen at the softening point temperature has been measured by Pfeiffer and Van Doormaal[(8)] in terms of penetration. Using a specially prepared, extra-long penetration needle they found a value of 800 pen for many, but not all bitumens. The exact value was found to vary with penetration index and wax content. It has also been demonstrated by direct measurement that the viscosity at the softening point temperature of the majority of bitumens is approximately 1200 Pa.s (12,000 poise).

5.2 The Fraass breaking point test

The Fraass breaking point test (IP 80) is one of very few tests which can be used to describe the behaviour of bitumens at very low temperature (as low as –30°C). The test was developed by A. Fraass[9] in 1937. It is essentially a research tool which determines the temperature at which the bitumen reaches a critical stiffness and cracks. However, a number of countries with very low winter temperatures, for example Canada, Finland, Norway and Sweden, have maximum allowable Fraass temperatures for individual grades of bitumen.

In the Fraass test (see photograph 5.3) a steel plaque 41 mm x 20 mm coated with 0.5 mm of bitumen is slowly flexed and released. The temperature of the plaque is reduced at 1°C per minute until the bitumen reaches a critical stiffness and cracks. The temperature at which the sample cracks is termed the breaking point and represents an equi-stiffness temperature. It has been shown that at fracture the bitumen has a stiffness of 2.1×10^9 Pa[10] which is approaching the maximum stiffness of 2.7×10^9 Pa (see section 5.6.4). The Fraass temperature can be predicted from penetration and softening point because it is equivalent to the temperature at which the bitumen has a penetration of 1.25.

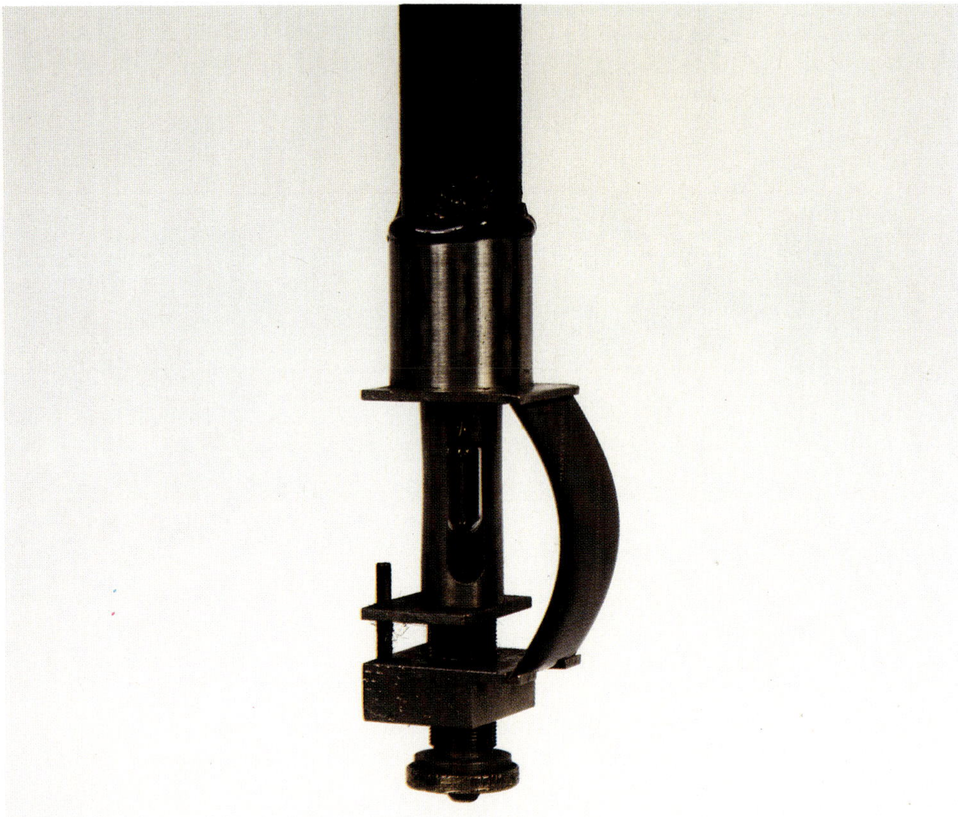

Photograph 5.3 – The Fraass breaking point test

Photograph 5.4 – A kinematic viscometer

5.3 Viscosity

This is a fundamental characteristic of a bitumen as it determines how the material will behave at a given temperature and over a temperature range. The basic unit of viscosity is the Pascal second (Pa.s). The absolute or dynamic viscosity of a bitumen, measured in Pascal seconds, is the shear stress applied to a sample of bitumen in Pascals divided by the shear rate per second; 1 Pa.s = 10 P (poise). The absolute viscosity of a bitumen can be measured using a sliding plate viscometer.

Viscosity can also be measured in units of m^2/s, or more commonly mm^2/s ($1\ mm^2/s$ = 1 centistoke (cSt)). These units relate to kinematic viscosity. Kinematic viscosity is measured using a capillary tube viscometer. Kinematic viscosity is related to dynamic viscosity by the expression:

$$\text{Kinematic viscosity} = \frac{\text{dynamic viscosity}}{\text{mass density}}$$

For many purposes it is usual to measure the viscosity of a bitumen by measuring the time required for a fixed quantity of material to flow through a standard orifice. These methods are useful for specification and comparative purposes and, if required, the results can be converted into more fundamental units of viscosity.

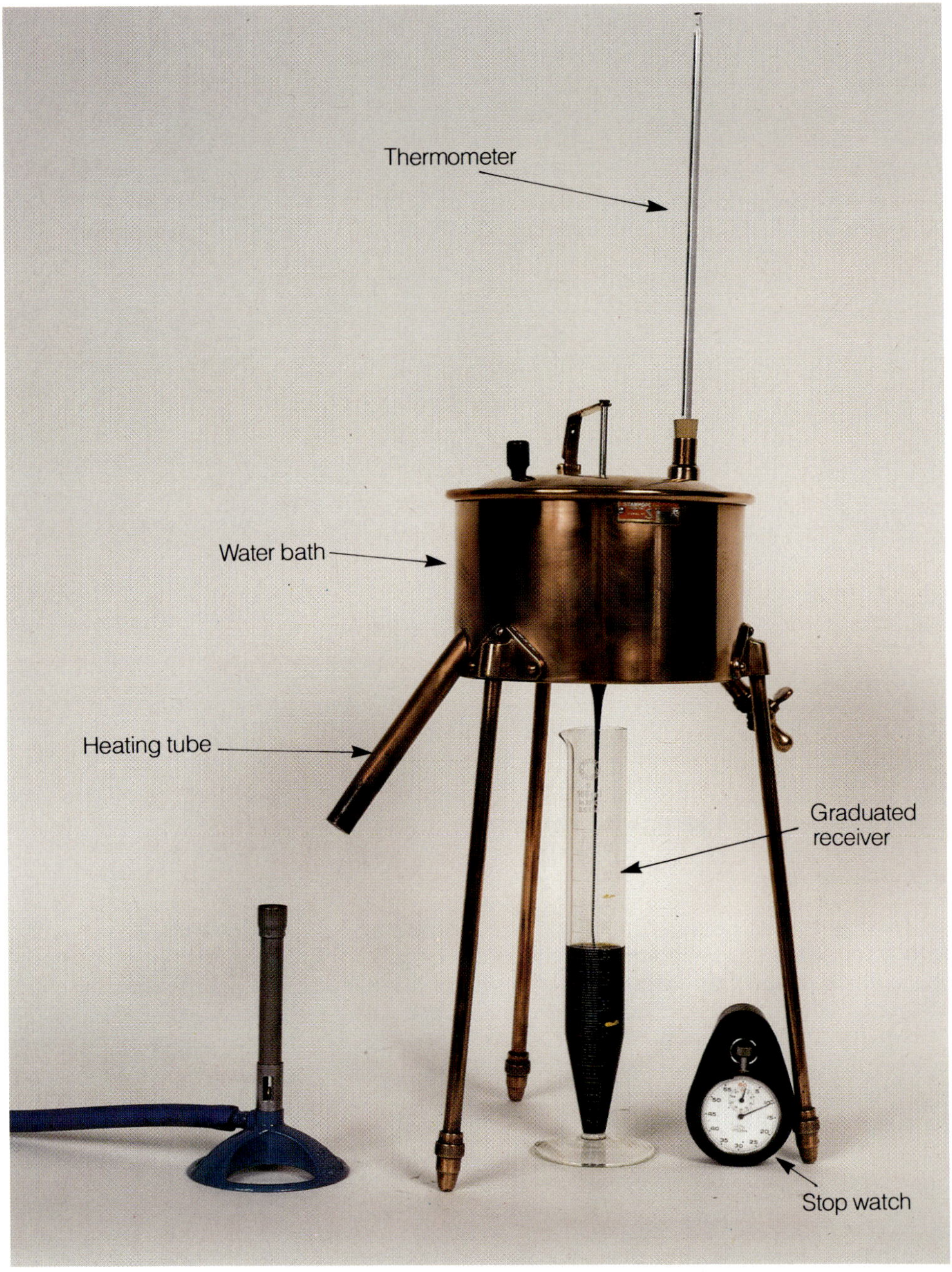

Photograph 5.5 – The standard tar viscometer

5.3.1 *The sliding plate viscometer*

A fundamental method of measuring viscosity is the sliding plate viscometer. This apparatus applies the definition of absolute or dynamic viscosity, ie it takes the shear stress (Pa) applied to a film of bitumen (5 to 50 microns thick) sandwiched between two flat plates and measures the resulting rate of strain (seconds^{-1}). The viscosity in Pascal seconds (Pa.s) is given by shear stress divided by rate of strain. The apparatus comprises a loading system which applies a uniform shear stress during a measurement, and a device to produce a record of the flow as a function of time. Generally the records show a curved toe due to visco-elastic behaviour, passing into a linear form whose slope is a direct measure of viscosity. Depending on the load and the size of the specimen, viscosities in the range of 10^5 to 10^9 Pa.s can be measured. A special feature is that the shear stress is the same throughout the specimen. Therefore this instrument is well suited to investigate the phenomenon of shear stress dependence. Shear stress dependence is the variation in measured viscosity with the magnitude of the applied shear stress.

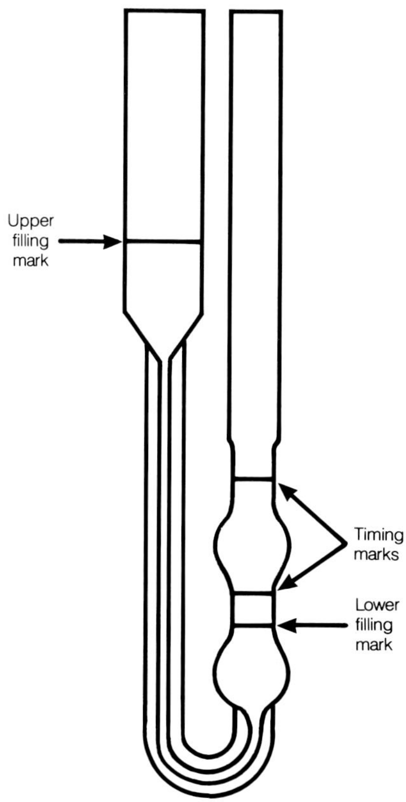

Figure 5.1 – U tube reverse flow viscometer

5.3.2 Capillary viscometers

Capillary viscometers are essentially narrow glass tubes through which the bitumen flows. The tube has narrower and wider parts and is provided with two or more marks to indicate a certain volume or flow (see figure 5.1 and photograph 5.4). The measurement of kinematic viscosity is made by timing the flow of bitumen through a glass capillary viscometer at a given temperature. Each viscometer is calibrated and the product of flow time and viscometer calibration factor gives the kinematic viscosity in mm²/s. Temperature/viscosity curves are obtained by measuring kinematic viscosity at a number of different temperatures, and plotting the viscosity on a logarithmic scale against temperature on a linear scale (see section 5.4 and figure 5.2).

5.3.3 Cup viscometers

Cutback bitumens and emulsions are both specified by viscosity. Cutback bitumens are specified using the standard tar viscometer[11] (STV, see photograph 5.5). The viscosity of emulsions can be determined using either the Engler viscometer or the Redwood II viscometer[12]. In these tests a metal cup is filled with cutback bitumen or emulsion at a standard temperature, and the time is recorded in seconds for a standard volume of material to flow out through the orifice in the bottom of the cup.

There are several cup type viscometers available which differ mainly in the size of opening through which the bitumen is drained. Since the stress is provided by gravity, the absolute viscosity, η, in Pa.s is given by:

$$\eta = \text{flow time x density x constant}$$

where the flow time is expressed in seconds

the density in g/ml

the constant depends on the instrument used, viz:

Viscometer	*Constant*
Saybolt Universal	0.000218
Redwood I	0.000247
Saybolt Furol	0.00218
Redwood II	0.00247
Engler	0.00758
Standard Tar Viscometer (4 mm)	0.0132
Standard Tar Viscometer (10 mm)	0.400

The test results may also be expressed as the kinematic viscosity, υ, in m²/s, or more commonly mm²/s, defined as:

$$\upsilon = \text{flow time x constant.}$$

5.4 The bitumen test data chart[13,14]

In the late sixties Heukelom developed a system to enable penetration, softening point, Fraass and viscosity data to be described as a function of temperature on one chart as shown in figure 5.2. This is known as the bitumen test data chart (BTDC). The chart consists of one horizontal scale for the temperature and two vertical scales for the penetration and viscosity. The temperature scale is linear,

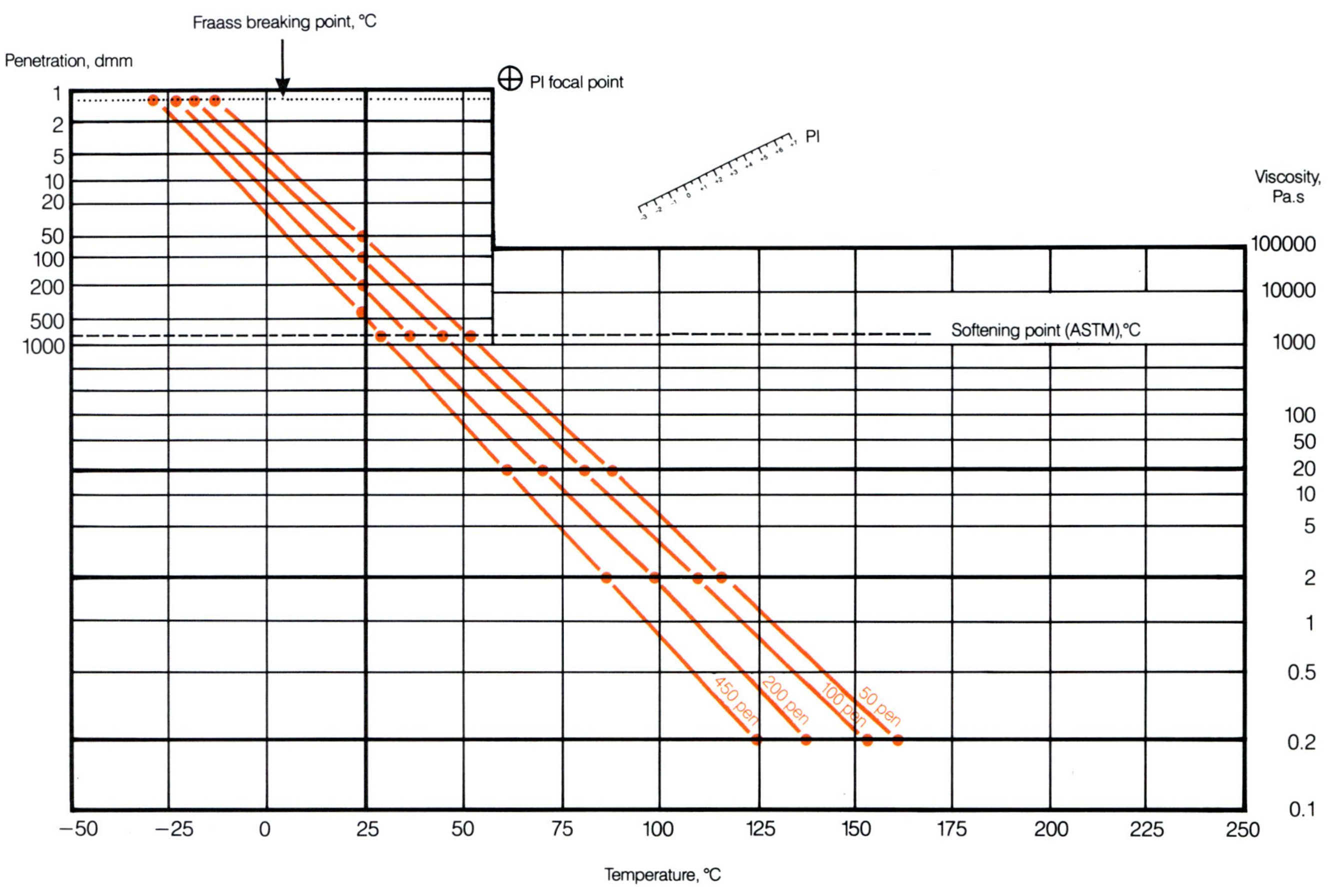

Figure 5.2 – A bitumen test data chart comparing penetration grade bitumens manufactured from one crude[14]

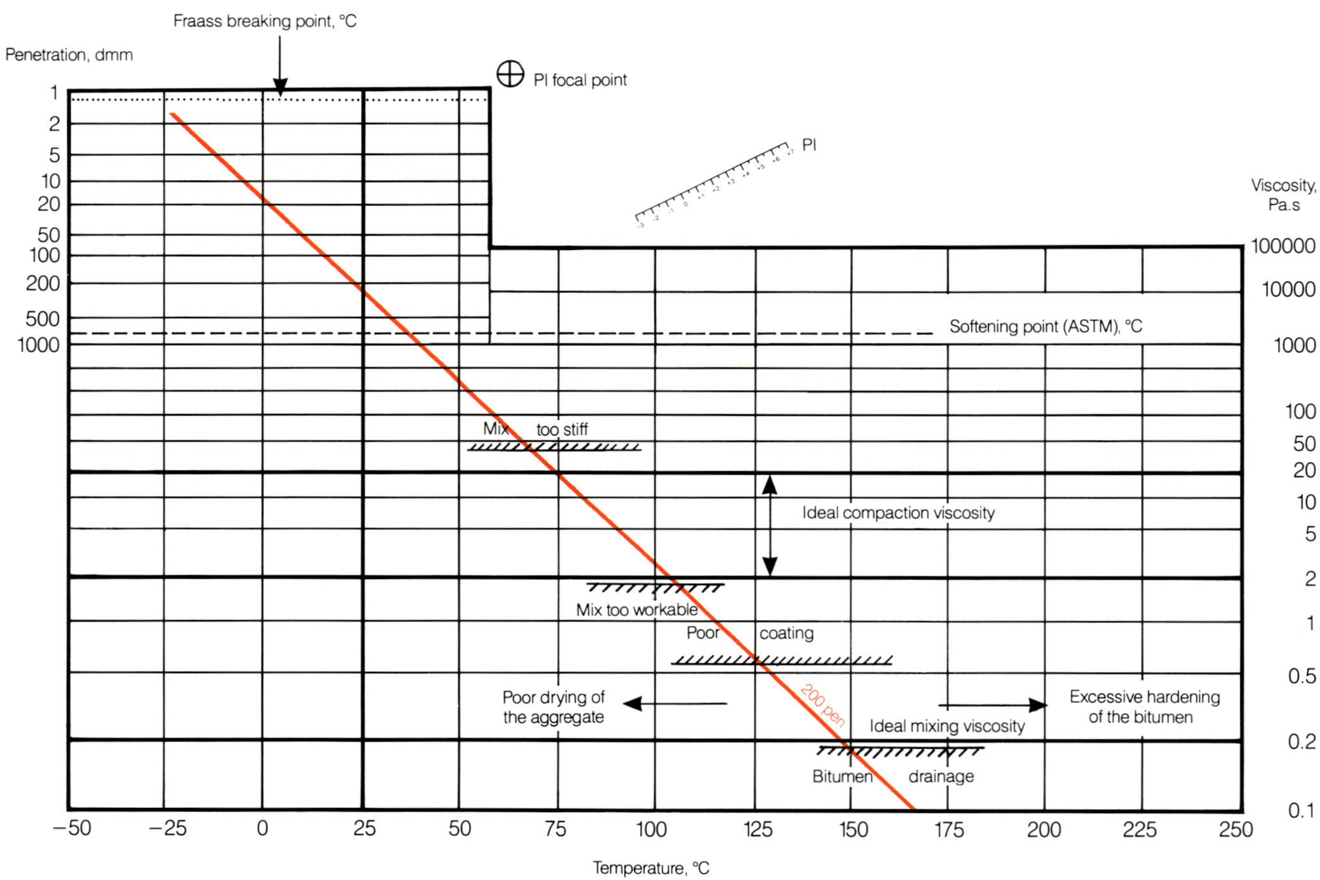

Figure 5.3 – A bitumen test data chart showing 'ideal' bitumen viscosities for optimal mixing and compaction of a dense bitumen macadam

the penetration scale is logarithmic and the viscosity scale has been devised so that penetration grade bitumens with 'normal' temperature susceptibility or penetration indices (PIs, see section 5.5) give straight-line relationships.

The BTDC shows how the viscosity of a bitumen depends on temperature, but it does not account for loading time. However, as the loading times for penetration, softening point and the Fraass breaking point tests are similar, these test data can be plotted with viscosity test data on the BTDC as shown in figure 5.2. As the test results on this chart form a straight-line relationship it is possible to predict the temperature/viscosity characteristics of a penetration grade bitumen, over a wide range of temperatures, from only the penetration and softening point.

During the manufacture and compaction of a bituminous mix there are optimum bitumen viscosities. This is illustrated in figure 5.3 for a dense bitumen macadam mix manufactured using 100 pen bitumen. If the viscosity of the bitumen is too high during mixing, the aggregate will not be properly coated and if the viscosity is too low, the bitumen will coat the aggregate easily but may drain off the aggregate during storage or transportation. For satisfactory coating the viscosity should be approximately 0.2 Pa.s.

During compaction, if the viscosity is too low, the mix will be excessively mobile, resulting in pushing of the material in front of the roller. High viscosities will significantly reduce the workability of the mix and little compaction will be achieved. It is widely recognised that the optimal bitumen viscosity for compaction is between 2 Pa.s and 20 Pa.s.

Thus the BTDC is a useful tool to ensure that appropriate operating temperatures are selected to achieve the appropriate viscosity for any grade of bitumen. Consideration of the viscosity requirements during manufacture and application has led to the operating temperatures given in figure 5.3. Particular mixes or circumstances may obviously dictate other operating temperatures.

The chart can also be used for comparing the temperature/viscosity characteristics of different types of bitumens. Three classes of bitumen can be distinguished using the BTDC:

5.4.1 Class S bitumens

The test data for a large group of bitumens can be represented by straight lines on the BTDC. This group, which has been designated Class S (S for straight line), comprises penetration bitumens of different origin with limited wax content. Figure 5.2 shows a chart with straight lines for a number of different penetration grade bitumens manufactured from one crude oil. The lines move towards the left of the chart as the bitumens become softer, but their slopes are equal, indicating that temperature susceptibilities are similar. Bitumens with the same penetration at 25°C, but from different origins, are shown in figure 5.4. The origin may influence the temperature susceptibility, which is reflected by the slope of the line. Therefore the temperature/viscosity characteristics of S type bitumens may be determined from their penetration and softening point only.

5.4.2 Class B bitumens

The test data of Class B (B for blown) bitumens give curves on the chart as shown in figure 5.5. The curves can be represented by two intersecting straight lines. The slope of the line in the high temperature range is about equal to that of an unblown bitumen of the same origin, but the line in the lower temperature range is less steep. Physically, there is no transition point, but it is very convenient that they are still straight lines in the penetration and viscosity regions; each of them can be characterised with two test values; thus, in all, four tests are required for a complete description: penetration, softening point and two high temperature viscosity measurements.

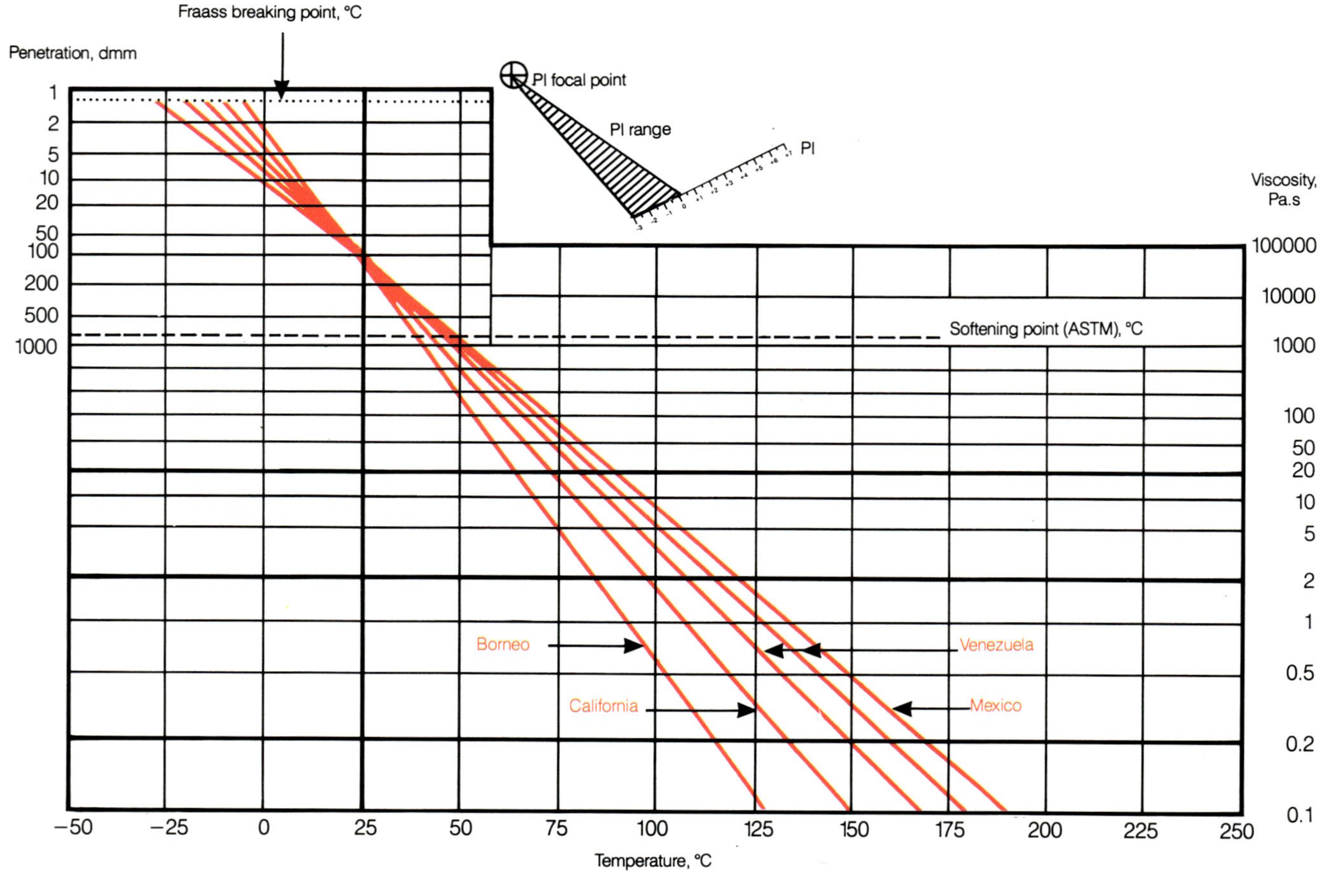

Figure 5.4 – A bitumen test data chart comparing several 100 pen bitumens manufactured from different crudes[14]

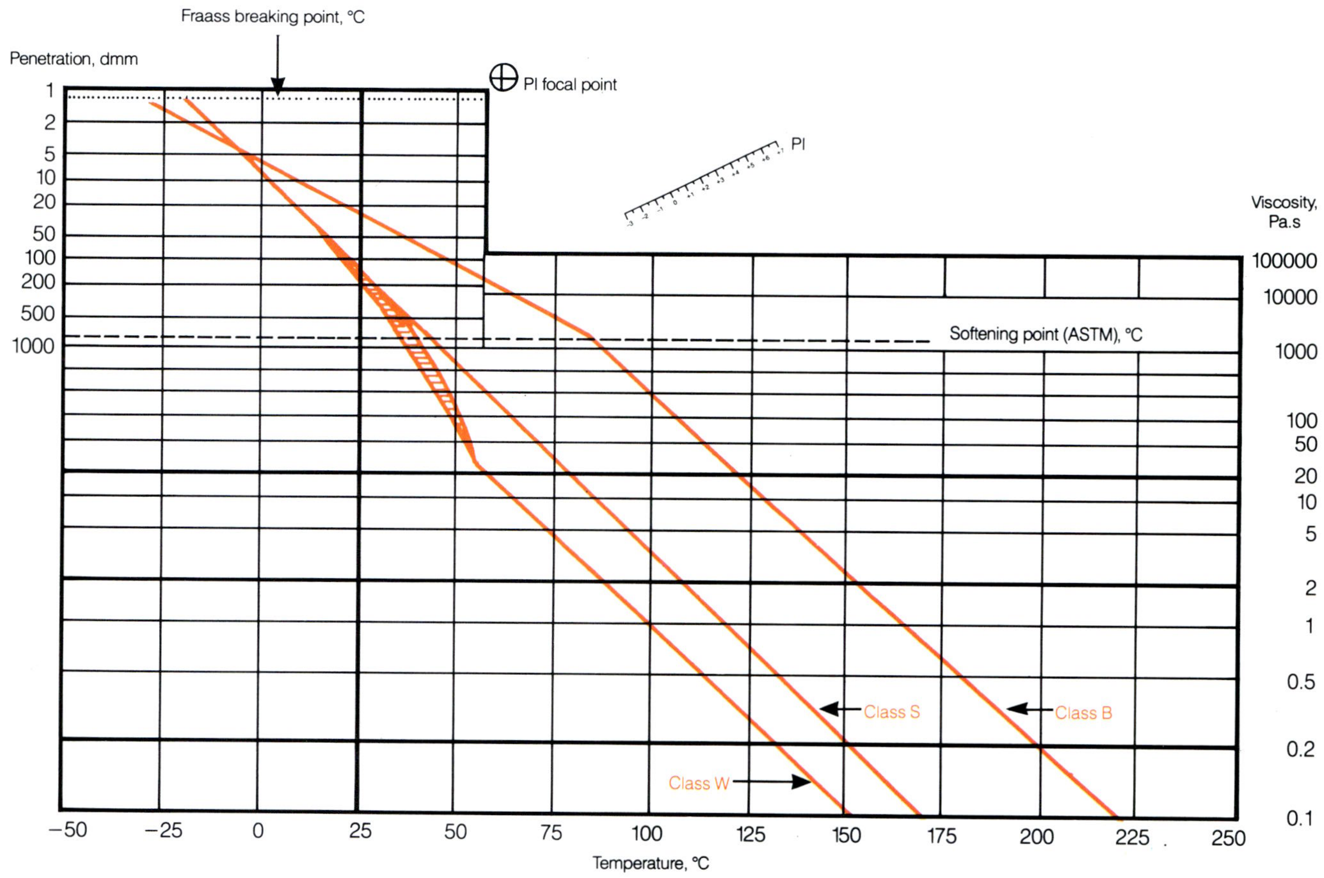

Figure 5.5 – A bitumen test data chart comparing class S, B and W type bitumens[14]

5.4.3 *Class W bitumens*

Class W (W for waxy) bitumens also give curves consisting of two straight lines; however, these are dissimilar to those of blown bitumens. The two branches of the curve give nearly equal slopes, but are not aligned. Figure 5.5 shows an example of an S type bitumen together with a curve for a similar bitumen with a wax content of 12 %. At low temperatures, when the wax is crystallised, there is hardly any difference between the two curves. At higher temperatures, when the wax is molten, the curve for the waxy bitumen is considerably lower down on the chart. Between the two straight branches there is a transition range in which the test data are scattered because the thermal history of the sample influences the viscosity result obtained over this range of temperatures.

5.5 Temperature susceptibility – penetration index (PI)

All bitumens display thermoplastic properties, ie they become softer when heated and harder when cooled. Several expressions exist to define the way the viscosity (or consistency) changes with the temperature. One of the best known is that developed by Pfeiffer and Van Doormaal(8). If the logarithm of penetration, P, is plotted against temperature, T, a straight line is obtained such that:

$$\log P = AT + K$$

where A is the temperature susceptibility of the logarithm of the penetration

K is a constant.

The value of A varies from about 0.015 to 0.06 showing that there may be a considerable difference in temperature susceptibility. Pfeiffer and Van Doormaal developed an expression for the temperature susceptibility which would assume a value of about zero for road bitumens. For this reason they defined penetration index (PI) as:

$$\frac{20 - PI}{10 + PI} = 50\,A$$

or explicitly

$$PI = \frac{20\,(1 - 25A)}{1 + 50A}$$

The value of PI ranges from about –3 for highly temperature susceptible bitumens to about +7 for highly blown low temperature susceptible (high PI) bitumens. The PI is an unequivocal function of A and hence it may be used for the same purpose. The values of A and PI can be derived from penetration measurements at two temperatures, T_1 and T_2, using the equation:

$$A = \frac{\log \text{pen at } T_1 - \log \text{pen at } T_2}{T_1 - T_2}$$

The consistency at the softening point can be expressed in terms of penetration, both by linear extrapolation of logarithm of pen vs temperature and by direct measurement with an extra long penetration needle at the ASTM softening point temperature. Pfeiffer and Van Doormaal found that at the ASTM softening point temperature most bitumens had a penetration of about 800. Replacing T_2 in the above equation by the ASTM softening point temperature and the penetration at T_2 by 800 they obtained the equation:

$$A = \frac{\log \text{pen } T_1 - \log 800}{T_1 - \text{ASTM softening point}}$$

Substituting this equation in the above PI equation and assuming a penetration test temperature of 25°C, we have:

$$PI = \frac{1952 - 500 \log pen - 20\, SP}{50 \log pen - SP - 120}$$

The assumption of a penetration of 800 at the softening point temperature is not valid for all bitumens. It is thus advisable to calculate the temperature susceptibility using penetration at two temperatures, T_1 and T_2.

The nomographs shown in figures 5.6 and 5.7 enable the PI to be deduced approximately from either the penetration at 25°C and the softening point temperature, or the penetration of the bitumen at two different temperatures. Due to the spread of the actual value of penetration at the softening point temperature, the value of PI calculated from one penetration and one softening point may vary from the precise value calculated from two penetrations. However, since penetration at 25°C and softening point are generally determined for bitumen specification control, it is normally the case that these figures are used.

One drawback of the PI system is that it uses the change in bitumen properties over a relatively small range of temperatures to characterise bitumen. Extrapolations to extremes of temperature can sometimes be misleading. PI can be used to give a good approximation of the behaviour to be expected but confirmation by stiffness or viscosity measurements is desirable.

5.6 Engineering properties of bitumen

Bitumen is widely used as a construction material in civil engineering, building and industrial fields, but its mechanical properties are more complex than materials such as steel or cement concrete. However, the mechanical properties of bitumens, which are of considerable value to both the civil engineer and the industrial bitumen user, can be defined in terms which are analogous to the elasticity moduli of solid materials.

5.6.1 The stiffness concept

Bitumens are visco-elastic materials and their deformation under stress is a function of both temperature and loading time. At high temperatures or long times of loading they behave as viscous liquids, whereas at very low temperatures or short times of loading they behave as elastic (brittle) solids. The intermediate range of temperature and loading times, more typical of conditions in service, results in visco-elastic behaviour.

In order to define the visco-elastic properties, Van der Poel[15], in 1954, introduced the concept of stiffness modulus as a fundamental parameter to describe the mechanical properties of bitumens, by analogy with the elastic modulus of solids. In this latter case, if a tensile stress, σ, is applied at loading time $t = 0$, a strain, ε, is instantly attained, which does not increase with loading time. The elastic modulus, E, of the material is expressed by Hooke's law:

$$E = \frac{\sigma}{\varepsilon}$$

In the case of visco-elastic materials such as bitumen, a tensile stress σ, applied at a loading time $t = 0$, causes a strain ε_t, which increases, but not proportionally with loading time. The stiffness modulus, S_t, at loading time t, is defined as the ratio between the applied stress and the resulting strain at loading time t:

$$S_t = \frac{\sigma}{\varepsilon_t}$$

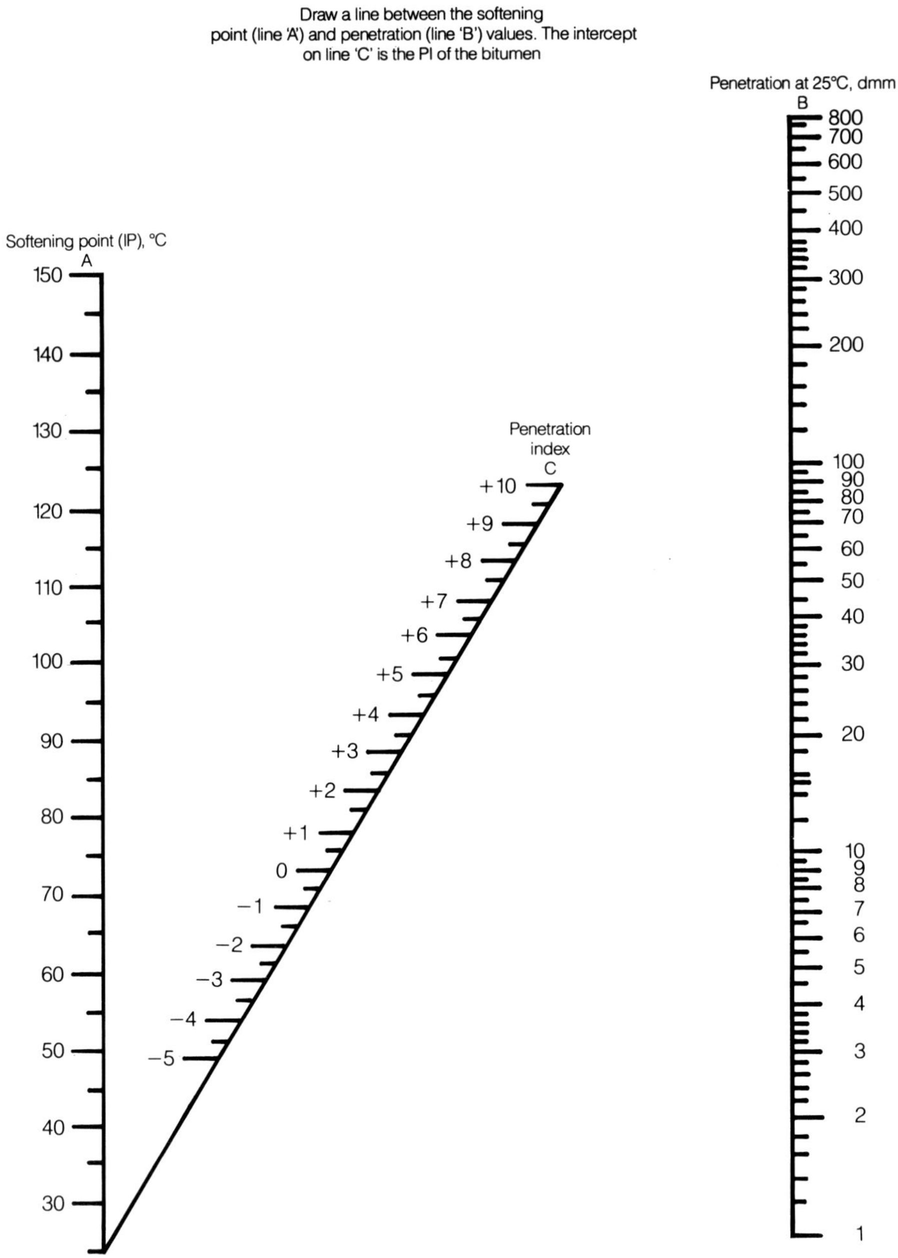

Figure 5.6 – Nomograph for penetration index (SP/Pen)

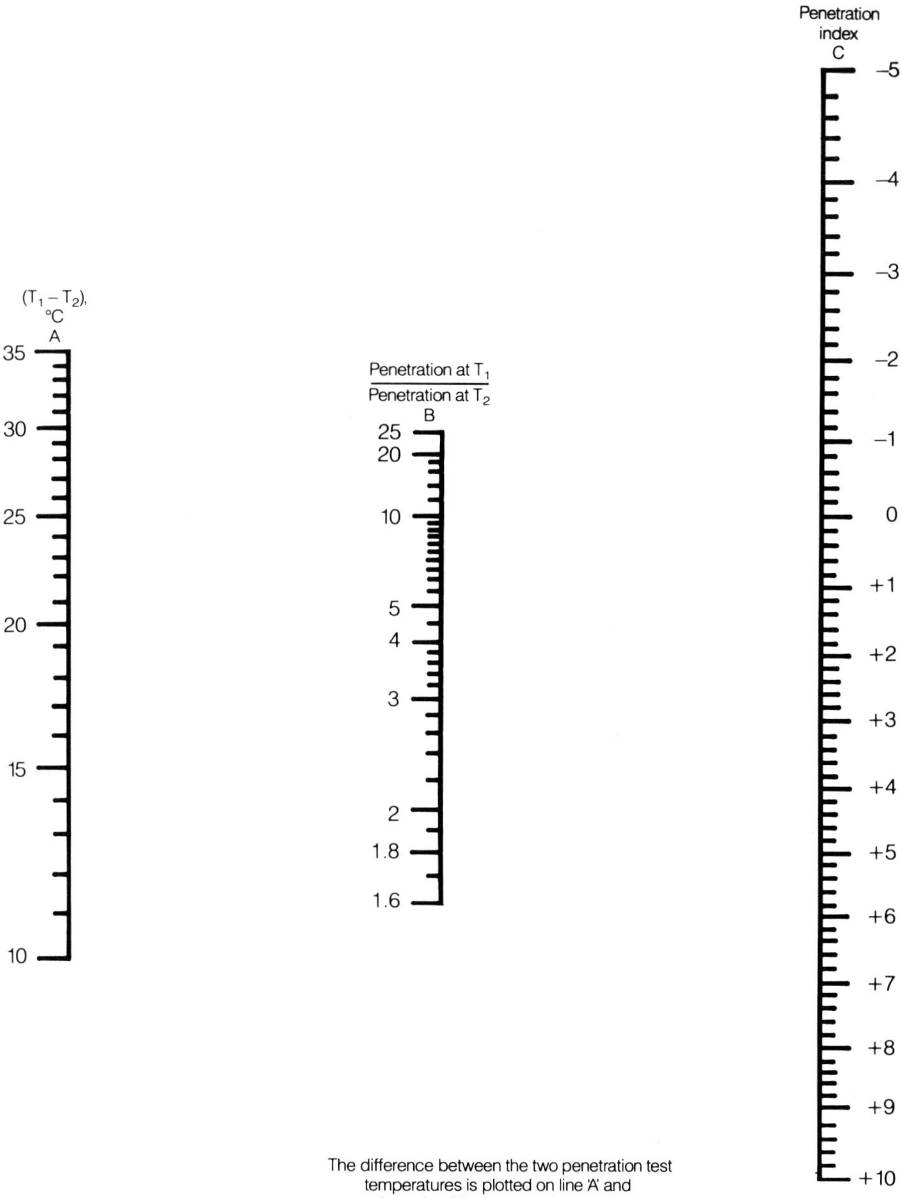

The difference between the two penetration test temperatures is plotted on line 'A' and the ratio of the two penetrations is plotted on line 'B'. A line is drawn through these two points and where it intercepts line 'C' is the PI of the bitumen.

Figure 5.7 – Nomograph for penetration index (Pen/Pen)

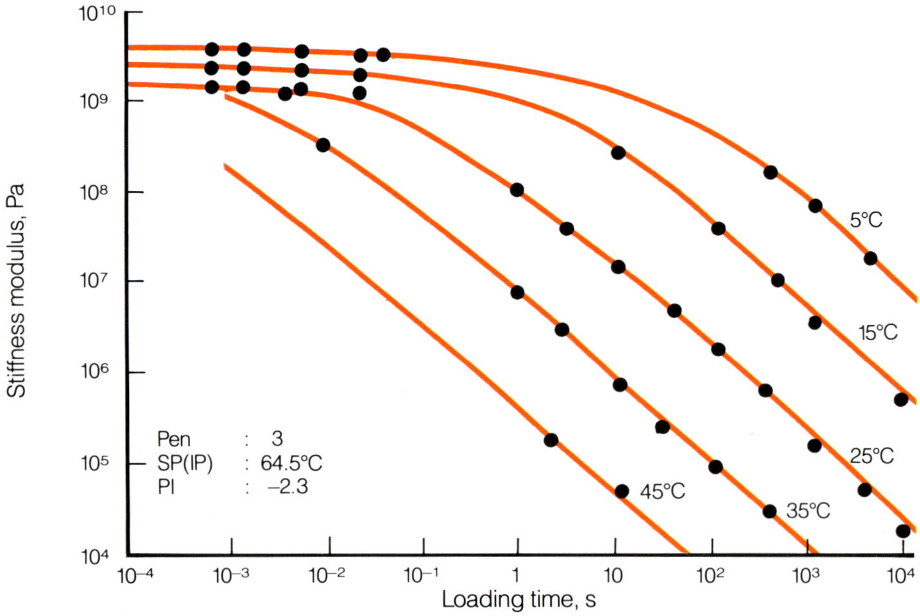

Figure 5.8a – The effect of temperature and loading time on the stiffness of a low PI bitumen

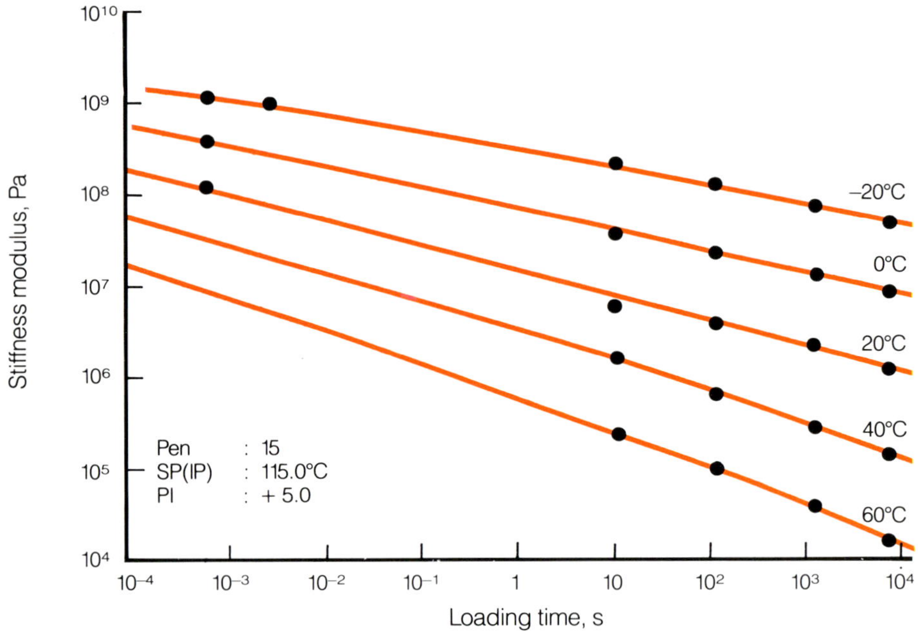

Figure 5.8b – The effect of temperature and loading time on the stiffness of 115/15

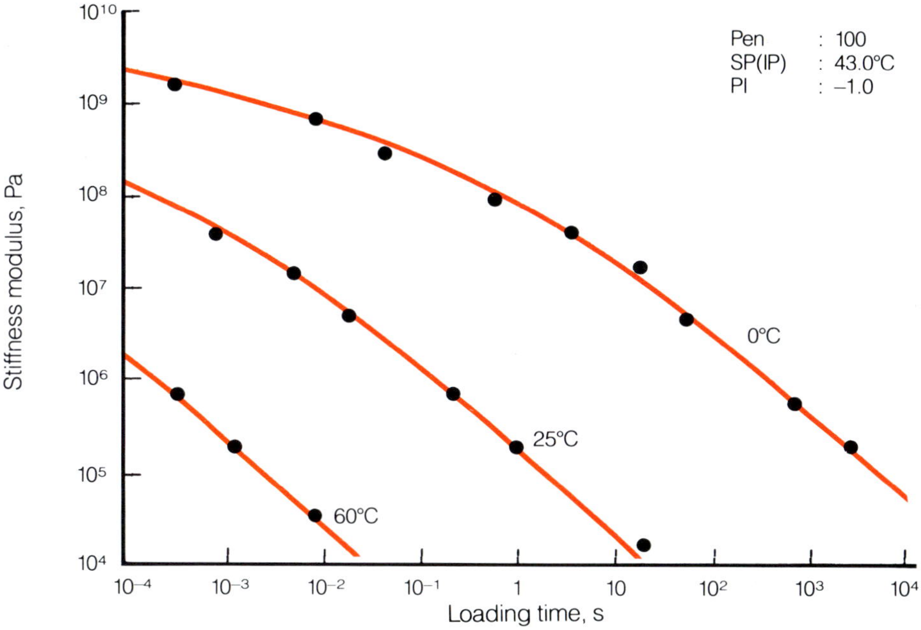

Figure 5.8c – The effect of temperature and loading time of the stiffness of 100 pen

It follows from the above that the value of the stiffness modulus is dependent on the loading time which is due to the special nature of bitumen. Similarly the stiffness modulus will also depend on temperature, T; consequently it is necessary to state both temperature and time of loading of any stiffness measurement:

$$S_{t,T} = \frac{\sigma}{\varepsilon_{t,T}}$$

The effect of changes in temperature and loading time on the stiffness modulus of three different bitumens are shown in figures 5.8a, 5.8b and 5.8c. Figure 5.8a shows a bitumen of low PI, –2.3. At very short loading times the stiffness modulus is virtually constant, asymptotic towards 2.5 to 3.0 x 10^9 Pa, and is, in this region, largely independent of temperature and loading time, ie S = E. The effect of PI is clearly illustrated by comparing figures 5.8a and 5.8b. The bitumen with the higher PI (+5) is considerably stiffer at higher temperatures and longer loading times, ie it is less temperature susceptible.

Figure 5.8c shows the relationship for a 100 pen bitumen. At a loading time of 0.02 second (which equates to a vehicle speed of ca 50 km/h) the stiffness modulus is ca 10^7 Pa at 25°C but falls to ca 5 x 10^4 Pa at 60°C. At low temperatures the stiffness modulus is high and therefore permanent deformation does not occur. However, at higher temperature or longer loading times (stationary traffic) the stiffness modulus is substantially reduced and under these conditions, therefore, permanent deformation of the road surface is much more likely to occur.

In order to appreciate fully the significance of the stiffness modulus and its measurement, it is necessary to consider the deformation under stress of simple solids and liquids. The deformation behaviour of visco-elastic materials can then be derived.

5.6.2 Determination of the stiffness modulus of bitumen

The methods to measure the stiffness modulus of bitumen are based on shear deformation, as shown in figure 5.9. The resistance to shear is expressed in terms of the shear modulus, G, which is defined as:

$$G = \frac{\tau}{\gamma} = \frac{\text{Shear stress}}{\text{Shear strain}}$$

The elastic modulus and shear modulus are related by the equation:

$$E = 2\,(1 + \mu)\,G$$

where μ = Poisson's ratio.

The value of μ depends on the compressibility of the material and may be assumed to be 0.5 for almost incompressible pure bitumens, while values of <0.5 have to be considered for asphalt mixes. Thus we have:

$$E \approx 3G$$

Shear stress can be determined statically in a creep test or dynamically by application of a sinusoidal load. In a creep test, the shear stress is applied from the starting time of the test. The deformation at loading times from about 1 second to 10^5 seconds or longer can be measured. In dynamic tests, the shear stress is usually applied as a sinusoidally varying stress of constant amplitude and fixed frequency. The deformation of the material under test also varies sinusoidally, with the same frequency as the applied stress. This is illustrated in figure 5.10.

The shear modulus, G_f, at frequency, f, is given by the ratio of the amplitudes of shear stress, τ, and shear strain, γ, according to:

$$G_f = \left(\frac{\tau}{\gamma}\right)_f$$

It follows that the stiffness modulus under dynamic conditions is:

$$S_f = 3\,G_f$$

Figure 5.10 shows the phase difference, φ, between shear stress and shear strain. This phase lag is a measure of the degree of elasticity of the bitumen under the test conditions, A purely elastic material would not show any phase difference between the stress and strain, whereas in the case of a purely viscous material there would be a phase lag of 90° or one-quarter cycle. With a visco-elastic material such as bitumen, the phase lag between stress and strain is between 0° and 90°, depending on the type and grade of bitumen, temperature and frequency. Small phase angles are found at low temperature and high frequency, and vice-versa, indicating that under these conditions the bitumen approximates respectively to elastic and viscous behaviour

By combining creep tests with dynamic tests, a considerable range of stiffness moduli and loading times can be covered. The stiffness modulus as a function of loading time is, therefore, often represented in a graph with logarithmic scales as illustrated in figure 5.11, in which the asymptotes represent the approximation of elastic and viscous response at short and long loading times, respectively.

Measurements of the stiffness as a function of loading time at various temperatures result in a graph as represented in figure 5.12. It appears that the stiffness/loading time curves obtained at different temperatures on one grade of bitumen all have the same shape and if shifted along the loading time

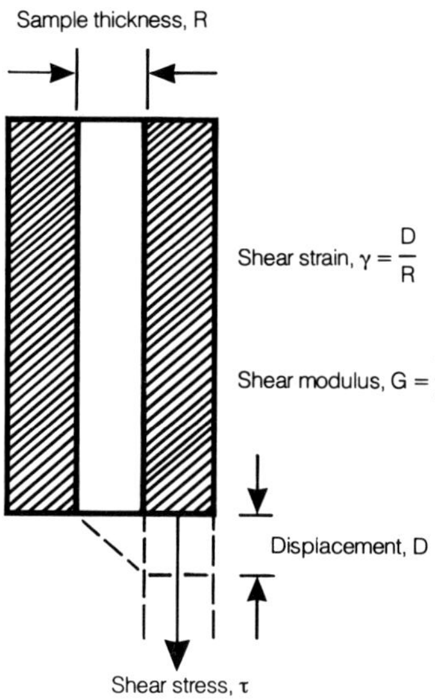

Figure 5.9 – A creep test in shear

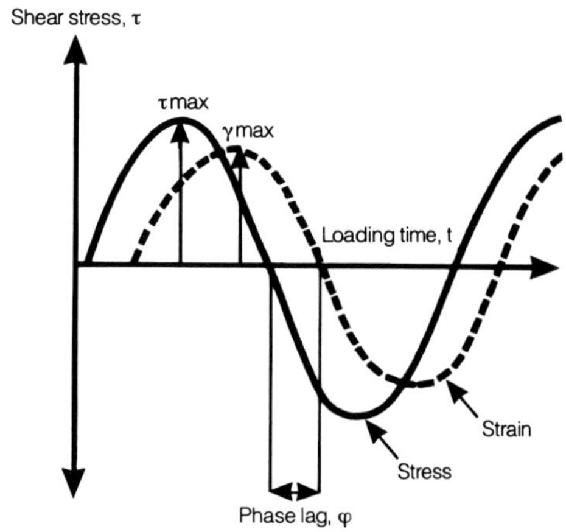

Figure 5.10 – A dynamic test in shear

Figure 5.11 – Stiffness modulus as a function of loading time or frequency

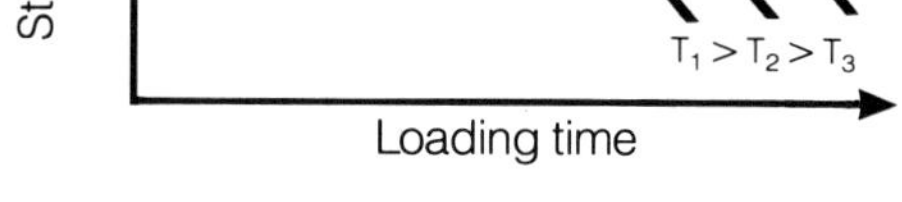

Figure 5.12 – Stiffness modulus as a function of loading time at different temperatures

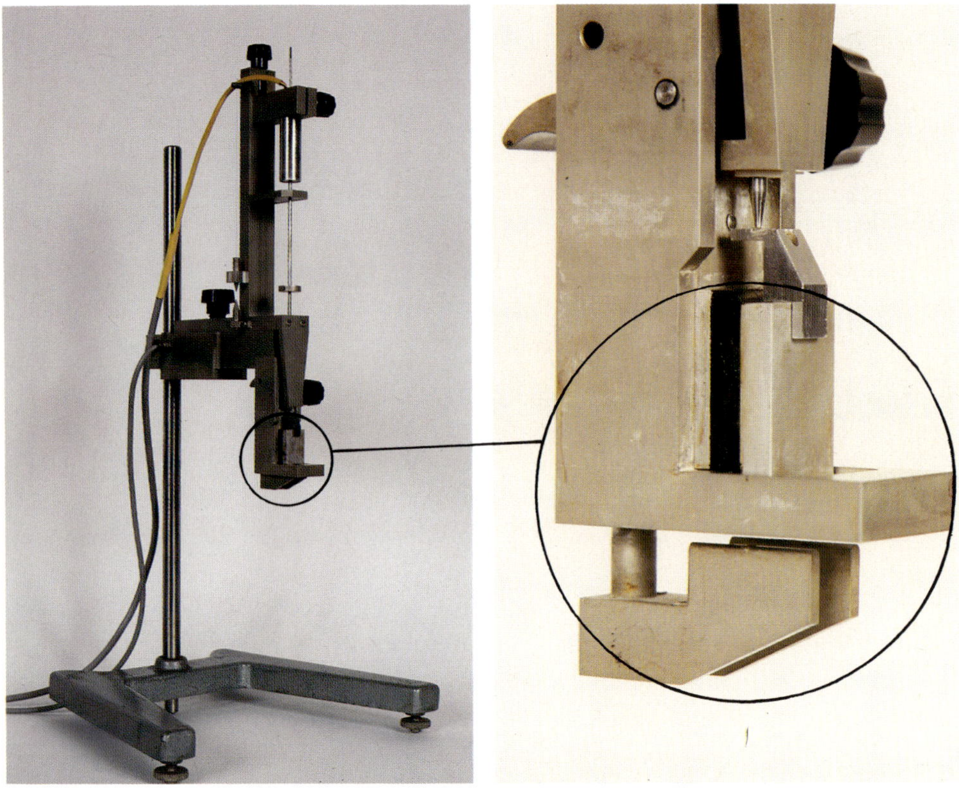

Photograph 5.6 – The sliding plate rheometer

axis would coincide. In this case the bitumen is said to be 'thermorheologically simple'. Most road and industrial grade bitumens belong to this category.

5.6.2.1 Measurement of stiffness modulus

The majority of instruments used to measure stiffness modulus are sophisticated and therefore expensive to purchase. However, in the late sixties the bitumen department at Koninklijke/Shell-Laboratorium Amsterdam (KSLA) developed the sliding plate rheometer[16], which combines simplicity and reliability. This test is similar to the sliding plate viscometer but uses thicker specimens between the plates. This instrument locks one plate in position and applies a load to the other plate (see photograph 5.6). The displacement (D) of this plate is recorded; knowing the shear stress, and the sample thickness (R), the stiffness modulus in the range 10^3 to 10^9 Pa can therefore be determined.

5.6.2.2 Prediction of stiffness modulus

If the direct measurement of stiffness modulus is not feasible, it can be predicted using the Van der Poel nomograph[15]. Van der Poel showed that two bitumens of the same PI at the same time of loading have equal stiffness at temperatures which differ from their respective softening points by

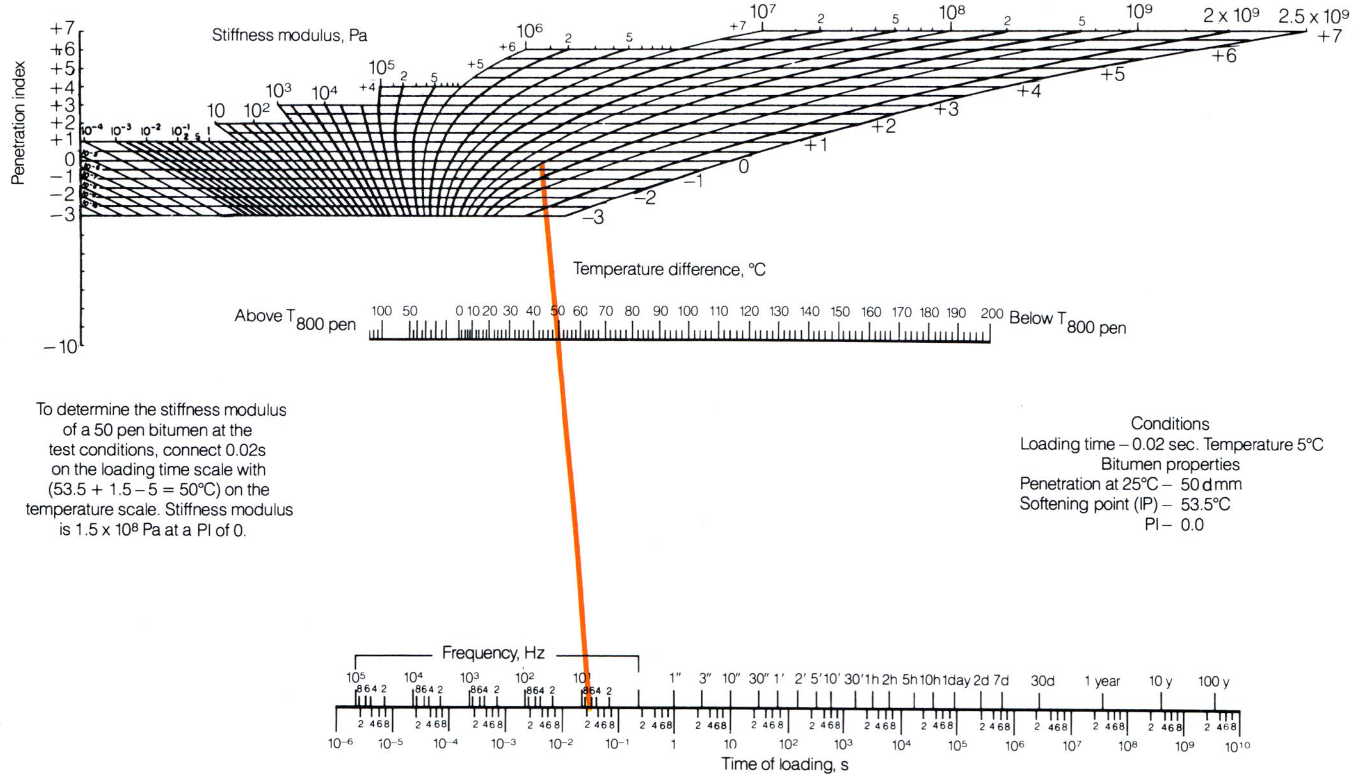

Figure 5.13 – Nomograph for determining the stiffness modulus of bitumens(15)

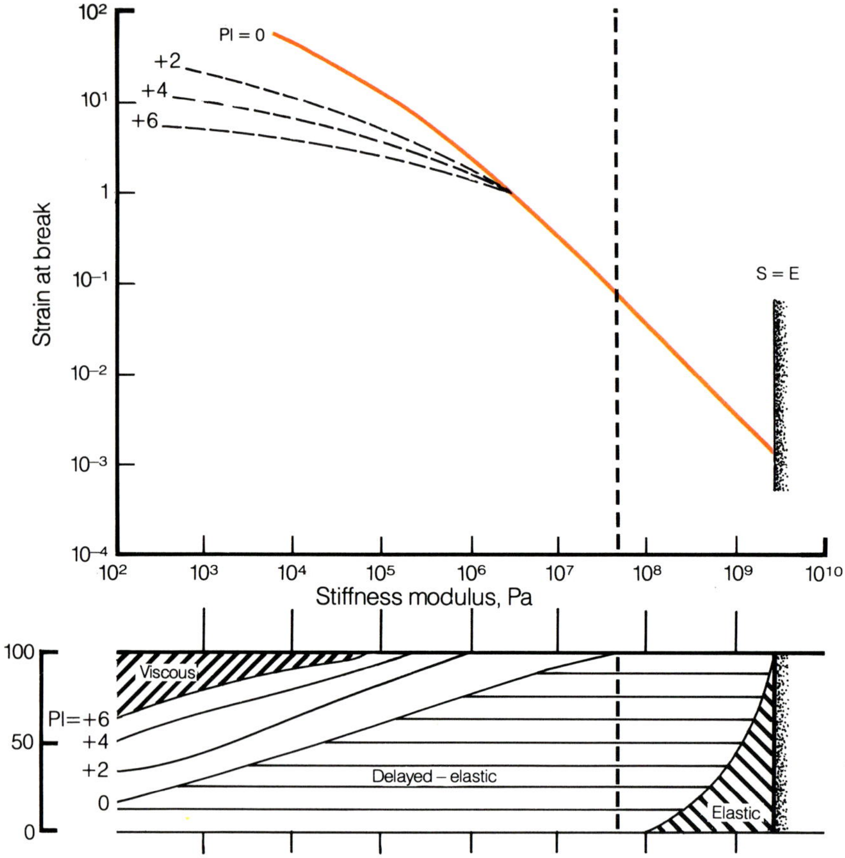

Figure 5.14 – Strain at break as a function of stiffness modulus and penetration index[17]

the same amount. He tested over forty bitumens, whose PIs varied from +6.3 to –2.3 at, many temperatures and frequencies, using both creep and dynamic tests. From the test data Van der Poel produced a nomograph, from which, using only penetration and softening point (ASTM D 36), it is possible to predict, to within a factor of 2, the stiffness modulus of a bitumen for any conditions of temperature and time of loading. Figure 5.13 shows the Van der Poel nomograph with the stiffness modulus determined for a 50 pen bitumen at a loading time of 0.02 seconds and a test temperature of 5°C.

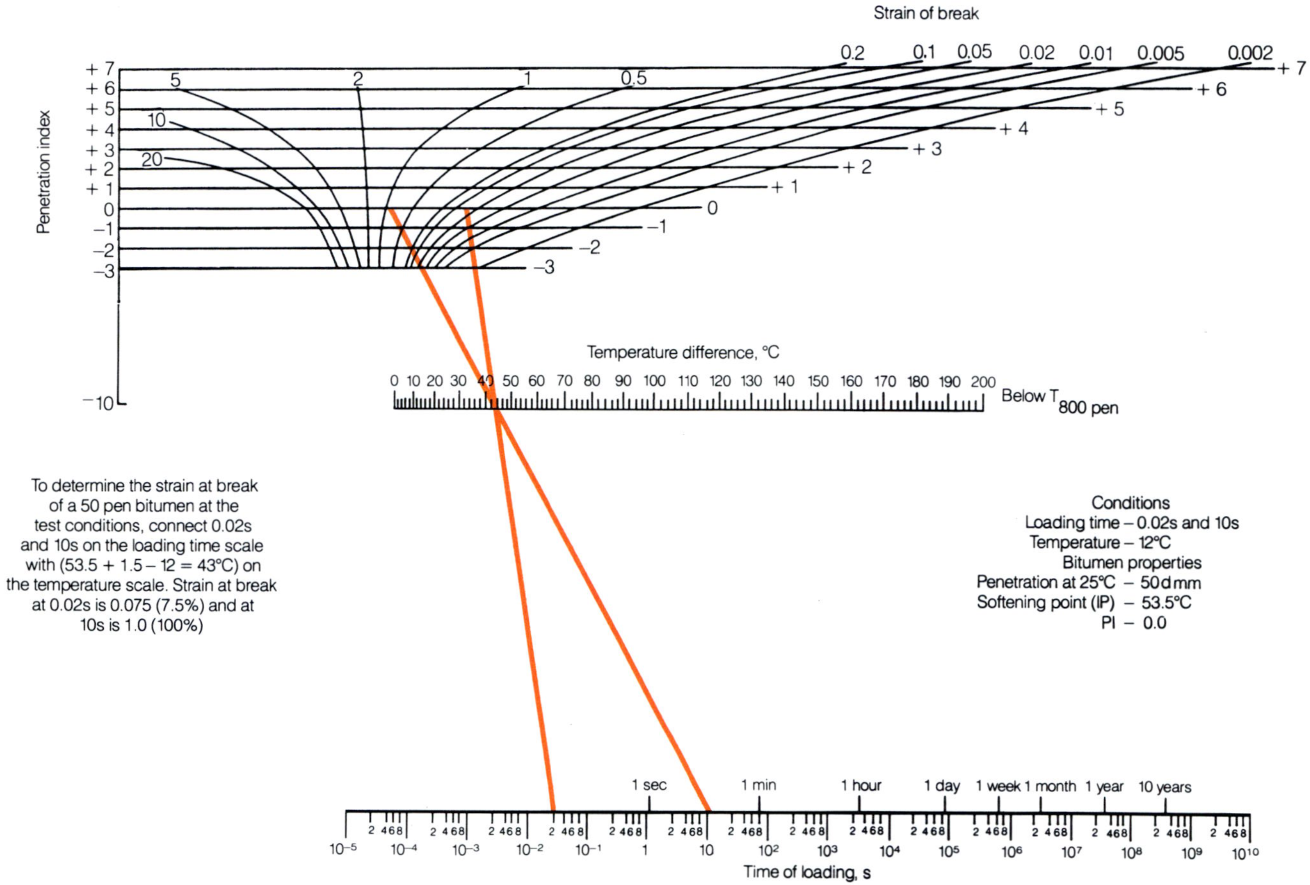

Figure 5.15 – Nomograph for estimating the strain at break of bitumens(18)

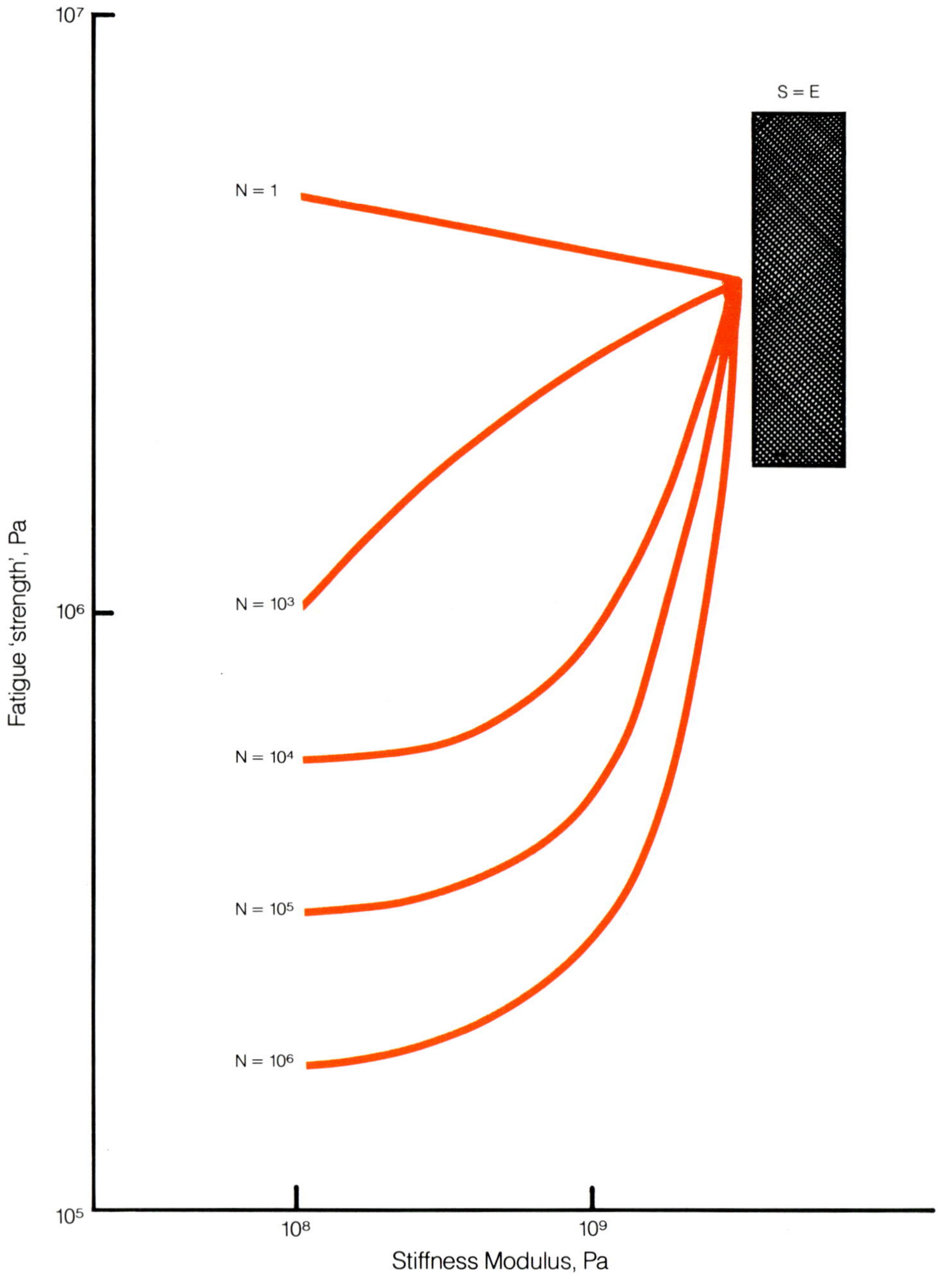

Figure 5.16 – Fatigue 'strength' as a function of stiffness modulus[18]

5.6.3 *Tensile strength*

Another important engineering property of bitumen is its breaking stress and/or tensile strain at break. The tensile properties of bitumen are, like stiffness modulus, dependent on both temperature and loading time. Assuming that the breaking properties of bitumen are dependent on its rheology at the instant of failure, its behaviour can be expressed in terms of the amounts of elastic, delayed elastic and viscous strain as shown in figure 5.14[17]. This shows, accepting the above assumption, that at stiffnesses greater than 5×10^6 Pa strain at break is a function of stiffness modulus. Only at low stiffness, where viscous deformation is large, is PI a significant factor. At high stiffness S = E. As the strain at break is defined as the ratio between change in the length and the instantaneous length, it will be equal to the elongation, ie increase in length divided by the initial length, as long as the increase in length remains low.

In practice break takes place under conditions of large stress. Large stresses normally occur at low temperature, ie at high stiffness; strain at break is then at its minimum. At higher temperatures, ie at low stiffness moduli, there will normally be no break at the same load but there will be deformation.

Strain at break can be predicted using the Heukelom nomograph[18], shown in figure 5.15, which has been derived from the Van der Poel nomograph. In the example strain at break of a bitumen with a softening point (IP) of 53.5°C and a PI of zero is given at 12°C. At a loading time of 0.02 seconds strain at break is 0.075 (7.5%) and at 10 seconds is 1.00 (100%).

5.6.4 *Fatigue 'strength'*

Like many other materials the strength of bitumen can be reduced by repeated loading, ie it fatigues. This is illustrated in figure 5.16 where the curves for fatigue 'strength' and breaking strength are shown as a function of stiffness modulus. All the tests have been carried out by bending at constant stress amplitude. The fatigue 'strength' is the stress which causes failure after 10^3, 10^4, 10^5, and 10^6 load cycles while the breaking strength corresponds to one cycle. The strength is reduced as the number of loadings increases. The figure shows that at high stiffness moduli the effect of load repetitions is significantly reduced and at the maximum stiffness modulus (S = E = 2.7×10^9 Pa) fatigue 'strength' is independent of the number of load repetitions. For a single bend the fatigue strength is virtually constant, ca 4×10^6 Pa for stiffness moduli greater than 5×10^7 Pa.

5.7 References

(1) Baven, H C: *School of Mines Quarterly*, vol 10, p 297, 1889.

(2) Institute of Petroleum: *Methods for analysis and testing*, part 1, volume 1.

(3) Annual book of the American Society for Testing and Materials Standards, section 4, volume 04.04: *Roofing, waterproofing and bituminous materials*, 1987.

(4) British Standard 2000: Part 49: 1983: *Penetration of bituminous materials*.

(5) British Standard 2000: Part 58: 1983: *Softening point of bitumen (ring and ball)*.

(6) Krom, C J: *Determination of the ring and ball softening point of asphaltic bitumens with and without stirring*. Journal of the Institute of Petroleum, vol 36, p 36, 1950.

(7) Pfeiffer, J Ph: *The properties of asphaltic bitumen*. Elsevier Publishing Company, 1950.

(8) Pfeiffer, J Ph and Van Doormaal, P M: *The rheological properties of asphaltic bitumens*. Journal of the Institute of Petroleum, vol 22, pp 414-440, 1936.

(9) Fraass, A: *Test methods for bitumen and bituminous mixture with specific reference to low temperature*. Bitumen, pp 152-155, 1937.

(10) Thenoux, G, Lees, G and Bell, C A: *Laboratory investigation of the Fraass brittle test*. Asphalt Technology, No 39, pp 34-46, July 1987.

(11) British Standard 2000: Part 72: 1982: *Viscosity of cutback bitumen and road oil*.

(12) British Standard 434: Part 1: 1984: *Bitumen road emulsions (anionic and cationic). Part 1. Specification for bitumen road emulsions*.

(13) Heukelom, W: *A bitumen test data chart for showing the effect of temperature on the mechanical behaviour of asphaltic bitumens*. Journal of the Institute of Petroleum Technologists, vol 55, pp 404-417, 1969.

(14) Heukelom, W: *An improved method of characterising asphaltic bitumens with the aid of their mechanical properties*. Proceedings of the Association of Asphalt Paving Technologists, vol 42, pp 62-98, 1973.

(15) Van der Poel, C: *A general system describing the visco-elastic properties of bitumen and its relation to routine test data*. Journal of Applied Chemistry, vol 4, pp 221-236, 1954.

(16) Fenijn, J and Krooshof, R C: *The sliding plate rheometer. A simple instrument for measuring the visco-elastic behaviour of bitumens and related substances in absolute units*. Proceedings of the fifteenth annual conference CTAA, vol 15, p 123, 1970.

(17) Heukelom, W and Wijga, P W O: *Bitumen testing*, Koninklijke/Shell-Laboratorium Amsterdam.

(18) Heukelom, W: *Observations on the rheology and fracture of bitumens and asphalt mixes*. Proceedings of the Association of Asphalt Paving Technologists, vol 35, p 358-399, 1966.

Chapter 6 – Bitumen constitution, structure and rheology

The rheology of a bitumen at a given temperature is determined by both the constitution (chemical composition) and structure (physical arrangement) of the predominantly hydrocarbon molecular structures in the material. Changes to either the constitution, structure or both will result in a change in the rheology. Thus to understand changes in bitumen rheology, it is essential to understand how the structure and constitution of a bitumen interact to influence rheology.

6.1 Bitumen constitution

The configuration of the internal structure of a bitumen is largely determined by the chemical constitution of the molecular species present. Bitumen is a complex chemical mixture of molecules of a predominantly hydrocarbon nature with a minor amount of structurally analogous heterocyclic species and functional groups containing sulphur, nitrogen and oxygen atoms[1,2,3]. Bitumen also contains trace quantities of metals such as vanadium, nickel, iron, magnesium and calcium which occur in the form of inorganic salts and oxides or in porphyrine structures. Elementary analysis of bitumens manufactured from a variety of crude oils show that most bitumens contain:

Carbon	82 – 88%
Hydrogen	8 – 11%
Sulphur	0 – 6%
Oxygen	0 – 1.5%
Nitrogen	0 – 1%

The precise composition varies according to the source of the crude oil from which the bitumen originates and the subsequent modification induced by semi-blowing and blowing during manufacture and subsequent ageing in service.

The chemical composition of bitumen is extremely complex; thus, a complete analysis of bitumen (if possible) would be extremely laborious and would produce such a large amount of data that correlation with the rheological properties would be impossible. However, it is possible to separate bitumen into two broad chemical groups called asphaltenes and maltenes. The maltenes can be further subdivided into saturates, aromatics and resins. The four groups are not well defined and there is inevitably some overlap between the groups. However, it does enable bitumen rheology to be compared with broad chemical composition.

The methods available for separating bitumens into fractions can be classified as follows:

- solvent extraction;
- adsorption by finely divided solids and removal of unadsorbed solution by filtration;
- chromatography;
- molecular distillation used in conjunction with one of the above techniques.

Solvent extraction is attractive because it is a relatively rapid technique[4], but the separation obtained is generally poorer than when using chromatography where a solvent effect is combined with selective adsorption. Similarly, simple adsorption methods[5] are not as effective as column chromatography in which the eluting solution is constantly re-exposed to fresh adsorbent and different equilibrium conditions as it progresses down the column. Molecular distillation is lengthy and has limitations in terms of how far type separation and distillation of high molecular weight components of bitumen can be effected.

Chromatographic techniques[6,7,8] have, therefore, been most widely used to define bitumen constitution. The basis of the method is to initially precipitate asphaltenes using n-heptane followed by chromatographic separation of the remaining material. Using this technique bitumens can be separated into the four groups: asphaltenes, resins, aromatics and saturates. Figure 6.1 shows a schematic representation of the chromatographic method.

The main characteristics of these four broad component groups are as follows:

6.1.1 Asphaltenes

These are n-heptane insoluble black or brown amorphous solids containing, in addition to carbon and hydrogen, some nitrogen, sulphur and oxygen. Asphaltenes are generally considered as highly polar and complex aromatic materials of fairly high molecular weight. Different methods of determining molecular weights have led to different values ranging widely from 600 to 300,000 depending on the separation technique employed. However, the majority of test data indicates that the molecular weights of asphaltenes range from 1,000 to 100,000; they have a particle size of 5 nm to 30 nm and a hydrogen/carbon (H/C) atomic ratio of about 1.1. The asphaltene content has a large effect on the rheological characteristics of a bitumen. Increasing the asphaltene content produces a harder bitumen with a lower penetration, higher softening point and consequently higher viscosity. Asphaltenes constitute 5% to 25% of the bitumen. Figure 6.2a shows a typical chemical structure of an asphaltene.

6.1.2 Resins

Resins are soluble in n-heptane; like asphaltenes they are largely composed of hydrogen and carbon, and contain small amounts of oxygen, sulphur and nitrogen. They are dark brown in colour, solid or semi-solid, and very polar in nature. This particular characteristic makes them strongly adhesive. They are dispersing agents or peptisers for the asphaltenes and the proportion of resins to asphaltenes governs to a degree the solution (SOL) or gelatinous (GEL) type character of the bitumen. Resins separated from bitumens are found to have molecular weights ranging from 500 to 50,000, a particle size of 1 nm to 5 nm and an H/C atomic ratio of 1.3 to 1.4.

6.1.3 Aromatics

Aromatics comprise the lowest molecular weight naphthenic aromatic compounds in the bitumen and represent the major proportion of the dispersion medium for the peptised asphaltenes. They constitute 40% to 65% of the total bitumen and are dark brown viscous liquids. The average molecular weight range is in the region of 300 to 2,000, they consist of non-polar carbon chains in which the unsaturated ring systems dominate, and they have a high dissolving ability for other high molecular weight hydrocarbons (see figure 6.2b).

6.1.4 Saturates

Saturates comprise straight and branch-chain aliphatic hydrocarbons, together with alkyl-naphthenes and some alkyl-aromatics. They are non-polar viscous oils which are straw or white in colour. The average molecular weight range is similar to that of aromatics and the components include both waxy and non-waxy saturates. This fraction forms 5% to 20% of the bitumen. Figure 6.2c shows a typical saturate structure.

An elemental analysis of the above four groups from a 100 pen bitumen is detailed in table 6.1[9].

6.2 Bitumen structure

Bitumen is traditionally regarded as a colloidal system[10] consisting of high molecular weight asphaltene micelles dispersed or dissolved in a lower molecular weight oily medium (maltenes). The micelles are considered to be asphaltenes together with an absorbed sheath of high molecular weight aromatic resins which act as a stabilising solvating layer. Away from the centre of the micelle there is a gradual transition to less polar aromatic resins, these layers extending outwards to the less aromatic oily dispersion medium.

In the presence of sufficient quantities of resins and aromatics of adequate solvating power the asphaltenes are fully peptised, and the resulting micelles have good mobility within the bitumen. These are known as solution or 'SOL' type bitumens, (see figure 6.3a). If the aromatic/resin fraction is not present in sufficient quantities to peptise the micelles, or has insufficient solvating power, the asphaltenes can associate further together. This can lead to an irregular open packed structure of linked micelles, in which the internal voids are filled with an intermicellar fluid of mixed constitution. These bitumens are known as gelatinous or 'GEL' type bitumens (see figure 6.3b), the best examples being the blown or oxidised grades used for roofing purposes. In practice most bitumens are of intermediate character.

The colloidal behaviour of the asphaltenes in bitumens results from aggregation and solvation. The degree to which they are peptised will considerably influence the resultant viscosity of the system. Such effects decrease with temperature and the GEL character of certain bitumens may be lost when they are heated to high temperatures. The viscosities of the saturates, aromatics, and resins depend on their molecular weight distributions. The higher the molecular weight the higher the viscosity. The viscosity of the continuous phase, ie the maltenes, imparts an inherent viscosity to the bitumen which is increased by the presence of the dispersed phase, ie the asphaltenes. The saturates fraction decreases the ability of the maltenes to solvate the asphaltenes, because high saturates contents can lead to marked agglomeration of the asphaltenes. Increase in GEL character and a lower temperature dependence for bitumens, therefore, not only results from the asphaltene content but also from the saturates content.

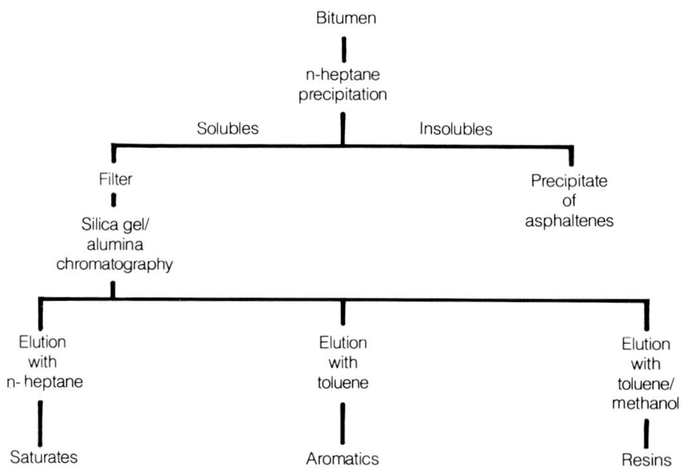

Figure 6.1 – Schematic representation of the analysis for broad chemical composition of bitumen

	Yield on bitumen, %w	Carbon, %w	Hydrogen, %w	Nitrogen, %w	Sulphur, %w	Oxygen, %w	Atomic ratio, H/C	Molecular weight
Asphaltenes (n-heptane)	5.7	82.0	7.3	1.0	7.8	0.8	1.1	11 300
Resins	19.8	81.6	9.1	1.0	5.2	–	1.4	1 270
Aromatics	62.4	83.3	10.4	0.1	5.6	–	1.5	870
Saturates	9.6	85.6	13.2	0.05	0.3	–	1.8	835

Table 6.1 – Typical elemental analysis of the four groups of a 100 pen bitumen[9]

6.3 The relationship between constitution and rheology

Systematic blending of saturates, aromatics, resins and asphaltene fractions separated from bitumen has demonstrated the dependence of rheology on constitution [11,12,13,14,15]. By holding the asphaltene content constant, and varying the concentration of the other three fractions, it has been demonstrated that:

increasing the aromatics content at a constant saturate to resins ratio has little effect on rheology, other than a marginal reduction in shear susceptibility;

maintaining a constant ratio of resins to aromatics and increasing the saturates content softens the bitumen;

the addition of resins hardens the bitumen, reduces the penetration index and shear susceptibility but increases the viscosity.

It has also been shown that the rheological properties of bitumens depend strongly on the asphaltene content. At constant temperature the viscosity of a bitumen increases as the concentration of the asphaltenes blended into the parent maltenes is increased. The increase in viscosity, however, is substantially greater than would be expected if the asphaltenes were spherical, non-solvated entities. This suggests that the asphaltenes can interact with each other and/or the solvating medium. Even in a dilute toluene solution the viscosity increase observed when increasing the asphaltene content corresponds to a concentration of non-solvated spheres, some five times higher than the amount of asphaltenes used. Bitumen asphaltenes are believed to be stacks of plate-like sheets formed of aromatic/naphthenic ring structures. The viscosity of a solution, in particular a dilute solution, depends on the shape of the asphaltene particles. Size is important only if shape changes significantly as size increases. At high temperatures the hydrogen bonds holding the sheets/stacks together are broken, resulting in a change in both the size and shape of the asphaltenes. Dissociation of the asphaltene entities continues until the limiting moiety, the unit sheet of condensed aromatic and naphthenic rings, is reached. Consequently, viscosity falls as temperature increases. However, as a hot bitumen cools, associations between asphaltenes occur to produce extended sheets. These, in turn, interact with other chemical types present (aromatics and resins) as well as stacking together to form discrete asphaltene particles.

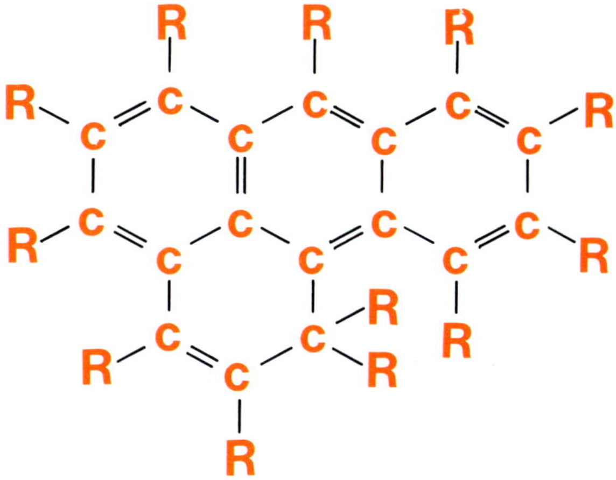

R – Aliphatic, napthenic or aromatic carbon chains

Figure 6.2a – Asphaltene structure

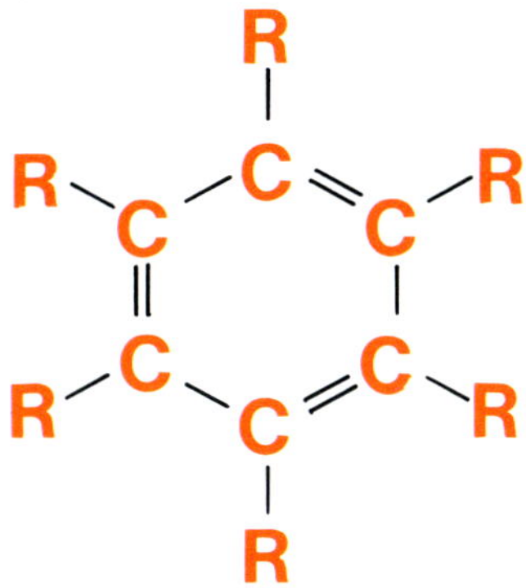

R – Aromatic and/or napthenic carbon chains

Figure 6.2b – Aromatic structure

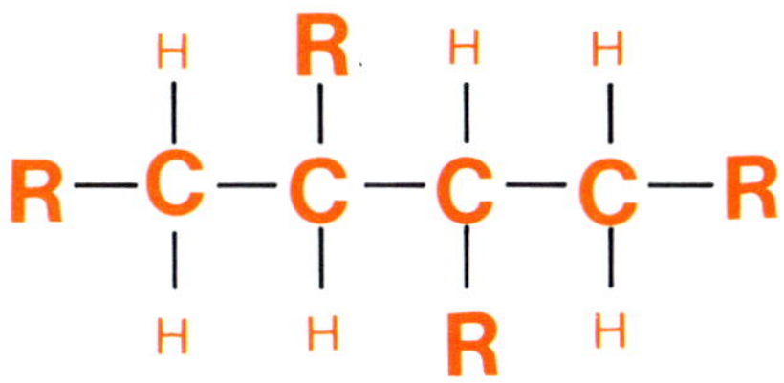

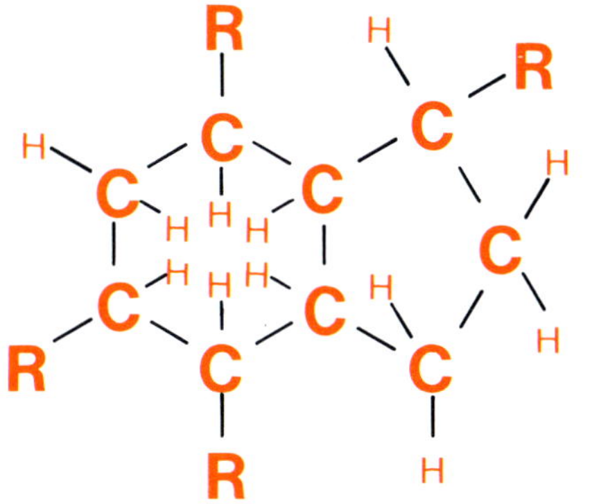

C – Carbon

H – Hydrogen

R – Aliphatic and/or napthenic carbon chains

Figure 6.2c – Saturate structures

	Vacuum residue	Distillation →				Blowing →		
Penetration at 25°C, dmm	285	185	99	44	12	84	46	9
Asphaltenes, %w	9.1	9.9	10.5	11.3	12.5	15.2	17.3	22.9
Resins, %w	18.6	16.7	18.2	17.7	21.3	21.0	22.1	21.5
Aromatics, %w	51.2	53.0	52.4	58.4	53.8	47.6	45.0	40.5
Saturates, %w	16.2	15.1	14.1	11.2	9.4	16.2	15.6	15.1

Note
The recovery of components is incomplete because of the techniques used.

Table 6.2 – Comparison of the broad chemical composition of distilled and blown bitumens manufactured from a single short residue [9]

The marked increase in non-Newtonian behaviour as a bitumen cools is a consequence of the inter and intra-molecular attractions between asphaltenes and other entities. Under shear these extended associations will deform or even dissociate, in a way not adequately described by classical Newtonian concepts. Consequently, at ambient and intermediate temperatures, it can be reasonably concluded that the rheology of bitumens is dominated by the degree of association of asphaltene particles and the relative amount of other species present in the system to stabilise these associations.

6.4 The relationship between broad chemical composition and physical properties

Atmospheric and vacuum distillation removes the lighter components from the bitumen feedstock. The loss of distillates leads to the preferential removal of saturates and concentration of asphaltenes. Air blowing of bitumens from a given vacuum residue or fluxed vacuum residue considerably increases the asphaltene content and decreases the aromatics content. Saturate and resin contents remain substantially of the same order as before commencing blowing; this is shown graphically in figure 6.4[16]. In table 6.2[9] a comparison of the chemical compositions is given for bitumens derived by distillation or blowing of the same feedstock.

Long term full-scale road trial studies have been undertaken to determine if the chemical composition of bitumen changes with time, using a full range of practical conditions, ie different mix types, different aggregates and bitumen contents[17]. The results of the studies are shown in figure 6.5 in terms of the ageing index (ratio of viscosity of recovered bitumen, η_r, to viscosity of the original bitumen, η_o, at 25°C) and also the chemical components. The major changes in viscosity are associated with the mixing and laying process. Changes in the viscosity of the binder are small with time. For the chemical composition the asphaltene content increased with mixing and has shown a gradual increase with time. The resin and aromatic content decreases with time. Although little change was expected in the saturates content, some increase has been noted, probably due to oil spillage from vehicles on the road. Overall changes after mixing are very small even though the initial void contents of the mixes studied were relatively high (5% to 8%), recovered bitumen samples were obtained from the top 3 mm of cores extracted from the test sections and service temperatures were high.

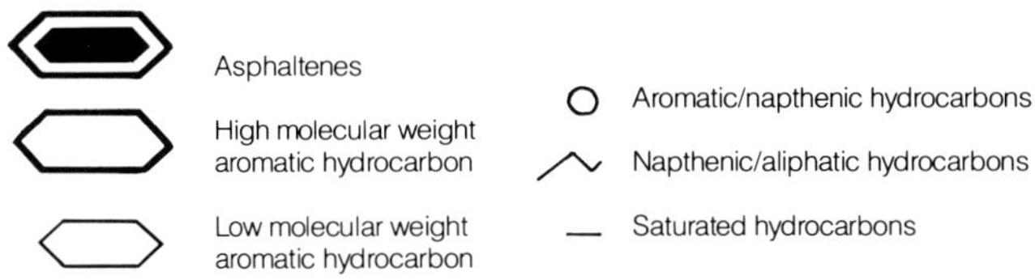

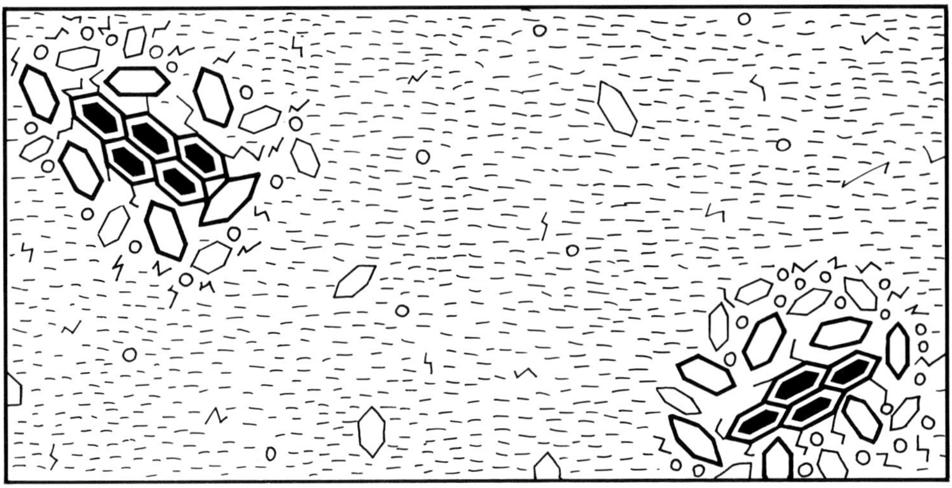

Figure 6.3a – Schematic representation of a 'SOL' type bitumen

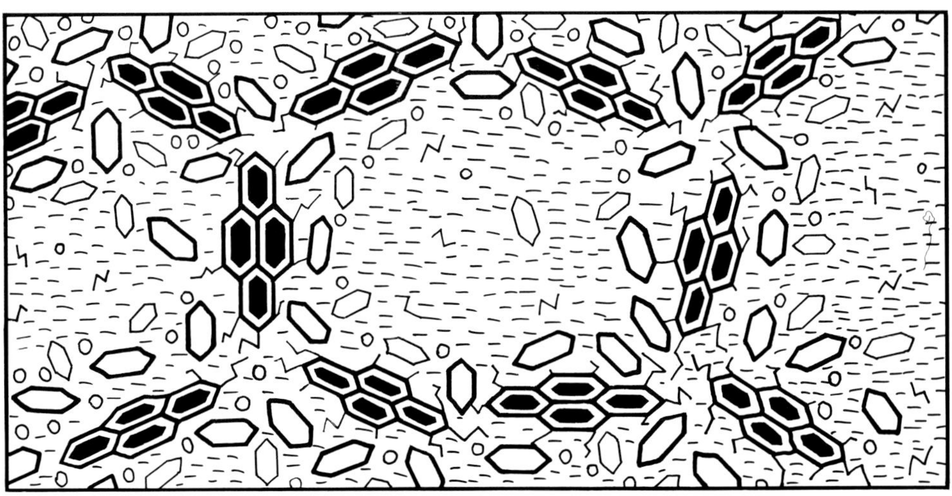

Figure 6.3b – Schematic representation of a 'GEL' type bitumen

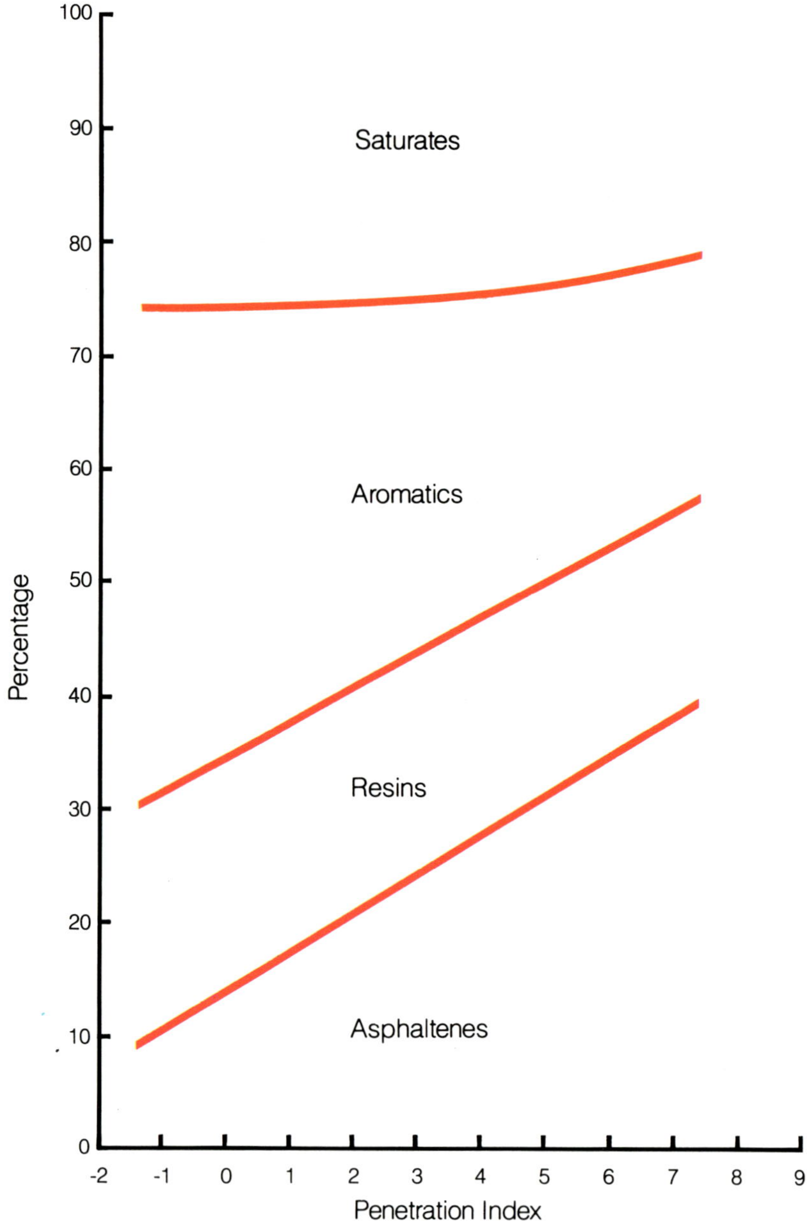

Figure 6.4 – Relationship between broad chemical composition and penetration index[16]

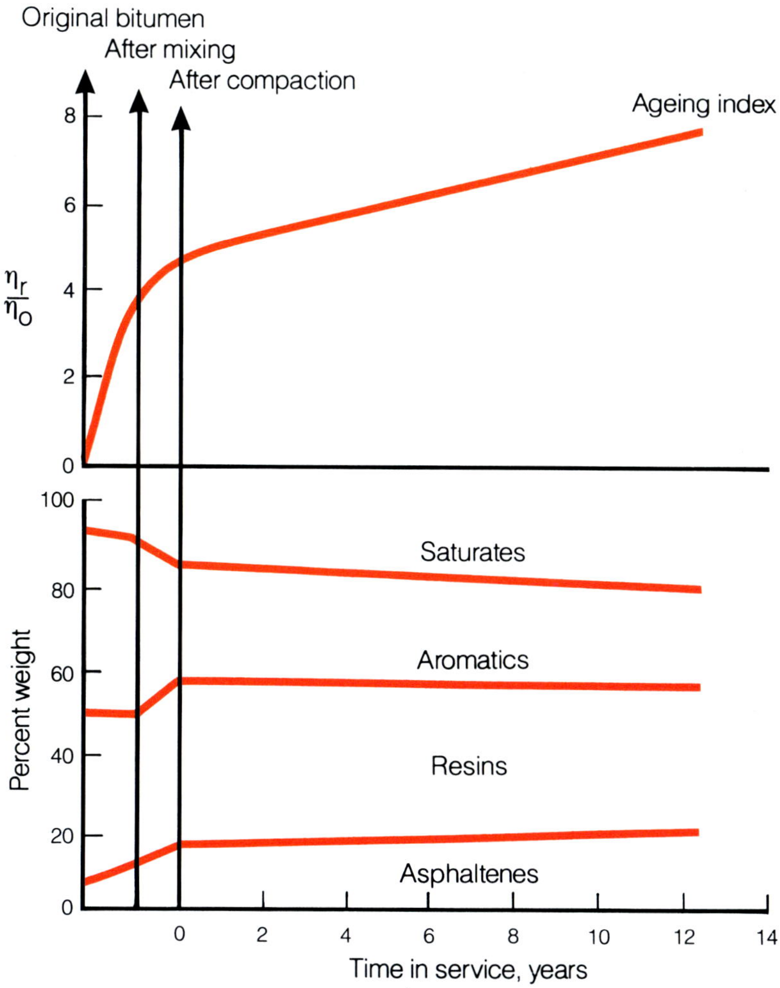

Figure 6.5 – Changes in bitumen composition during mixing, laying and in service[17]

Whilst chemical composition can be related to physical properties given specific components, it should be noted that bitumens of very different chemical compositions can have very similar physical properties[18]. It is, therefore, impossible to describe bitumens generally in terms of chemical component concentrations, and defining individual components, such as a minimum asphaltene content, has little, if any, relevance.

6.5 References

(1) Traxler, R N: *The physical chemistry of asphaltic bitumen.* Chemical Review, vol 19, No 2.

(2) Romberg, J W, Nesmitts, S D and Traxler, R N: *Some chemical aspects of the components of asphalt.* Journal of Chemical and Engineering Data, vol 4, No 2, April 1959.

(3) Traxler, R N and Coombs, C E: *The colloidal nature of asphalt as shown by its flow properties.* Thirteenth colloid symposium, St. Louis, Missouri, June 1936.

(4) Traxler, R N and Schweyer, H E: *Oil and Gas Journal*, vol 52, p 158, 1953.

(5) Marcusson, J and Agnew, Z: Chem., vol 29, p 346, 1916.

(6) Schweyer, H E, Chelton, H and Brenner, H H: *A chromatographic study of asphalt.* Proceedings of the Association of Asphalt Paving Technologists, vol 24, p 3, 1955.

(7) Middleton, W R: American Chemistry Society Symposium, vol 3, A-45, 1958.

(8) Corbett, L W and Swarbrick, R E: *Clues to asphalt composition.* Proceedings of the Association of Asphalt Paving Technologists, vol 27, p 107, 1958.

(9) Chipperfield, E H: *Bitumen production, properties and uses in relation to occupational exposures.* IARC review on bitumen carcinogenicity. Institute of Petroleum, Report IP 84-006, 1984.

(10) Girdler, R B: *Constitution of asphaltenes and related studies.* Proceedings of the Association of Asphalt Paving Technologists, vol 34, p 45, 1965.

(11) McKay, J R et al.: *Petroleum asphaltenes: chemistry and composition.* Advances in Chemistry Series No 170, Paper 9.

(12) Reerink, H: *Size and shape of asphaltene particles in relationship to high temperature viscosity.* Ind. Eng. Chem. Prod. Res. Deve., vol 12, No 1, 1973.

(13) Griffin, R L and Miles, T K: *Relationship of asphalt properties to chemical constitution.* Journal of Chemical and Engineering Data, vol 6, July 1961.

(14) Griffin, R L et al.: *Influence of composition of paving asphalt on viscosity, viscosity/temperature, susceptibility and durability.* Journal of Chemical and Engineering Data, vol 4, No 4, October 1959.

(15) ASTM Special Publication: *The influence of asphalt composition on rheology.* No 294, 1960.

(16) Lubbers, H E: *Bitumen in de weg- en waterbouw.* Nederlands Adviesbureau voor Bitumentoepassingen, April 1985.

(17) Chipperfield, E H, Duthie, J L and Girdler, R B: *Asphalt characteristics in relation to road performance.* Proceedings of the Association of Asphalt Paving Technologists, vol 39, pp 575-613, 1970.

(18) Chipperfield, E H, Duthie, J L and Girdler, R B: *Research on road bitumen.* Journal of the Institute of Municipal Engineers, pp 216-220, July 1968.

Chapter 7 – The influence of bitumen properties on performance in practice

The performance of bitumen-bound mixes in practice is significantly influenced by the rheological (or mechanical) properties and to a lesser extent the chemical constitution of the bitumen. The latter is particularly important at the road surface because the constitution of the bitumen influences the rate of oxidation and thereby how rapidly the bitumen is eroded by traffic. These factors are, in turn, influenced by changes due to the effect of air, temperature and water on the bitumen. There are, of course, many other factors influencing behaviour, including the nature of the aggregate, mix composition, bitumen content (ie bitumen film thickness), degree of compaction, etc, all of which influence long term durability.

Bitumens are visco-elastic materials and their behaviour varies from purely viscous to elastic, depending upon time of loading and temperature. During the mixing and compaction of bituminous mixes, and at high service temperatures, the properties can be considered in terms of viscosity but for most service conditions bitumens behave visco-elastically and their properties can be expressed through a stiffness modulus.

The rheological requirements for the bitumen during mixing, compaction and in service are illustrated in figure 7.1 and the critical requirements are summarised in table 7.1.

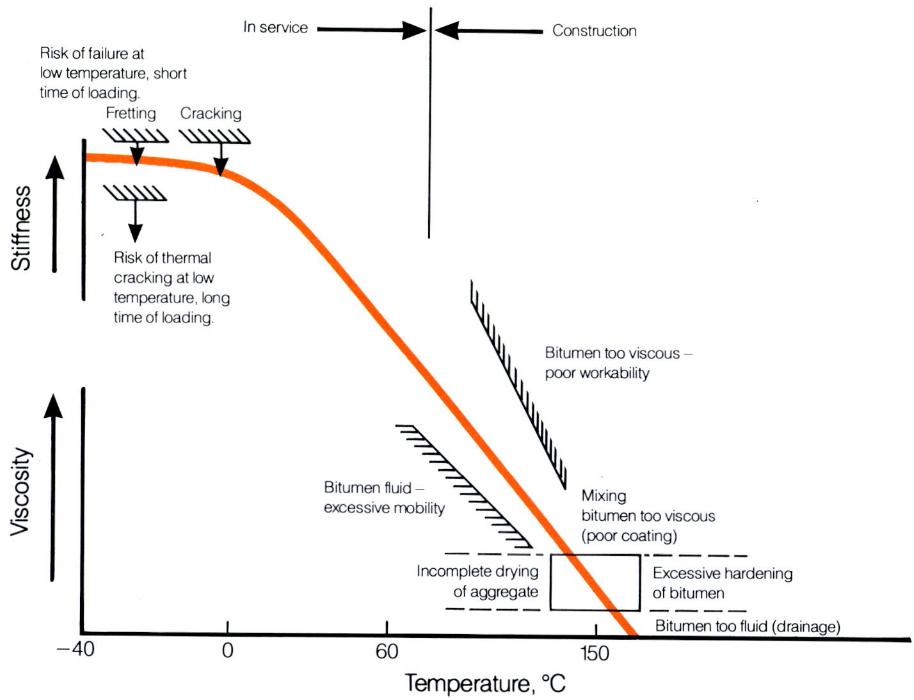

Figure 7.1 – Properties of penetration grade bitumens during construction and subsequently in service

Behaviour during	Condition		Significant property of the bitumen in the mix
	Temperature, °C	Time of loading, s	
Application			
Mixing	High (>100°C)	–	Viscosity, approximately 0.2 Pa.s
Laying	High	–	Viscosity
Compaction	High	–	Viscosity, minimum 5 Pa.s, maximum 30 Pa.s
In service			
Permanent deformation	High road temperature (30-60°C)	Long $>10^{-2}$	Minimum viscosity determined by penetration index and the softening point of the bitumen
Fatting up	High road temperature (30-60°C)	Long $>10^{-2}$	
Cracking			
– Traffic stresses	Low road temperature	Short (10^{-2})	Maximum bitumen stiffness
– Thermal stresses	Low road temperature	Long	
Fretting	Low road temperature	Short (10^{-3})	

Table 7.1 – Engineering requirements of bitumens during application and in service[4]

7.1 The influence of bitumen properties during construction

7.1.1 Mixing and transport

During mixing the hot bitumen must be readily able to coat the dried and heated mineral aggregate, given the shearing conditions employed, in a relatively short period of time (typically 30 to 90 seconds); this determines the lowest mixing temperature. Whilst the mixing temperature must be sufficiently high to allow rapid distribution of the bitumen on the aggregate, the use of the minimum mixing time at the lowest temperature possible should be advocated. The higher the mixing temperature the greater will be the oxidation of the bitumen exposed in thin films on the aggregate surface. This is illustrated in figure 7.2 where an increase of 5.5°C in mixing temperature, for a standard mixing time of 30 seconds, results in an increase of 1°C in the softening point of the bitumen[1]. There are, therefore, upper and lower limits to mixing temperature. If macadams are mixed too hot, drainage of bitumen off the aggregate may occur during hot storage or transport to site, leading to variations in bitumen content. Here the filler has an important function. As it is added to the mix after coating of the aggregate is complete, it 'stabilises' or increases the apparent viscosity of the bitumen, reducing drainage.

These different considerations combine to give an optimal bitumen viscosity of 0.2 Pa.s (2 poise) at mixing temperatures. It has been shown[2] that the temperature required to achieve a viscosity of 0.2 Pa.s can be crudely estimated by simply adding 110°C to the softening point of the bitumen (see figure 7.3). The drawback with this method is that it does not take account of the penetration index (PI) of the bitumen. A more precise estimate can be determined by plotting the penetration and softening point of the bitumen on the bitumen test data chart (see chapter 5, figure 5.3). The various conditions described above are summarised in figure 7.1.

When materials are being laid at low ambient temperatures, or if haulage over long distances is necessary, mixing temperatures are often increased to offset these two factors. However, increasing the mixing temperature will considerably accelerate the rate of bitumen oxidation which will increase the viscosity of the bitumen. Thus a significant proportion of the reduction in viscosity achieved by increasing the mixing temperature will be lost because of additional oxidation of the bitumen. If bituminous materials are transported in properly sheeted, well insulated vehicles, the loss in temperature is low, about 2°C per hour.

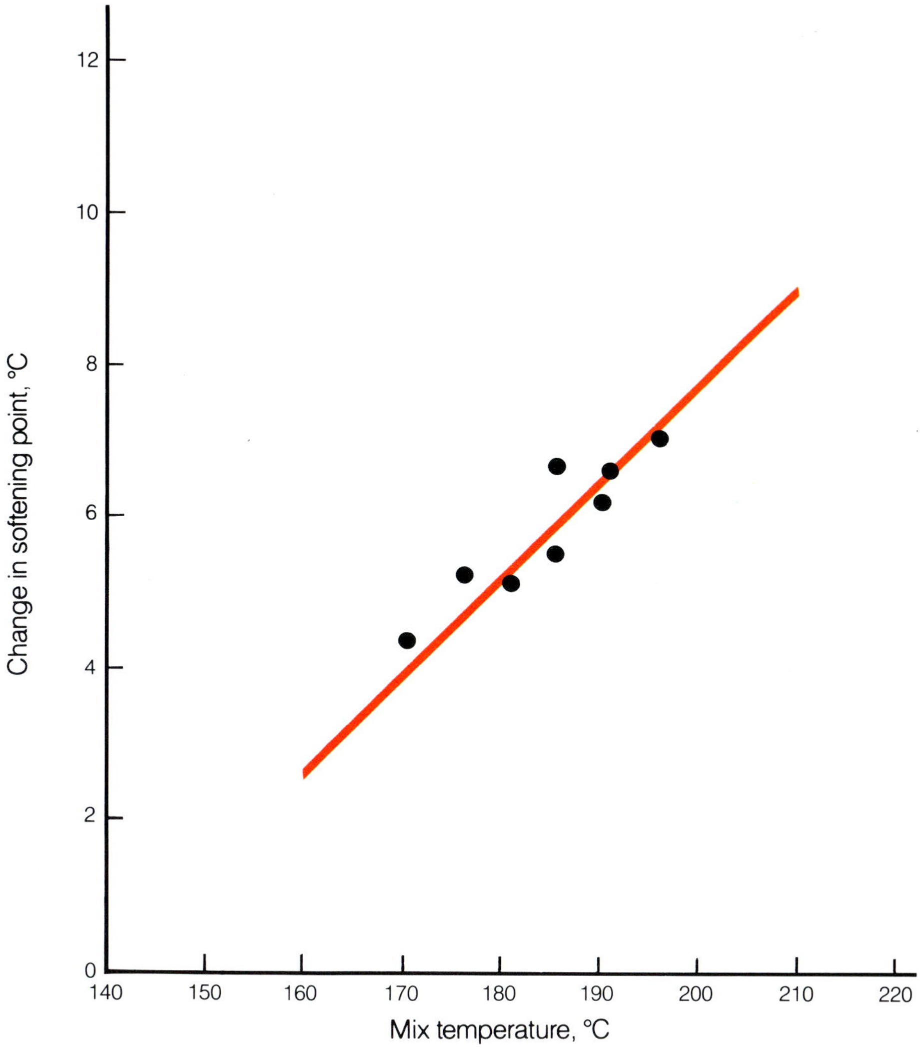

Figure 7.2 – Relationship between the temperature of the mix and change in softening point[1]

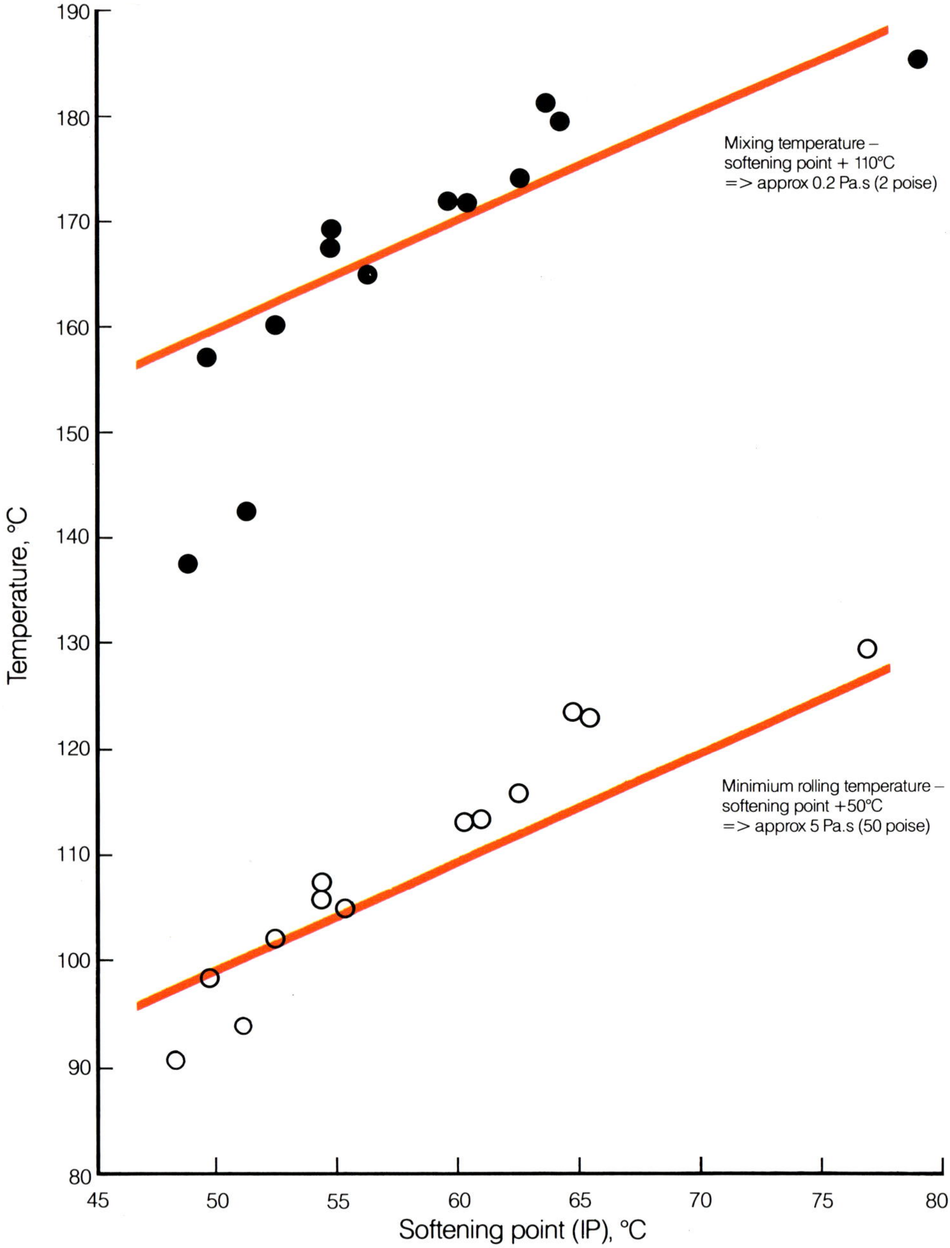

Figure 7.3 – The relationship between the softening point of the bitumen and equi-viscous temperatures for mixing and rolling hot rolled asphalt wearing course[(2)]

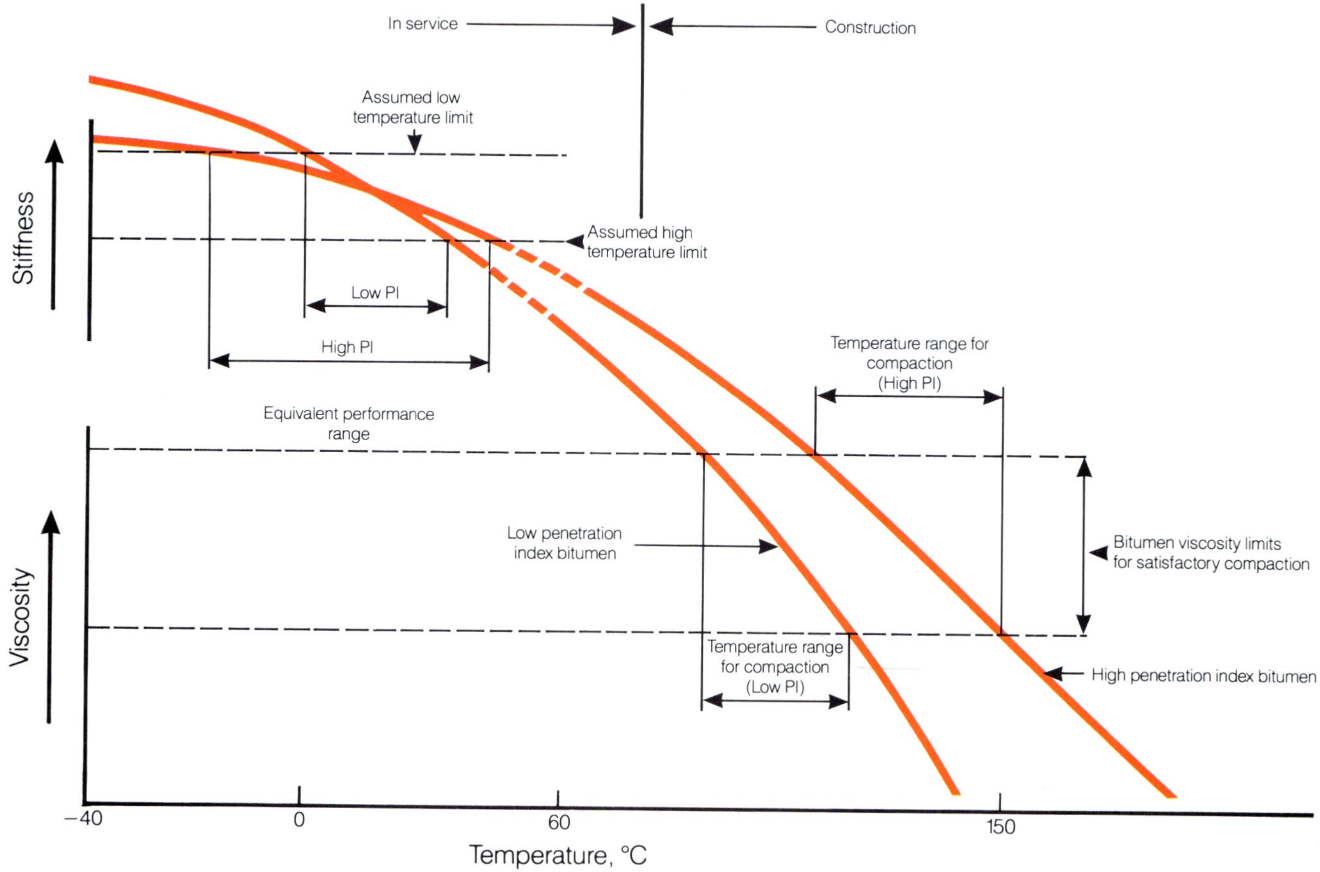

Figure 7.4 – The influence of penetration index on behaviour during construction and subsequent performance in practice

7.1.2 Laying and compaction

In discharging the mix into a machine-paver a loss of temperature of up to 20°C will be experienced and spreading the material into a mat will again reduce the temperature. The temperature loss will be dependent on the thickness of the layer, ambient temperature, wind speed and the temperature of the substrate on which the new material is being placed. The two most crucial factors influencing the cooling of the layer are wind speed and layer thickness(3). Once the mat has been spread it must still be sufficiently workable to enable the material to be satisfactorily compacted with the available equipment. For compaction the viscosity of the bitumen should be between 5 Pa.s (50 poise) and 30 Pa.s (300 poise). At viscosities lower than 5 Pa.s the material will probably be too mobile to compact and at viscosities greater than 30 Pa.s the material will be too stiff to allow any further compaction. The minimum rolling temperature can be estimated by adding 50°C to the softening point of the bitumen(2) (see figure 7.3).

The slope of the viscosity/temperature relationship or PI in this temperature region is particularly important because it determines the temperature range over which the viscosity of the bitumen remains at a suitable level for compaction. Figure 7.4 clearly shows that bitumens with a high temperature susceptibility, ie a low penetration index, PI, have a much narrower temperature 'window' in which satisfactory compaction of the material can be achieved.

7.2 The influence of bitumen properties on performance of bituminous mixes in service

After the mix has been manufactured, laid and compacted, its behaviour in service must be considered. The critical conditions for the performance of bituminous mixes are generally associated with either high service temperatures (30°C to 60°C) or low service temperatures (<5°C). At high temperatures the problems are deformation and fatting-up and at low temperatures cracking and surface fretting.

7.2.1 Performance at high service temperatures

7.2.1.1 Permanent deformation

Theoretically, deformation occurs thoughout the pavement structure, (see chapter 15, sections 15.2 and 15.2.1, and chapter 17, section 17.3.2). However, as asphalt mixes have a low thermal conductivity, in practice the majority of the deformation is plastic deformation of the wearing course (see photograph 7.1). This can occur under moving or stationary traffic, and particularly under the high shearing stresses imposed by braking, accelerating or turning traffic. The primary factor influencing plastic deformation is the mix composition, but for a given composition behaviour will be governed by the viscosity of the bitumen. Plastic deformation is greatest at high service temperatures, for which 60°C may be taken as a maximum *in-situ* temperature. At such temperatures, the cumulative effect of repeated loadings of short duration will be determined by bitumen viscosity. It has been estimated that during the long hot summer experienced in the U.K. during 1976, deformation in the wheel-tracks of hot rolled asphalt wearing courses was between two and four times the rate of an average summer.

The significance of the PI of the bitumen was confirmed by an extensive full-scale road trial carried out on the Colnbrook bypass(4). Bitumens of widely differing rheological characteristics were employed in hot rolled asphalt wearing course mixes. Figure 7.5 shows the relationship between rut depth on the road after 8 years and the PI of the bitumen; the advantage to be gained from using higher PI bitumens is obvious.

Photograph 7.1 – Permanent deformation

Work in several laboratories has shown that deformation occurring in a wheel-tracking test can be predicted using the uniaxial unconfined creep compression test (see chapter 14, section 14.1.2 and chapter 15, section 15.2). Creep curves of mix stiffness (S_{mix}) plotted against bitumen stiffness (S_{bit}) can be used to compare different mixes, and to predict their relative deformations, if subjected to the same temperature and loading conditions (see chapter 15, section 15.2).

The shape of a creep curve for a typical hot rolled asphalt wearing course is such that it can be approximated for most of its length by a straight line having a slope of 0.25, thus:

$$S_{mix} = k(S_{bit})^{0.25}$$

for two mixes having S_{bit} values in a known ratio:

$$S_{mix} \text{ ratio} = (S_{bit} \text{ ratio})^{0.25}$$

for bituminous mixes in similar road structures subjected to the same traffic:

$$\text{Deformation} = k \times \frac{1}{S_{mix}}$$

for two mixes having S_{mix} values in a known ratio:

$$\text{Deformation ratio} = \frac{1}{S_{mix}} \text{ ratio}$$

The above relationships can be used to compare the effect of changing the penetration of the bitumen and/or PI.

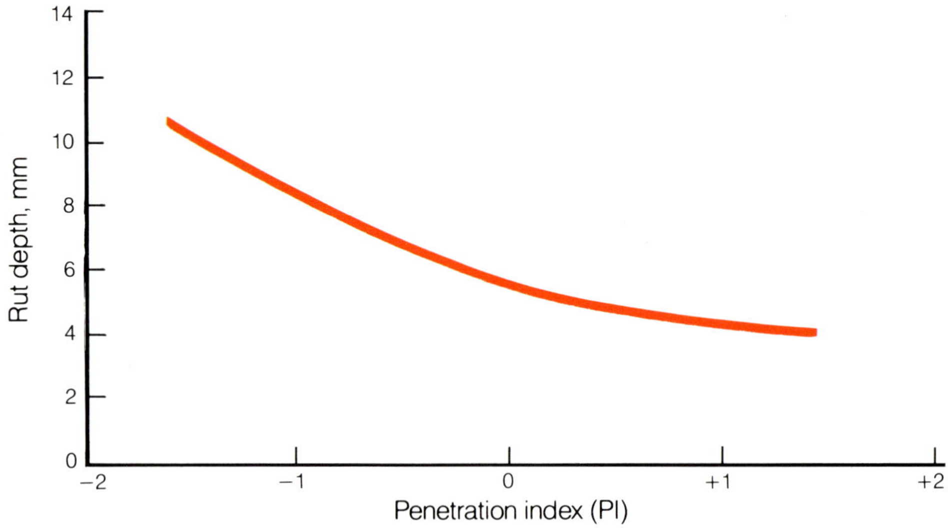

Figure 7.5 – Rut depth as a function of in-situ PI measured on the Colnbrook bypass after eight years' service(4)

Bitumen penetration	PI	Viscosity at 40°C Pa.s	Relative S_{bit}	Relative S_{mix}	Relative deformation
100	−0.8	6×10^3	0.2	0.67	1.5
60	−0.6	18×10^3	0.6	0.88	1.1
50	−0.5	30×10^3	1	1	1
40	−0.5	40×10^3	1.3	1.07	0.94
30	−0.4	90×10^3	3.0	1.32	0.75

Table 7.2a – Relative deformation of mixes at 40°C, effect of bitumen penetration

Bitumen penetration	PI	Viscosity at 40°C Pa.s	Relative S_{bit}	Relative S_{mix}	Relative deformation
40	−0.5	4×10^4	1	1	1
40	+0.5	1.5×10^5	3.8	1.41	0.71
40	+2.0	6×10^5	15	1.97	0.50

Table 7.2b – Relative deformation of mixes at 40°C, effect of bitumen penetration index

Calculations made on this basis show that if the deformation of a 50 pen bitumen is regarded as unity, a mix manufactured using a 30 pen bitumen subjected to the same temperature and loading regime will deform by 0.75 units. This assumes that the loading time for determination of viscosity and stiffness is the same. Conversely, if a 100 pen bitumen is used, 1.5 units of deformation will occur (see table 7.2a). Clearly little improvement in deformation resistance is achieved by using a harder bitumen.

Increasing the PI of the bitumen significantly improves resistance to deformation. For example, at 40°C a bitumen with a penetration of 40 and PI of −0.5 has a viscosity of 4 x 10^4 Pa.s, whereas a bitumen with the same penetration but having a PI of +2.0 has a viscosity of 6 x 10^5 Pa.s at 40°C. This gives a factor of 15 in the viscosity and hence in the stiffness of the two bitumens. Thus, applying the above 0.25 power formula, gives an increase of approximately 2 in mix stiffness. That is, the deformation of a mix made with a bitumen having a PI of +2.0 would be half of one made with a standard bitumen (see table 7.2b). Figure 7.6 shows the above theoretical relationship for relative deformation plotted as a function of PI. This clearly shows that the theoretical relationship is supported by both simulative laboratory wheel-tracking tests and full-scale road trials.

Correlation of bitumen properties with both Marshall and wheel-tracking tests(2) shows that the correlation between bitumen penetration and Marshall stability is poor, whereas softening point correlates very well with both Marshall stability and deformation in the wheel-tracking test (see figures 7.7a, 7.7b and 7.7c). For every 5°C increase in softening point the Marshall stability was increased by over 1.3 kN and the wheel-tracking rate almost halved.

7.2.1.2 Fatting-up

Fatting up occurs because consolidation of the aggregate in the mix by traffic reduces the void content, eventually forcing bitumen to the surface. This will be exacerbated if the bitumen content is too high or if the void content is too low. Migration of bitumen to the surface results in a smooth, shiny surface which has poor resistance to skidding in wet weather. Fatting-up is most likely to occur at high service temperatures; thus, limiting the softening point or viscosity of the bitumen at 60°C will limit fatting-up.

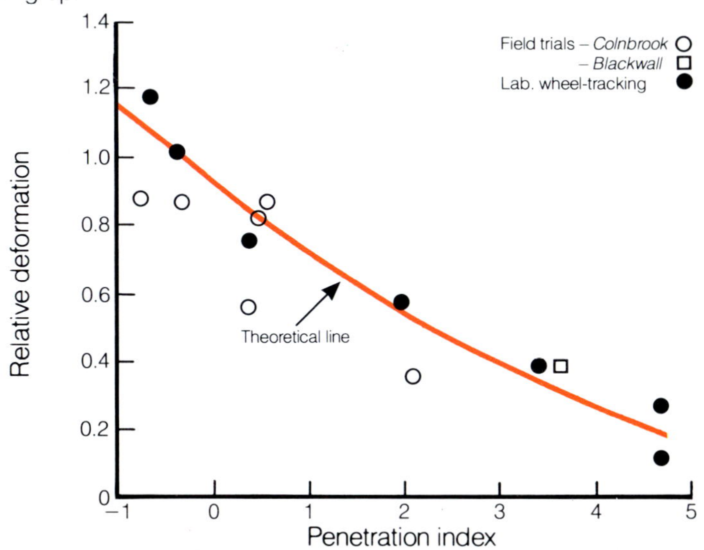

Figure 7.6 – Relative deformation as a function of penetration index

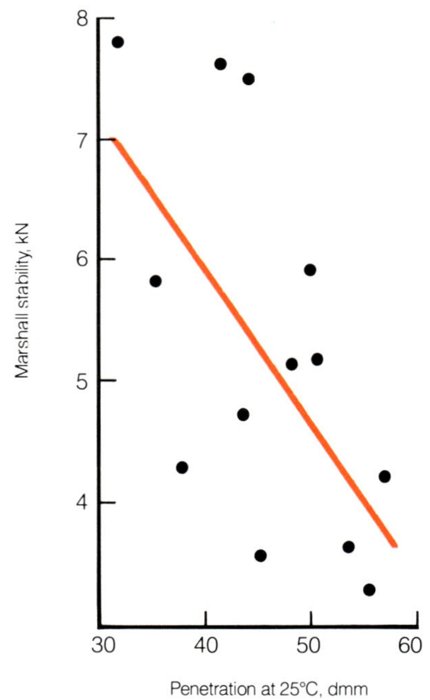

Figure 7.7a

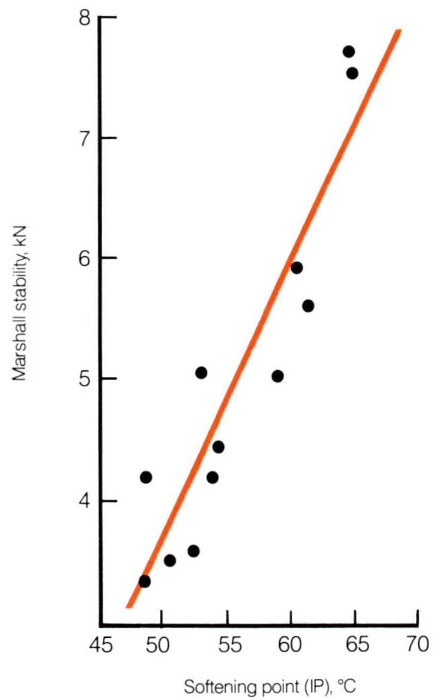

Figure 7.7b

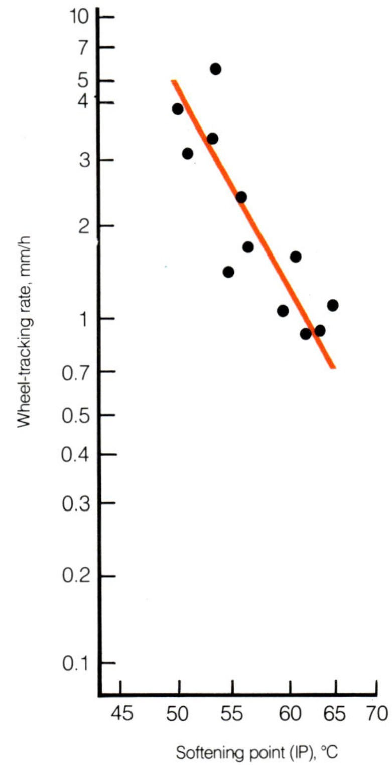

Figure 7.7c

The relationship between bitumen properties and Marshall stability and wheel-tracking rate[1]

7.2.2 Performance at low service temperatures

Low-temperature failure is associated with either cracking of the mix or with fretting of individual aggregate particles from the surface of the mix. An extensive study of the rheology and fracture of bitumens and bituminous mixes has shown that their fracture behaviour is directly related to their stiffness modulus(5). This resulted in the nomograph shown in chaper 5, figure 5.15 from which the strain or elongation at break of a bitumen under any specific conditions of temperature and time of loading can be estimated from the penetration and softening point of the bitumen.

7.2.2.1 Cracking

An asphalt mix will crack when the repeated stress and resulting strain imposed by traffic and/or temperature changes exceed the breaking or fatigue 'strength' of the mix. Figure 7.8 shows that the critical conditions occur when the stiffness of the bitumen is high, ie at low temperature and at short times of loading. In many countries very severe conditions of this kind may occur when temperatures rise just above 0°C after a sustained period below freezing point. Under these conditions the stability of the subgrade reduces dramatically, leading to high tensile stresses in the asphalt under traffic loads with the consequent danger of cracking. This is colloquially known as 'spring break-up'.

7.2.2.2 Thermal cracking

During periods of cooling, asphalt mixes try to contract; however, asphalt within a road structure is effectively constrained and therefore cannot change its length. As a result thermal tensile stresses develop which, together with stresses caused by traffic, may exceed the breaking strength of the asphalt, resulting in a transverse thermal crack. This is a problem particularly in the Mid-West of North America, Canada and in Siberia where temperatures can fall as low as −40°C.

From the stiffness and fracture properties of both the bitumen and the asphalt it is possible to calculate the conditions under which thermal fracture is liable to occur, allowing for stress relaxation. Laboratory studies have shown(6,7,8) that the temperature at which cracking occurs is related to the low-temperature stiffness of the bitumen. Bitumens of lower temperature susceptibility, ie high PI, reduce the 'cracking' temperature. Softer bitumens provide some improvement, but use of these is limited because of performance criteria at high service temperatures. Failure temperatures from laboratory tests of U.K. bitumen grades are in the order of −30°C which suggests that thermal cracking should not occur in the U.K. Until recently, there has been a paucity of evidence of thermal cracking in the U.K. However, investigations of cracking on the M4 motorway in Berkshire(9) found cracking which was initiated at the surface progressing downwards through the bituminous layers. It was hypothesised that diurnal thermal changes and sustained low temperature during winter were a contributory factor to the initiation and growth of the cracks.

7.2.2.3 Load associated cracking

When a bituminous pavement is loaded tensile stresses and strains are induced at the underside of the bitumen-bound layer, (see chapter 17, figure 17.9b). If the structure is inadequate for the imposed loading or if the characteristics of the subgrade change through ingress of water, the tensile strength of the material will be exceeded and a crack at the bottom of the bituminous layer will result. Repeated loads inducing tensile stresses above the tensile 'strength' of the mix will cause the crack to propagate upwards towards the road surface, ie the material fatigues. The measurement and prediction of fatigue characteristics are discussed in chapter 15, sections 15.3 and 15.3.1.

7.2.2.4 Reflective cracking

Until recently it was believed that reflective or reflection cracking occurred as a result of diurnal expansion and contraction of cementitious layers. These movements induced high tensile strains in the material immediately above cracks or joints, as shown in figure 7.9a. As a result a crack develops

Photograph 7.2 – Load associated cracking

Photograph 7.3 – Reflective cracking

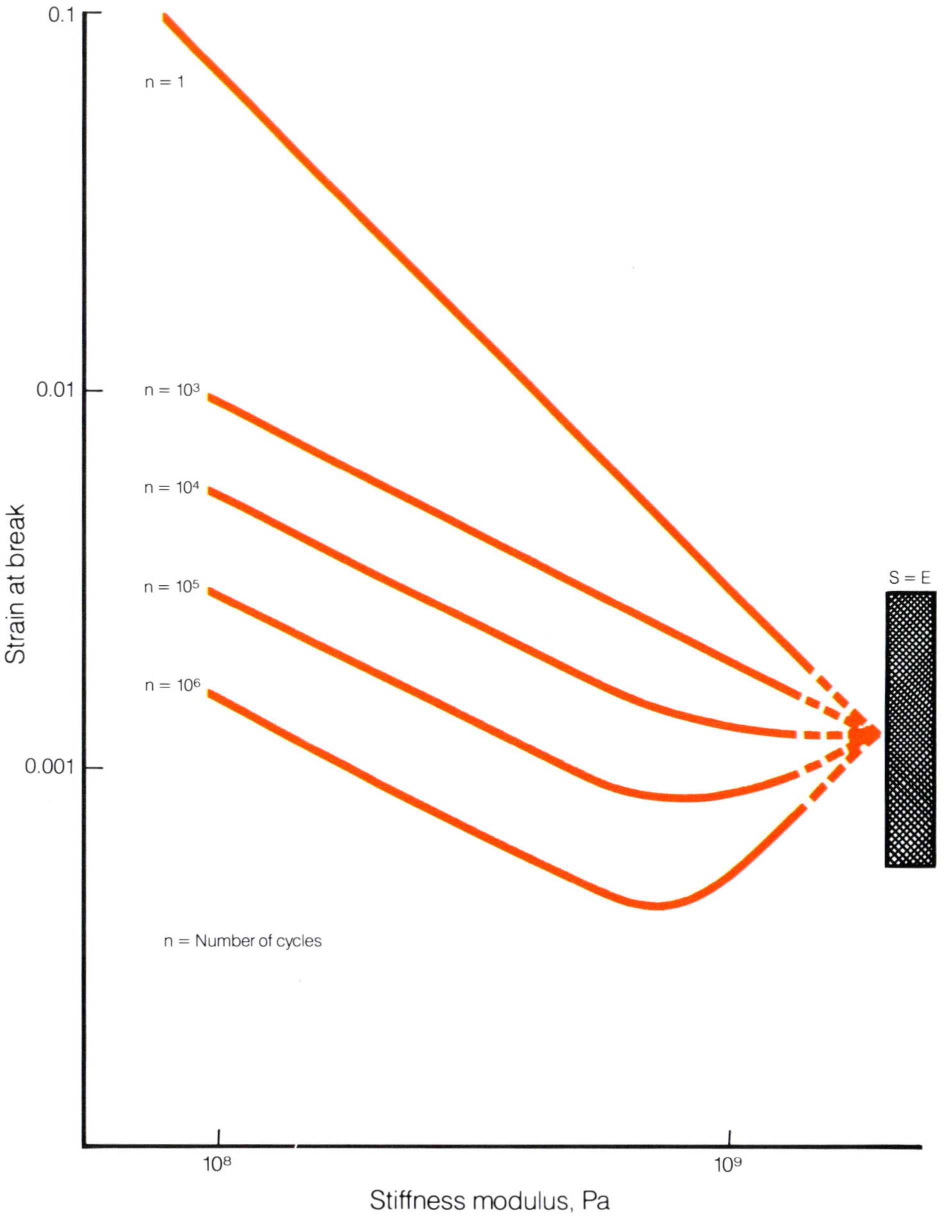

Figure 7.8 – Strain at break as a function of stiffness modulus[5]

which is progressively propagated towards the surface. However, examination of reflective cracking on the road[9,10] has shown that cracks can also be initiated at the road surface and propagate downwards to meet the cracks in the lean concrete. In addition to thermally induced stresses and strains, vertical movements between adjacent cementitious slabs induced by traffic (see figure 7.9b) will result in shear stresses. The magnitude of this shear stress will depend on the thickness of the bituminous and cementitious layers, the subgrade support and the amount of interlock between the slabs. Generally the magnitude of the traffic induced shear stresses is too small to initiate a crack but, if the mechanical interlock across the existing crack breaks down,the shear stresses can increase and accelerate propagation of the crack.

A third mechanism has been proposed[10] which suggests that cracking at the surface of the pavements can be induced by a combination of thermal contraction and warping of the pavement under cold winter conditions, (see figure 7.9c). Under these conditions the bituminous surfacing is very stiff and is least able to accommodate tensile stresses and strains caused by thermal contraction. In cold weather the upper layers are at a lower temperature than the lower layers and the consequent differential contraction with depth causes the concrete slab to warp, inducing a tensile stress at the road surface, which may result in surface cracking which propagates down to the crack in the lean concrete.

7.2.2.5 Fretting

Fretting is the progressive loss of interstitial fines from the road surface. It occurs when small movements of individual aggregate particles under the action of traffic are sufficient to develop tensile stresses or strains which exceed the breaking strength of the bitumen. Fretting is most likely to occur at low temperatues and at short loading times when the stiffness of the bitumen is high; fretting is also often associated with wet weather.

The major factors influencing fretting are the bitumen content of the mix and the degree of compaction. Loss of aggregate can be due to either loss of adhesion between the aggregate and the bitumen, or brittle fracture of the bitumen film connecting particles of aggregate. The first condition should not arise if suitable aggregates are selected. In the majority of cases fretting is associated with a low degree of compaction or too low a bitumen content. The choice of bitumen grade may control the degree of compaction which can be achieved initially, especially if operations are being conducted in adverse weather conditions and a suitable compaction temperature is not used. The higher the penetration of the bitumen or the lower the stiffness at low temperatures, the greater will be the resistance to fretting.

7.2.2.5.1 Fretting of macadam wearing course

For bitumen macadam mixes three different forms of fretting are identifiable:

- *Superficial fretting*
  This involves the loss of interstitial fines from the road surface. The material subsequently closes up under traffic without further deterioration or detrimental effect on performance.
- *Severe fretting*
  If superficial fretting does not close up under traffic, the loss of interstitial fines results in lack of mechanical interlock which is likely to lead to loss of coarse aggregate on the surface and loss of internal cohesion due to ingress of water. The presence of water breaks down the adhesive bond between the bitumen and the aggregate, resulting in stripping of the bitumen.In addition the uncoated fines and water form a slurry which is pumped through the material, as a result of pore pressures induced by moving traffic, abrading the coarse aggregate and exacerbating the problem. This will eventually lead to collapse of the material and lack of internal stability, often visible on the road as slight rutting.

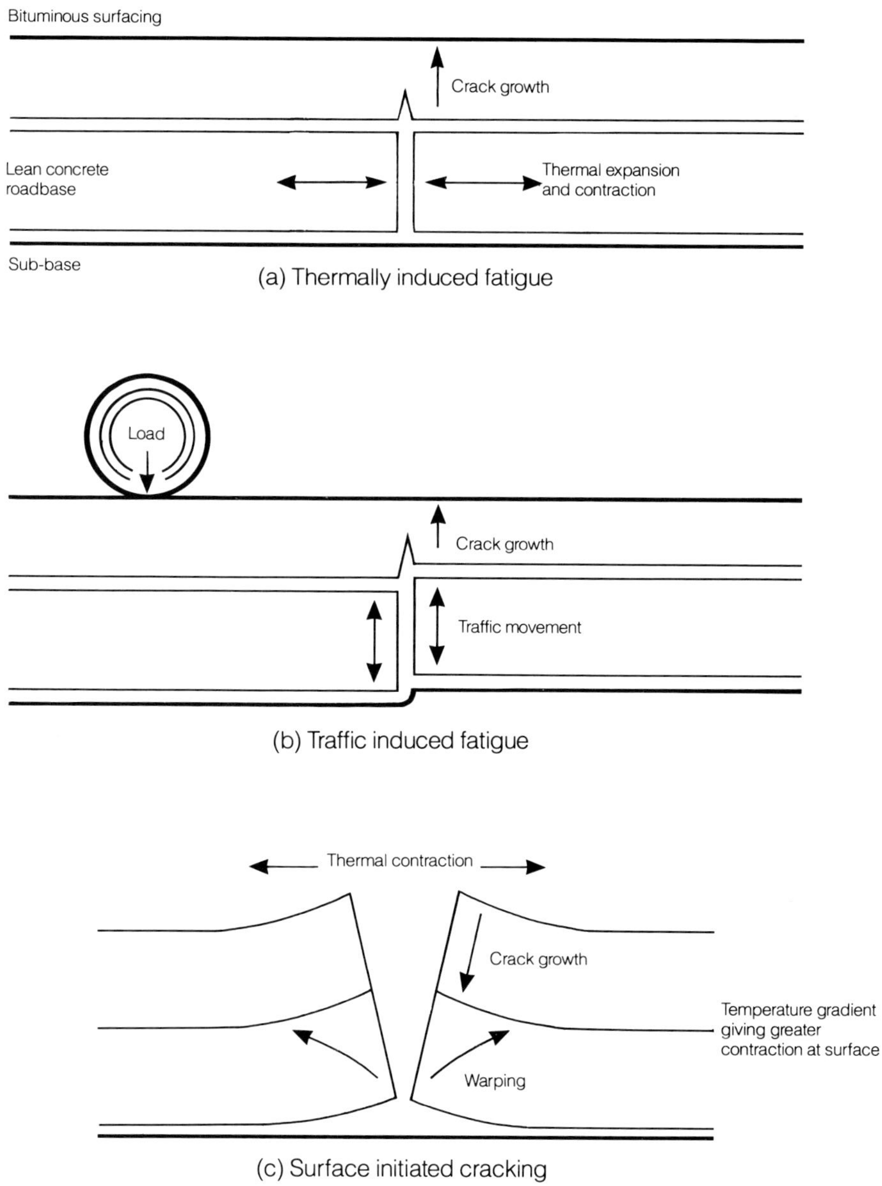

Figure 7.9 – Mechanisms of reflective cracking[10]

Bitumen/aggregate adhesion and the effect of prolonged water contact are discussed in depth in chapter 9.

- *Ravelling*

  Ravelling differs from the above in that it involves plucking out of surface aggregate by traffic without any loss of cohesion of internal fines.It occurs when individual aggregate particles move under the action of traffic. If the tensile stress, induced in the bitumen as a result of the movement, exceeds the breaking strength of the bitumen, cohesive fracture of the bitumen will occur and the aggregate particles will be detached from the road surface. Thus ravelling is most likely to occur at low temperatures and at short loading times when the stiffness of the bitumen is high.

In the autumn/winter of 1976/77, the incidence of fretting failures on newly-laid wearing course was substantially larger than observed in previous years. As a result, in 1977 the then Asphalt and Coated Macadam Association (ACMA) carried out a survey of its members as part of an investigation to determine the root cause of the failures.

The survey highlighted the following factors which were considered to significantly influence the fretting of dense bitumen macadam:

- Inadequate initial compaction due to:

  inadequate or irregular layer thickness;

  inadequate support by the substrate;

  the mixture being too stiff.

  Inadequate initial compaction, caused for whatever reason, will result in permeable materials that allow ingress of water which may result in 'severe fretting' as described above.

- Inadequate secondary compaction due to low traffic volumes.

  'Superficial fretting' can be arrested by the beneficial kneading of the surface by traffic. However, without secondary compaction, 'superficial fretting' may progress to 'severe fretting'

- The prolonged cold and wet weather conditions experienced during autumn/winter 1976/77.

  Paving before and during wet weather often results in insufficient compaction unless the utmost care is taken, viz:

  the time available for effective compaction is severely limited; with thin layers during cold and wet weather the time available for satisfactory compaction can be as little as four minutes;

  thin layers laid on cold substrates cool very rapidly; consequently the material stiffens prematurely before satisfactory compaction can be achieved; in addition, thin layers do not provide sufficient depth for particles to orientate and pack down effectively.

- The use of lower bitumen contents and harder grades of bitumen.

  Low bitumen contents and harder bitumens increase the stiffness of the mix, making the material more difficult to compact, thus significantly increasing the likelihood of inadequate compaction. The choice of bitumen grade in particular will strongly influence the workability of the mix. Correct grade selection is essential, especially if operations are conducted in adverse weather conditions.

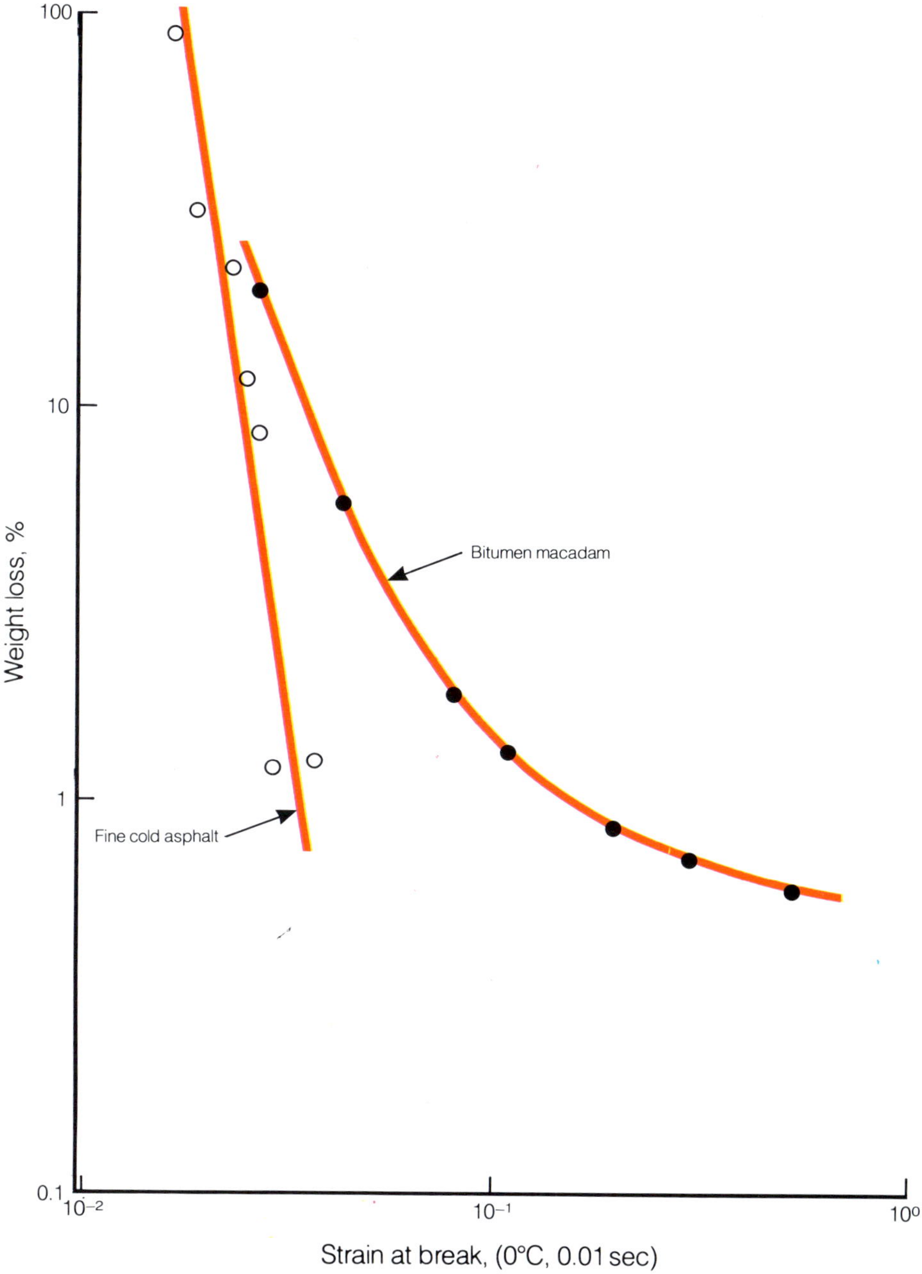

Figure 7.10 – Relationship between fretting and strain at break for fine cold asphalt and bitumen macadam(11)

Simulative abrasion wheel-tracking tests[11] at 0°C on fine cold asphalt and bitumen macadam, manufactured using different grades of bitumen, showed that the grade used significantly influenced resistance to fretting. Figure 7.10 shows the strain at break for each grade, determined using the Heukelom nomograph (see chapter 5, figure 5.15), at 0°C and a loading time of 0.01 seconds, plotted against percentage loss of material. The amount of material fretted from the surface increased dramatically as harder grades of bitumen were used, ie as strain at break decreased. It is interesting to note that the fine cold asphalt was more sensitive to bitumen grade than the bitumen macadam.

7.2.2.5.2 Fretting of hot rolled asphalt wearing course

In hot rolled asphalt mixes there are also three forms of fretting:

- *Superficial fretting of asphalt mortar*

  As with macadams this process involves the loss of interstitial fines. A small quantity of this fine material is eroded from the exposed asphalt mortar by traffic. The surface of the mortar subsequently closes up under traffic without further deterioration or detrimental effect on performance.

- *Severe fretting of asphalt mortar and loss of chippings*

  This process involves severe erosion of the asphalt mortar between the chippings. Eventually the support for the chipping is removed and it is plucked out (see photograph 7.4). Figure 7.11 shows a profile of void content in hot rolled asphalt wearing course compacted under adverse weather conditions. This shows that although the mean void content of the wearing course is satisfactory the void content of the critical top 5 mm to 10 mm has an unacceptably high void content. This material is permeable to air and water and is vulnerable to premature fretting.

Photograph 7.4 – Fretting of hot rolled asphalt

- *Loss of chippings and subsequent fretting of asphalt mortar*
  To achieve a texture depth of 1.5 mm relatively heavy rates of spread of pre-coated chippings are applied. Inevitably, a small proportion of these chippings will not be adequately embedded into the asphalt and they are plucked-out during early trafficking. This is acknowledged in the specification of this material and up to 5% chipping loss is considered acceptable[12,13].

  Where greater loss of chippings occurs it is likely to be the result of:

  inappropriate specification of texture depth;

  inadequate embedment of chippings;

  adhesive failure between the pre-coated chipping and the asphalt.

In 1984 a survey was carried out by British Aggregate Construction Materials Industries (BACMI) as a part of a wider survey into chipping loss by the Institution of Highways and Transportation (IHT)[14,15]. BACMI's survey showed[16] that over half the chipping loss problems were at high stress, low-speed sites such as roundabouts and junctions where a minimum 1.5 mm texture depth had been erroneously specified (see chapter 13, section 13.5).

Failure to achieve adequate chipping embedment usually occurs as a result of one or more of the following factors:

the asphalt is unworkable, ie too stiff;

the temperature of the asphalt during compaction is too low;

the asphalt cooled prematurely due to poor weather conditions during laying;

excessive application of pre-coated chippings;

inadequate compaction of the material, (see figure 7.11).

One method of treating patches where the chippings are inadequately embedded is to reheat the surface with an infra-red heater and, when the asphalt is sufficiently hot, embed the chippings further into the asphalt. Used judiciously this technique should not significantly harden the bitumen except at the surface; however, this will be rapidly eroded by traffic to expose the aggregate at the surface of the asphalt.

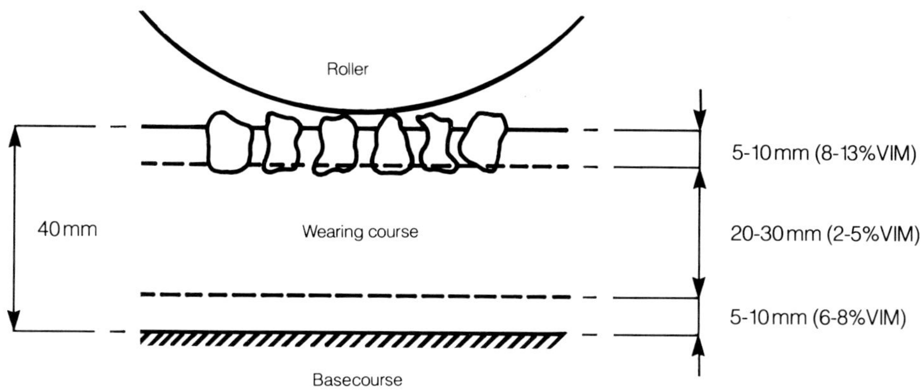

Figure 7.11 – Compaction of hot rolled asphalt wearing course under adverse weather conditions

Some chipping loss may occur as a result of failure to achieve or maintain bond between the pre-coated chipping and the asphalt. It has been found that the bitumen on pre-coated chippings can be 'coked' and made non-adhesive by storage of the chippings in large stockpiles immediately after manufacture. The adhesivity of the bitumen coating can be tested by the hot sand test which is described in BS598: Part 3: 1985[17].

When the chippings have been plucked out of the asphalt, asphalt mortar is exposed to traffic. It is likely that this material will be poorly compacted and as a result will fret rapidly.

7.2.3 The influence of the bitumen on the development of surface texture in hot rolled asphalt wearing course

It has long been accepted that to maintain a satisfactory non-skid surface on hot rolled asphalt it is necessary to have adequate surface texture obtained by rolling chippings into the surface, and by using a sand/filler/bitumen mortar that will 'weather' to allow new aggregate particles to be exposed.

Full-scale road trials carried out jointly by the Transport and Road Research Laboratory (TRRL)[18,19] and Shell International Petroleum Company Limited in the early sixties clearly showed that the origin or type of the bitumen was a primary factor in the 'weathering' of the surface of the mortar. Sideway-force coefficient (SFC) measurements over a period of eight years (normalising the results to main summer SFC values) clearly showed that the bitumen significantly influenced SFC. Bitumens incorporating a proportion of Trinidad Lake Asphalt (TLA) had high SFC values, whereas Venezuelan-based bitumens had significantly lower SFC values. This difference in performance was attributed to differences in the rates of hardening of the binders by atmospheric oxidation. The Venezuelan bitumen was not being eroded from the surface of the mortar and as a result was tending to produce a shiny, smooth surface, ie the bitumen was too durable. However, the soluble portion of the TLA at the road surface was very much harder than the original soluble binder, despite the softening effect of oil droppings.

It was thought, therefore, that if the susceptibility to oxidation of the Venezuelan bitumen could be increased to that of TLA, then, if oxidation was the principal factor influencing 'weatherability', this blended product should behave on the road similarly to TLA. Full-scale road trials and pressure-oxidation tests on Venezuelan/TLA and Venezuelan/coal-tar pitch blends showed that these mixtures had significantly higher rates of hardening, in some cases similar to TLA[20].

As a consequence of these findings, the first Ministry of Transport specification in 1963 required a proportion of vertical-retort coal-tar pitch in bitumen. This was subsequently relaxed to permit first the use of coke-oven pitch and later any coal-tar pitch carbonised at a temperature of over 600°C.

The Specification for Road and Bridge Works[21] published by the Ministry of Transport in 1969 specified that only bitumen/TLA or pitch/bitumen blends were acceptable for hot rolled asphalt surfacings for trunk roads and motorways. However, in the late sixties the disadvantages of pitch-bitumen were becoming more evident. Pitch-bitumen was showing a greater tendency to harden at high mixing temperatures than bitumen. At the same time its higher temperature susceptibility was resulting in greater deformation. In addition the cost of blending pitch with the bitumen was significantly increasing the cost of the product.

During the same period, the sources of crude oil suitable for the manufacture of bitumen had widened and more bitumen was being refined from Middle East crude oils. However, although bitumens manufactured from Middle East crude had weathering properties at least as good as pitch-bitumen mixtures they were still excluded by the Ministry of Transport specification.

There was therefore an urgent need for a laboratory test which would measure a property of the binder which would correlate with performance in practice. Considerable efforts were directed by the TRRL to discover the exact mechanisms involved in giving high skid resistance. Many different bitumen properties were investigaged in an attempt to discover a laboratory test which discriminated between bitumens of differing 'weatherability' and thereby provide a test which could be used to reject bitumens which were too durable. Green(22) showed that the ability of bitumens to weather under the combined effects of ultra-violet light in sunlight, oxidation, rain and oil deposition from traffic was related to the permittivity or dielectric constant of the bitumen.

7.2.3.1 Specification of permittivity

Road trials carried out by the TRRL on unchipped cores indicated that 50 pen bitumen required a minimum permittivity value of 2.650 to achieve a satisfactory surface texture on hot rolled asphalt surfacings. Cores made with 35 pen and 70 pen bitumens confirmed that the development of skid resistance was related to permittivity but was independent of the penetration of the bitumen. Consequently 50 pen bitumen was included in the Department of Transport specification(23) provided its permittivity at 25°C and 1.59 kHz was at least 2.650. A value of 2.630 for 35 pen bitumen was subsequently included in the specification(24) and values for 35, 50, and 70 pen bitumen were included in the revised edition of British Standard (BS) 3690(25) published in 1982. Since the introduction of the permittivity specification there has been a number of developments in hot rolled asphalt technology which question the relevance of the permittivity criteria specified in BS 3690 : Part 1 : 1982, viz:

- the use of the Marshall test to design hot-rolled asphalt mixes has tended to result in a small reduction in the bitumen content;
- the rates of spread of pre-coated chippings have increased, and it is now common to apply them at a rate that covers over 70% of the surface in order to achieve texture depth specifications; thus, the surface area of the mortar in contact with the vehicle tyre is substantially reduced;
- the rate of deposition of oil from constant loss lubrication systems has virtually disappeared as more efficient systems are now being fitted to commercial vehicles.

A full-scale road trial was laid on the M4(26) in 1982, using bitumens covering a wide range of permittivities, to re-assess the permissible level of bitumen permittivity in the light of the above changes. Measurements of texture depth on this trial showed that there would be no loss in developed texture if the permittivity for all grades of bitumen was reduced to a minimum value of 2.630. Consequently a uniform permittivity of 2.630 has been applied to 35, 50, 70 and 100 pen bitumens(27,28).

7.3 References

(1) Whiteoak, C D and Fordyce, D: *Asphalt workability. Its measurement, and how it can be modified.* Shell Bitumen Review 64, pp 14-17, September 1989.

(2) Jacobs, F A: *Hot rolled asphalt: Effect of binder properties on resistance to deformation.* Transport and Road Research Laboratory, Report LR 1003, 1981.

(3) Daines, M E: *Cooling of bituminous layers and time available for their compaction.* Transport and Road Research Laboratory, Report RR 4, 1985.

(4) Dormon, G M: *Some observations on the properties of bitumen and their relation to performance in practice and specifications.* Shell Construction Service Report.

(5) Heukelom, W: *Observations on the rheology and fracture of bitumens and asphalt mixes.* Proceedings of the Association of Asphalt Paving Technologists, vol 35, pp 358-399, 1966.

(6) Burgess, R A et al: *Relationships between predicted fracture temperatures and low temperature field performance.* Proceedings of the Association of Asphalt Paving Technologists, vol 40, February 1971.

(7) Hills, J F and Brien, D: *The fracture of bitumens and asphalt mixes by temperature induced stresses.* Proceedings of the Association of Asphalt Paving Technologists, vol 35, pp 292, 1966.

(8) Fabb, T R J: *The influence of mix composition, binder properties and cooling rate on asphalt cracking at low temperatures.* Proceedings of the Association of Asphalt Paving Technologists, vol 43, pp 285-331, February 1974.

(9) Burt, A R: *M4 motorway, a composite pavement. Surface cracking.* Highways and Transportation, vol 34, pp 16-19, December 1987.

(10) Nunn, M E: *An investigation of reflection cracking in composite pavements.* Proceedings of the conference on reflective cracking in pavements, pp 146-153, Liège, March 1989.

(11) Dormon, G M and Snashall, H M: *The mechanical properties of bitumen in relation to road performance.* Society of Chemical Industry, March 1971.

(12) Department of Transport: *Texture depth of bituminous surfacings for high speed roads.* Departmental Standard HD/3/80.

(13) Department of Transport: *Texture depth of bituminous surfacings for high speed roads.* Advice Note HA/8/80.

(14) Johnson, W M and Salt, G F: *Loss of chippings from rolled asphalt wearing courses.* Paper presented at the Institution of Highways and Transportation National Conference, Keele University, September 1985.

(15) Johnson, W M and Salt, G F: *Loss of chippings from rolled asphalt wearing courses.* Highways and Transportation, vol 32, pp 22-27, July 1985.

(16) White, M J: *Flexible pavements and bituminous materials.* Residential course at Newcastle University, section G, *Bituminous materials – special requirements, problems and processes,* pp G1-G16, September 1989.

(17) British Standard 598: Part 3: 1985: *Sampling and examination of bituminous mixtures for roads and other paved areas. Part 3. Methods for design and physical testing.*

(18) Please, A and Lamb, D R: *The role of binder in the maintenance of resistance to skidding of dense asphalt surfacings under heavy traffic.* Transport and Road Research Laboratory, Report LR 319, 1970.

(19) Please, A and Lamb, D R: *Binder properties and the texture of asphalt surfacings.* Proceedings of the Association of Asphalt Paving Technologists, vol 40, 1971.

(20) Lee, A R and Dickinson, E J: *Improvements in bituminous road surfacing materials.* British Patent No 714,153, 1954.

(21) Ministry of Transport: *Specification for Road and Bridge Works.* London, 1969. (HMSO).

(22) Green, E H: *An acceptance test for bitumen for rolled asphalt wearing course.* Transport and Road Research Laboratory, Report LR 777, 1977.

(23) Department of Transport: *Specification for road and bridge works. Amendment to specification for bitumen binder in rolled asphalt wearing course.* Technical Memorandum H1/77, London, 1977.

(24) Department of Transport: *Specification for road and bridge works. Rolled asphalt wearing course*. Departmental Standard HD/2/79, London, 1979.

(25) British Standard 3690: Part 1: 1982. *Bitumens for building and civil engineering. Part 1. Specification for bitumens for road purposes.*

(26) Daines, M E: *Bitumen permittivity and texture depth of rolled asphalt: An experiment on motorway M4*. Transport and Road Research Laboratory, Supplementary Report 816, 1983.

(27) British Standard 3690: Part 1: 1982: *Bitumens for building and civil engineering. Part 1. Specification for bitumens for road purposes*. Draft amendment number 1. Document number 88/13214 DC, July 1988.

(28) British Standard 3690: Part 1: 1989. *Bitumens for building and civil engineering. Part 1. Specification for bitumens for road purposes.*

Photograph 7.5 – Application of chippings to hot rolled asphalt. *(Photograph reproduced by kind permission of Redland Aggregates Limited)*

Chapter 8 – Durability of bitumen

Long-term studies have shown that if a bituminous surfacing is to achieve its design life it is important that the bitumen is not excessively hardened during storage, during the manufacturing process or in service on the road. Bitumen in common with many organic substances is affected by the presence of oxygen, ultra-violet radiation and by changes in temperature. In bitumen these external influences cause it to harden, resulting in a decrease in penetration, an increase in softening point and an increase in penetration index (PI).

8.1 Bitumen hardening

The tendency for bitumen to harden under atmospheric influences has been known and studied for many years. Four principal mechanisms of bitumen hardening have been identified:

- oxidation;
- loss of volatiles;
- physical hardening;
- exudative hardening.

8.1.1 Oxidation

Like many organic substances, bitumen slowly oxidises when in contact with air. The polar oxygen-containing groups formed tend to associate into micelles of higher molecular weight and thereby increase the viscosity of the bitumen. Polar hydroxyl, carbonyl and carboxylic groups are formed resulting in larger and more complex molecules which make the bitumen harder and less flexible. The degree of oxidation is highly dependent on the temperature, time and the thickness of the bitumen film. Oxidation rate doubles for each 10°C increase in temperature above 100°C. Hardening due to oxidation has long been held to be the main cause of ageing, to the extent that other factors have hardly been considered. However, it has been shown that although other factors are generally less important than oxidation they are certainly measurable.

8.1.2 Loss of volatiles

Evaporation of volatile components depends mainly upon temperature and the conditions of exposure. Penetration grade bitumens are relatively involatile and therefore the amount of hardening resulting from loss of volatiles is usually fairly small.

8.1.3 Physical hardening

Physical hardening occurs when the bitumen is at ambient temperature and is usually attributed to reorientation of bitumen molecules and the slow crystallisation of waxes. Physical hardening is reversible in that upon reheating the original viscosity of the bitumen is obtained.

This phenomenon can be easily reproduced in the laboratory by measuring the penetration of a sample of bitumen (without reheating) over a period of time. As time progresses the penetration of the sample will fall.

8.1.4 Exudative hardening[1]

Exudative hardening results from the movement of an oily component which exudes from the bitumen into the mineral aggregate. It is a function of both the exudation tendency of the bitumen (see chapter 3, section 3.5.3.2) and the porosity of the aggregate.

8.2 Hardening of bitumen during storage, mixing and in service

The circumstances under which hardening occurs vary considerably. During storage the bitumen is in bulk at a relatively high temperature for a period of days or weeks; during mixing, hot storage, transport and application it is a thin film at high temperature for a relatively short period, and in service it is a thin film at a low or moderate temperature for a very long time. In service access of air is important and is dependent on the void content of the mix; little change normally occurs in dense, well compacted mixes.

8.2.1 Hardening of bitumen in bulk storage

When bitumen is stored in bulk at high temperature very little hardening occurs. This is because the surface of the bitumen exposed to the oxygen is very small in relation to the large volume. However, if the bitumen is being circulated and is falling from the pipe entry at the top of the tank to the surface of the bitumen, significant hardening may occur. This is because the surface area of the bitumen, falling from the entry pipe will be relatively large, providing the opportunity for oxygen to react with the bitumen (see chapter 2, section 2.2.1).

8.2.2 Hardening of bitumen during mixing with aggregate

During the mixing process all the aggregate and filler is coated with a thin film of bitumen usually between 5 microns and 15 microns thick. If each coated aggregate surface could be spread out into a continuous flat surface, the bitumen from 1 tonne of a dense bitumen macadam would produce a film of bitumen approximately 10,000 m^2 in area, ie over 1.5 average football pitches. Thus when bitumen is mixed with hot aggregate and spread into thin films in an asphalt pugmill, conditions are very favourable for oxidation and loss of volatiles. Hardening of bitumen during this process is well known and is taken into account when selecting the grade of bitumen to be used. As a *very* rough rule-of-thumb, during mixing bitumens harden by approximately one grade, eg 100 pen to 70 pen, 70 pen to 50 pen, etc. However, the amount of hardening depends on a number of factors such as temperature, duration of mixing, bitumen content, ie bitumen film thickness, etc. The prevention of detrimental hardening during mixing depends on careful control of all these factors. Temperature control and bitumen content in particular are critical. Figure 7.2 clearly shows the additional bitumen hardening, measured by increasing softening point, as mixing temperatures are increased. Similarly, figure 8.1 shows that reducing the thickness of the bitumen film significantly increases the ageing index (ratio of the viscosity of the aged bitumen, η_a, to the viscosity of the virgin bitumen, η_o) of the bitumen [2].

The type of mixer used also affects the amount of hardening during mixing. It has been recognised for some time that the amount of hardening in a drum mixer is often less than in a conventional batch mixer[3]. This is probably the result of shorter bitumen/aggregate mixing times and the presence of steam in the drum, limiting the availability of oxygen. However, the multiplicity of different designs of drum mixers means that variation in the amount of hardening between different designs of plant is almost inevitable. But, accepting this, a study carried out by Shell Bitumen U.K. on two different drum mixers has shown that, for equivalent mixing temperatures, the overall reduction in penetration is 5% to 15% lower and the increase in softening point is up to 4°C lower in a drum mixer than a conventional batch mixer.

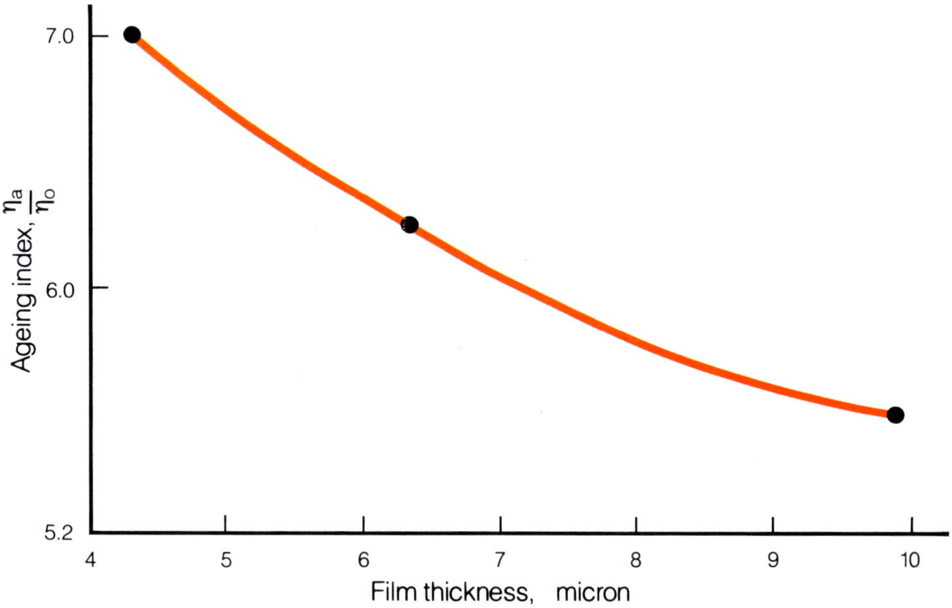

Figure 8.1 – Effect of film thickness on the ageing index(2)

8.2.3 Hardening of bitumen in a mix during hot storage, transport and laying

The introduction into the U.K. of high throughput drum mixers has necessitated the use of hot silo storage for mixed materials, to provide a cushion between continuous production and peaks and troughs in demand. Thus, mixes are not only being stored in the delivery vehicle during transport but also in hot silo storage.

Some hardening of the bitumen will take place during hot storage, be it in a silo or in a vehicle. The amount of hardening will, as discussed previously, depend principally upon duration of exposure to oxygen, the thickness of the bitumen film and the temperature of the mix. When a mix is discharged into a storage silo, air enters with the mix and some is trapped in the voids of the material. During storage some of the oxygen in this entrained air will react with the bitumen. If no additional air enters the silo, oxidation of the bitumen will cease.

To ensure that no additional air enters the silo, it is important that the entry and discharge gates are air tight and there are no other openings where air can enter the silo. If the discharge gate is not air tight the silo may behave like a chimney drawing air in at the discharge gate (which exits the loading gate) resulting in oxidation and cooling of the stored material. In addition the silo should be as full as practicable to eliminate as much free air from the top of the silo as possible. Air remaining at the top of the silo will react with the top surface of the mix. However, this reaction forms carbon dioxide, which, because it is heavier than air, tends to blanket the surface of the mix protecting it from further oxidation. In the U.S.A. some silos have the facility to be pressurised with exhaust gases, containing no oxygen, from a burner. The exhaust gases purge the silo of entrained air and provide a slight positive pressure preventing air entering the silo.

Studies carried out in the U.S.A.[4] suggest that if oxidation in the silo is limited to that induced by entrained air, then little or no additional oxidation will occur during transport and laying. This, it was hypothesised, was because no significant quantity of fresh air is entrained in the mix during vehicle loading. Thus little or no additional oxygen was available for oxidation. In fact, it was observed that if the mix was discharged directly from the pugmill into the delivery vehicle, the amount of hardening during transport was very similar to that occurring during silo storage.

If materials are being laid at low ambient temperatures or if the mix may have to be retained for a period in hot storage, there is a temptation to increase mixing temperatures to offset these two factors. However, increasing the mixing temperature will considerably accelerate the rate of bitumen oxidation, resulting in an increase in bitumen viscosity. Thus a significant proportion of the reduction in viscosity achieved by increasing the mixing temperature will be lost because of additional oxidation of the bitumen, which may ultimately adversely affect long-term performance of the material.

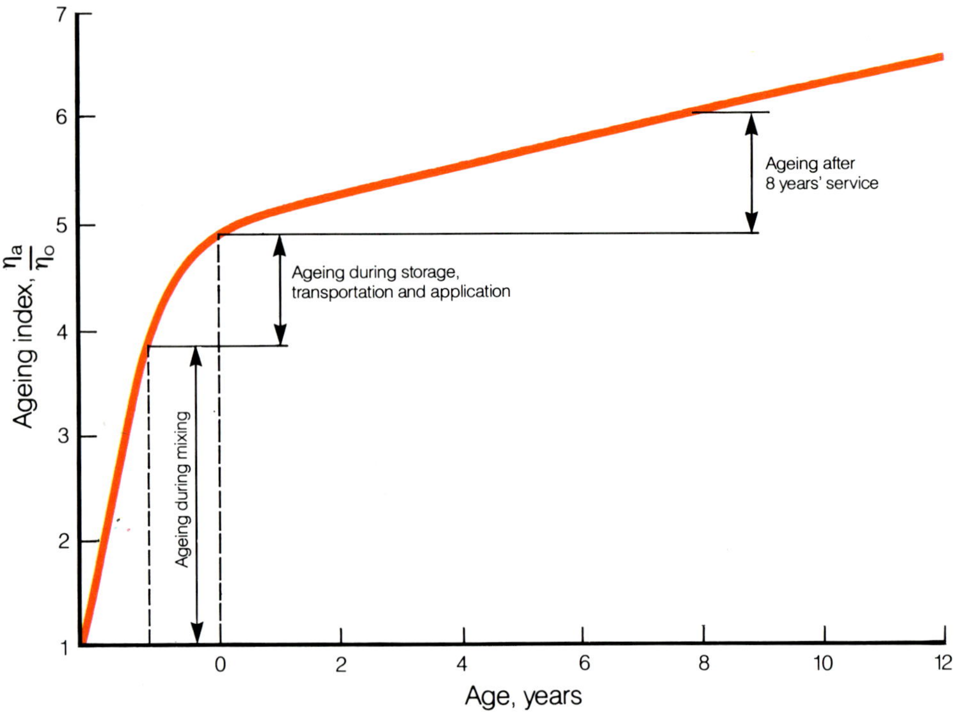

Figure 8.2 – Ageing of bitumen during mixing, subsequently during storage, transportation and application and finally in service

8.2.4 Hardening of bitumen on the road

The majority of bitumen hardening occurs during mixing and to a lesser extent during hot storage/transport. However, under adverse conditions hardening of the bitumen can also occur on the road. This is illustrated in figure 8.2 which shows the ageing index of the bitumen after mixing, storage, transport, application and subsequently in service.

The main factor which influences bitumen hardening on the road is the void content of the mix. Table 8.1 shows bitumens recovered from three asphaltic concrete mixes after 15 years service[5]. The bitumen recovered from the mix with the lowest void content had hardened very little. However, where the void content was high, allowing constant ingress of air, substantial hardening had occurred. The PI of the material with the highest void contents has increased appreciably; limiting the amount of stress relaxation which can occur, and may result in cracking of the compacted material. Similarly, figure 8.3 shows the *in-situ* bitumen properties of five year old asphaltic concrete with void contents ranging from 3% to 12%[6]. At void contents less than 5% very little hardening in service has occurred. However, at void contents greater than 9% the *in-situ* penetration has fallen from 70 to less than 25.

Road	A	B	C
Voids in mix, %	4	5	7
Properties after mixing and laying			
Softening point (IP), °C	64	63	66
Penetration at 25°C, dmm	33	33	30
Penetration index (Pen, SP)	+0.7	+0.7	+0.9
Stiffness (S_0), Pa (calc)			
10^4 secs, 25°C	1.4×10^3	1.4×10^3	2.5×10^3
10^4 secs, 0°C	5.0×10^5	5.0×10^5	7.0×10^5
10^{-2} secs, 25°C	2.5×10^7	3.0×10^7	3.0×10^7
10^{-2} secs, 0°C	3.0×10^8	3.0×10^8	3.0×10^8
Properties after 15 years service:			
Softening point (IP), °C	68	76	88
Penetration at 25°C, dmm	24	15	11
Penetration index (Pen, SP)	+0.8	+1.1	+2.1
Stiffness (S_{15}), Pa (calc)			
10^4 secs, 25°C	4×10^3	20×10^3	150×10^3
10^4 secs, 0°C	13×10^5	40×10^5	80×10^5
10^{-2} secs, 25°C	4×10^7	7×10^7	8×10^7
10^{-2} secs, 0°C	4×10^8	6×10^8	6×10^8
Ageing index (S_{15}/S_0)			
10^4 secs, 25°C	2.8	14	60
10^4 secs, 0°C	2.6	8	11
10^{-2} secs, 25°C	1.6	2.3	2.7
10^{-2} secs, 0°C	1.3	2.0	2.0

Table 8.1 – Hardening of bitumen in service[5]

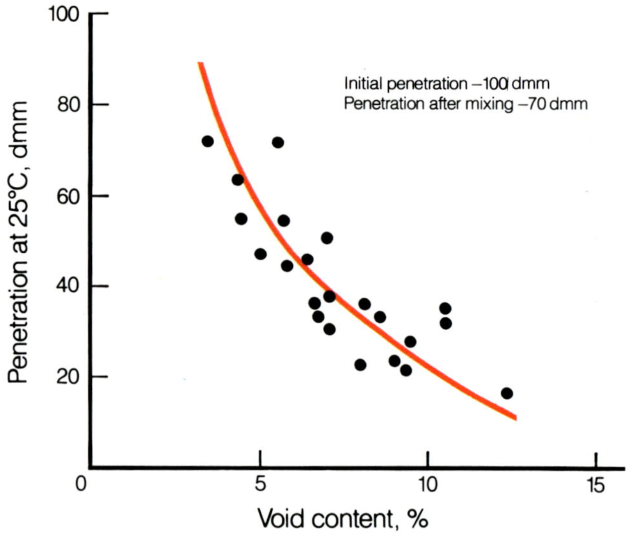

Figure 8.3 – The effect of void content on the hardening of bitumen on the road[6]

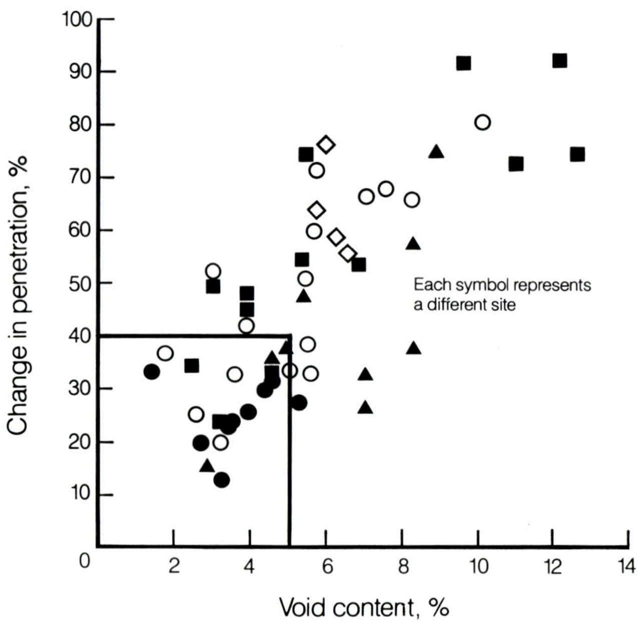

Figure 8.4 – The effect of void content on the hardening of high penetration index bitumen

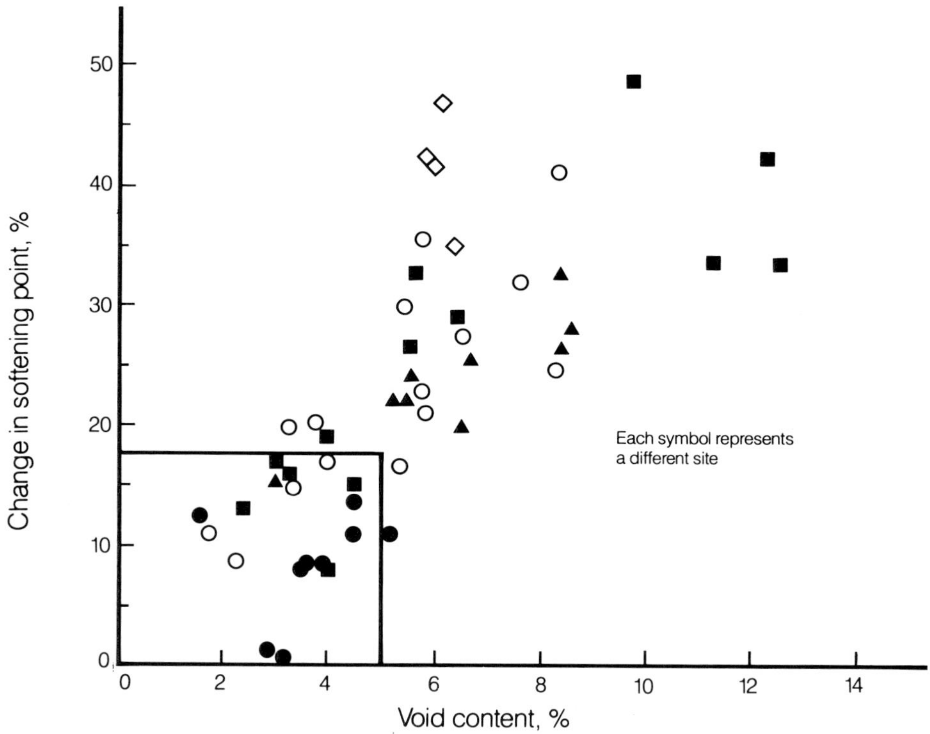

Figure 8.5 – The effect of void content on the hardening of high penetration index bitumen

In 1982 an extensive study was carried out by Shell Bitumen U.K. on hot rolled asphalt wearing course cores from sites where high PI bitumen had been used. The cores were taken from sites of varying performance and various ages and the important properties influencing the performance of the asphalt in practice were determined, viz:

level of compaction achieved;

mix composition;

in-situ bitumen properties.

In general the mix analyses showed no major deviation from the specified aggregate grading and bitumen content. However, trends did emerge relating void content and change in bitumen properties (including hardening during mixing and transport). Where good performance was reported by the local authority, the level of compaction was generally good, less than 5%, and no abnormal hardening of the bitumen *in-situ* was found. However, on sites where poor performance had been observed, more variable compaction was found with void contents up to 12%, associated with severe hardening of the bitumen. Figures 8.4 and 8.5 show the percentage change in penetration and softening point respectively. In both cases the trend is seen for the greatest hardening in mixes where compaction was poor, and where good compaction was achieved, ie void contents less than 5%, the amount of hardening was usually acceptable.

Figure 8.6 shows data on hot rolled asphalt wearing course, manufactured using 50 pen bitumen, from the Colnbrook bypass experiment(7). This shows that after eight years in service little additional compaction of the asphalt had occurred and the void content remained at about 5%. In this well compacted material no significant bitumen hardening had occurred.

The bitumen at the surface of the road hardens much faster than the bitumen in the bulk of the asphalt. The reasons for this are threefold:

- a constant supply of fresh oxygen;
- higher temperatures at the road surface;
- photo-oxidation of the bitumen by ultra-violet radiation.

Photo-oxidation causes a skin 4 microns to 5 microns thick to be rapidly formed on the surface of the bitumen film. This is induced by natural ultra-violet radiation which, it is believed, is absorbed in the upper 10 microns of a film of bitumen. It has been shown that the skin formation can retard oxygen absorption and loss of volatiles. However, the oxidised material is highly soluble in rainwater and is easily abraded away, so exposing fresh bitumen(8).

Although both continuously graded and gap-graded mixes are regarded as dense, gap-graded materials are considered to be more durable. They are generally less permeable to air than continuously graded mixes for similar air void contents(9). This is because the voids in gap-graded mixes are discrete and not interconnecting as is usually the case with continuously graded mixes.

The speed of surface oxidation is illustrated in figure 8.7 in which penetration, softening point, and PI of bitumens recovered from the surface and bulk of an asphaltic concrete mix over a number of years are plotted(6). After seven years the penetration at the surface is 25, whereas the bitumen in the bulk of the mix has a penetration of 45. Similarly there is a substantial difference in the PI. In the bulk of the mix the PI has only marginally changed, whereas the PI of the bitumen at the surface has increased to over four. Surface oxidation is, in fact, desirable because it enables bitumens to be eroded away relatively quickly, exposing new aggregate surfaces and thereby improving the skid resistance of the road surface. This is discussed in depth in chapter 7, section 7.2.3.

Bitumen content (bitumen film thickness) also plays a very important role in the speed of bitumen hardening on the road. Laboratory work has shown(10) that the oxidation of a film of bitumen, at high ambient temperatures, ie between 40°C and 60°C, is limited to a depth of about 4 microns as shown in figure 8.8. The significance of this will depend on the average bitumen film thickness, which as stated earlier can range from 5 microns to 15 microns. It has been suggested(11) that a minimum bitumen film thickness of 6 microns and 8 microns is required for satisfactory performance of continuously graded and gap-graded mixes respectively. Thus in general the lower the bitumen content, the thinner the bitumen film and the faster hardening will occur. The bitumen film thickness can be determined using the method given in chapter 20, section 20.4. Typical bitumen film thicknesses are as follows:

dense bitumen macadam	$>$5 micron
hot rolled asphalt	$>$7 micron
pervious macadam	$>$12 micron

This method of calculating bitumen film thickness determines surface area factors for the aggregate down to 75 micron. Clearly some of the material passing this sieve will be substantially finer than 75 micron, particularly for materials where limestone filler is added to the mix. Therefore, for high filler content materials such as hot rolled asphalt the literal bitumen film thickness will be lower than the above figure. However, if we consider that the bitumen and filler (the binder), coat the coarse and fine aggregate then the above figure will underestimate the *binder* film thickness.

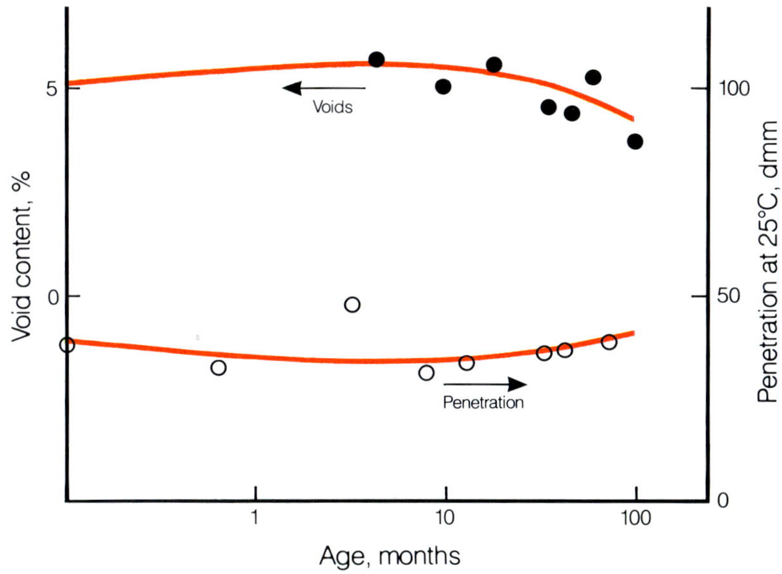

Figure 8.6 – Changes in void content and bitumen penetration, Colnbrook bypass experiment[4]

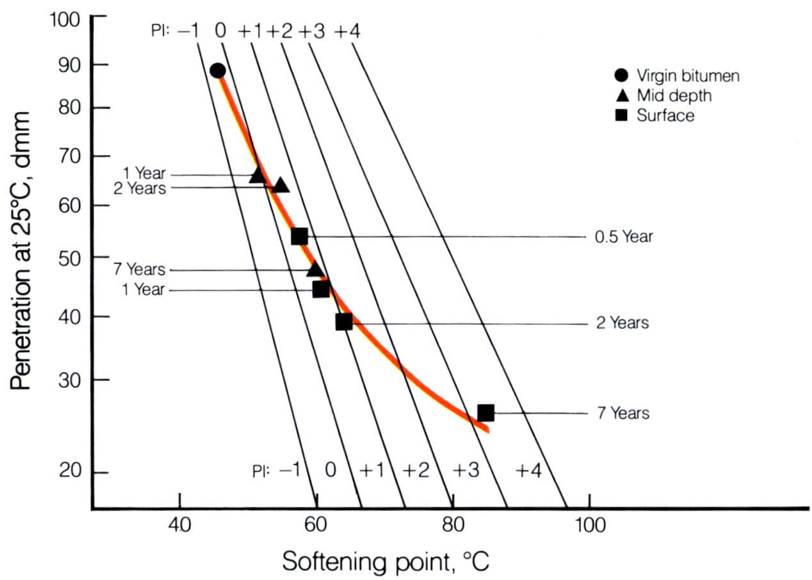

Figure 8.7 – Hardening of bitumen at the surface and at mid-depth of an asphaltic concrete wearing course[6]

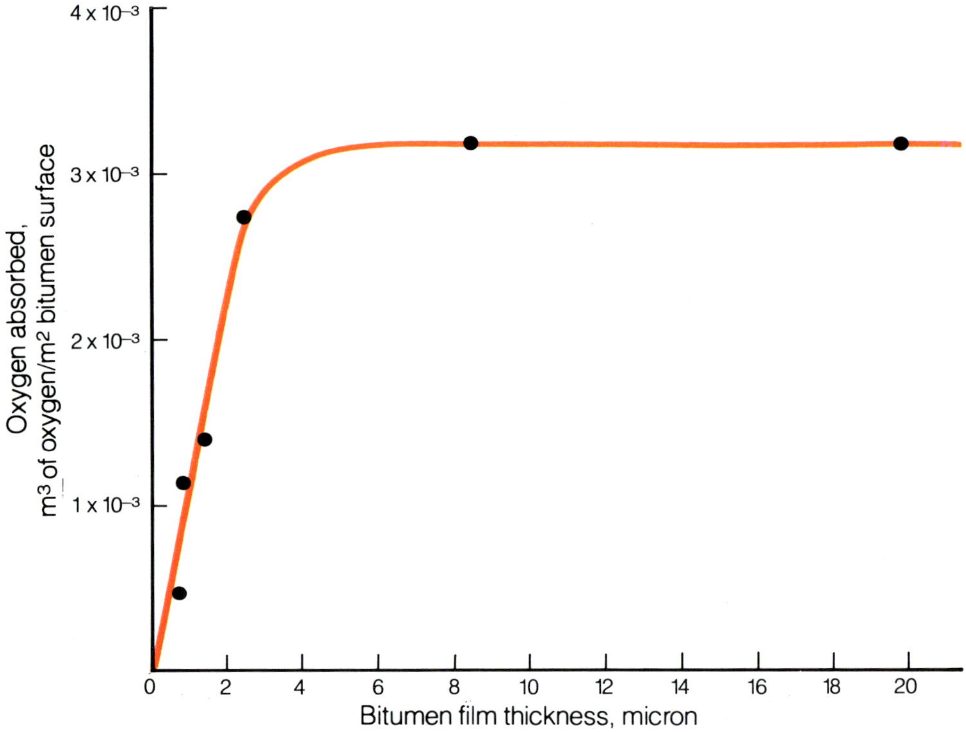

Figure 8.8 – Relationship between oxygen absorbed at 50°C and film thickness[9]

8.3 Ageing tests

It is clearly desirable that there should be laboratory tests that determine quantatively the resistance of bitumens to hardening at the various stages during the production process. A number of tests already exist to measure the effect of heat and air on bitumen. The main aim of these tests is to identify bitumens which are too volatile or are too susceptible to oxidation to give good performance in service.

The loss on heating test (BS 2000 : Part 45, ASTM D 6, IP 45) is a part of many bitumen specifications. This test is suitable for ranking bitumens according to their tendency to harden and indicates if a material has been contaminated with light oils. The ageing conditions in the test are akin to those in bulk storage but are far removed from those during mixing. The main disadvantages of the test are that the surface-to-volume ratio of the sample is too low and an oxidised skin tends to form quickly on the surface of the bitumen. The skin hampers further oxidation and evaporation, whereas under practical mixing conditions homogeneous hardening of the bitumen in thin films occurs on the aggregate.

The thin film oven test (TFOT) (ASTM D 1754) simulates practical conditions somewhat better. In this test the bitumen is stored at 163°C for five hours in a layer 3.2 mm thick. It is claimed that in this test the amount of hardening that takes place is about the same as that obtained in practice. However, in this test diffusion in the bitumen film is also limited and it is not possible to obtain homogeneous hardening or ageing, and therefore the test is far from ideal.

In 1963 the State of California Department of Public Works, Division of Highways, developed a test which simulates far better what happens to a bitumen during mixing – the rolling thin-film oven test (RFTOT) [12,13]. In this test eight cylindrical glass containers containing 35 g of bitumen are fixed in a vertically rotating shelf (see photograph 3.3). During the test the bitumen flows continuously around the inner surface of each container in a relatively thin film with pre-heated air periodically blown into each glass jar. The test temperature is normally 163°C for a period of 75 minutes. The method ensures that all the bitumen is exposed to heat and air, and continuous movement ensures that no skin develops to protect the bitumen. A homogeneously aged material is obtained, similar to that produced during full-scale mixing. Clearly the conditions in the test are not identical to those found in practice but experience has shown that the amount of hardening in the RTFOT correlates reasonably well with that observed in a conventional batch mixer.

The RTFOT was accepted in 1970 by the American Society for Testing Materials as method ASTM D 2872.

8.4 References

(1) van Gooswilligen, G, de Bats, F Th and Harrison, T: *Quality of paving grade bitumen – A practical approach in terms of functional tests.* Proceedings of the fourth Eurobitume symposium, pp 290-297, Madrid, October 1989.

(2) Griffin, R L, Miles, T K and Penther, C J: *Microfilm durability test for asphalt.* Proceedings of the Association of Asphalt Paving Technologists, vol 24, p 31, 1955.

(3) Haas, S: *Drum-dryer mixers in North Dakota.* Proceedings of the Association of Asphalt Paving Technologists, vol 43, p 417, 1974.

(4) Brock, J D: *Oxidation of asphalt.* Astec Industries technical bulletin T-103, 1986.

(5) Edwards, J M: *Dense bituminous mixes.* Paper presented to the conference on road engineering in Asia and Australasia, Kuala Lumpur, June 1973.

(6) Lubbers, H E: *Bitumen in de weg- en waterbouw.* Nederlands Adviesbureau voor Bitumentoepassingen, April 1985.

(7) Please, A and Lamb, D R: *The role of the binder on the maintenance of resistance to skidding of dense asphaltic surfacings under heavy traffic.* Road Research Laboratory, Report LR 319, 1970.

(8) Dickinson, E J, Nicholas, J H and Boas-Traube, S: *Physical factors affecting the absorption of oxygen by thin films of bituminous binders.* Journal of Applied Chemistry, vol 8, p 673, 1958.

(9) Marais, C P: *Gap-graded asphalt surfacings – the South African scene.* Proceedings of the second conference on asphalt paving for Southern Africa, vol 3, 1974.

(10) Blokker, P C and Van Hoorn, H: *Durability of bitumen in theory and practice.* Proceedings of the fifth World Petroleum Congress, paper 27, p 417, 1959.

(11) Campen, W H, Smith, J R, Erickson, L G and Mertz, L R: *The relationship betweeen voids, surface area, film thickness and stability in bituminous paving mixtures.* Proceedings of the Association of Asphalt Paving Technologists, vol 28, p 149, 1959.

(12) Hveem, F N, Zube, E and Skog, J: *Proposed new tests and specifications for paving grade asphalts*. Proceedings of the Association of Asphalt Paving Technologists, vol 32, 1963.

(13) *The rolling thin-film oven test*. Shell Bitumen Review 42, pp 18-19, 1973.

Chapter 9 – Adhesion of bitumen

One of the principal functions of bitumen is to act as an adhesive to bind together aggregate particles. This it has been doing with outstanding success on roads in the U.K. for over 70 years and the incidence of adhesion failure in well compacted dense bituminous mixes is rare. There are occasions, however, when the adhesive bond may fail and the durability of the material may be seriously reduced.

The adhesion of bitumen to the vast majority of aggregates presents few problems in the absence of water (although excessive dust tends to mop up bitumen, resulting in thinner bitumen films making the material less durable). However, because aggregate is wetted more easily by water than by bitumen, the presence of water results in difficulties, either in the initial coating of the aggregate or maintaining an adequate bond between the bitumen and the aggregate. Failure of the bond already formed is commonly referred to as 'stripping', which is brought about by the displacement of the bitumen from the aggregate surface by water.

The nature of the aggregates exerts a considerable influence on bitumen adhesion due to differences in their affinity for bitumen. The majority of aggregates are classified as 'hydrophilic' (water loving) or 'oleophobic' (oil hating). Aggregate types with a high silicon oxide content, eg quartz and granite, ie acidic aggregates, are more difficult to coat with bitumen than basic aggregates such as basalt or diabase.

9.1 Adhesion in the absence of water

Bitumen adheres well to most types of road aggregate provided that they are clean, dry and not too dusty. To obtain an adhesive bond it is necessary for the bitumen to 'wet' the aggregate; this process is largely controlled by the viscosity of the bitumen. The higher the viscosity, the longer it will take the bitumen to wet the aggregate. For practical purposes it is necessary that wetting should occur almost instantaneously, and therefore, too high a viscosity could result in inadequate coating. This is particularly important if the aggregate is dusty because there is always a tendency for the layer of dust on the aggregate to be preferentially coated by the bitumen, increasing its viscosity and possibly resulting in inadequate coating of the large aggregate. It is unusual for the bond between the bitumen and the aggregate to fail once good adhesion has been established, except by the action of water.

9.2 Adhesion in the presence of water

There would be practically no adhesion problems in bitumen-bound roads were it not for a characteristic feature of the British climate – rain. In any one year roads in the U.K. are wet for about the number of hours equal to 100 times the rainfall in inches[1]. Effectively this means that they are wet for a quarter to half the year, depending upon location.

Adhesion problems in the presence of water arise in two ways: firstly due to the aggregate being wet before mixing and secondly due to the effect of rain on the material after it has been laid. It is normally not possible to coat a cold wet aggregate with bitumen and therefore the water film is removed by heating the aggregate, thereby keeping the viscosity of the bitumen low during the mixing process and enabling the bitumen to spread over the surface of the aggregate unimpeded by water.

9.2.1 Adhesion failures caused by the presence of water

Detachment of bitumen off the aggregate (or stripping) is associated with mixes which are permeable to water. There is little risk of stripping in low void content dense asphalt or macadam. In materials that are permeable to water, even those that are relatively dense, there is a risk of stripping, resulting in a loss of internal cohesion and possibly disintegration of the surfacing. The risk of stripping occurring is, as stated earlier, a function of the affinity between the aggregate and the bitumen and its consequent ability to resist the displacing effect of water. Most aggregates offer good resistance; some, unfortunately, do not. This does not mean that wet weather inevitably brings problems with certain aggregates; it merely implies a greater risk. Practical experience over many years has shown that surfacings which are manufactured using aggregates which have a mediocre affinity for bitumen often perform satisfactorily in normal conditions; the majority of failures of these materials are associated with abnormally adverse weather and traffic.

A national survey of the problem of premature deterioration of bitumen macadam wearing courses(2), showed that a number of cases of premature deterioration were attributable to adhesion and stripping problems. Results from the survey and observations from a number of sites(3) showed that two types of deterioration, which were dependent on material properties, were recognised:

- deterioration of the weaker types of coarse aggregate in carpets where the aggregate was of variable quality (although giving acceptable test results); initially pits were formed on the road surface which eventually coalesced to form potholes;
- breakdown of the aggregate/bitumen bond leading to lack of adhesion and stripping, producing a more generalised fretting of the surface; it occurred with both weak and hard aggregates which possess a poor affinity for bitumen.

The nature of the mineral aggregate, particularly the filler, is the dominant factor influencing adhesion. Reported differences in bitumens are largely due to differences in rheological properties. Maintenance of bitumen adhesion is governed by the minimum viscosity of the bitumen in service. This can be illustrated by comparing the Marshall properties of asphalt before and after soaking in water for 4 days at 60°C(4). These tests showed that differences between different bitumens were primarily due to differences in void content of the specimens tested. The variation in void content was the result of the specimens being compacted at the same temperature and not at an equi-viscous temperature. Differences between bitumens were eliminated when specimens were compacted at equi-viscous temperatures.

Figure 9.1 shows the retained Marshall stability plotted against the amount of water absorbed during 4 days soaking at 60°C. The graph clearly shows the importance of absorbed water and the dominance of aggregate and filler type. The volume of water absorbed depends on the void content of the mix and, for a particular void content, the water absorbed is significantly lower if the filler is limestone, and even lower if hydrated lime is present. For retained stabilities less than 30%, the amount of water absorbed was greater than the void content of the mixes, showing that considerable swelling must have taken place. This suggests that some failures which appear to be due to loss of adhesion between the aggregate and the bitumen, may, in fact, be due to a breakdown of the structure of the mix.

9.3 Fundamental properties of the bitumen/aggregate/water system

The displacement of one liquid by another on a solid surface arises from the physio-chemical forces acting on the system. The surfaces of most aggregates are electrically charged; for example, silica,

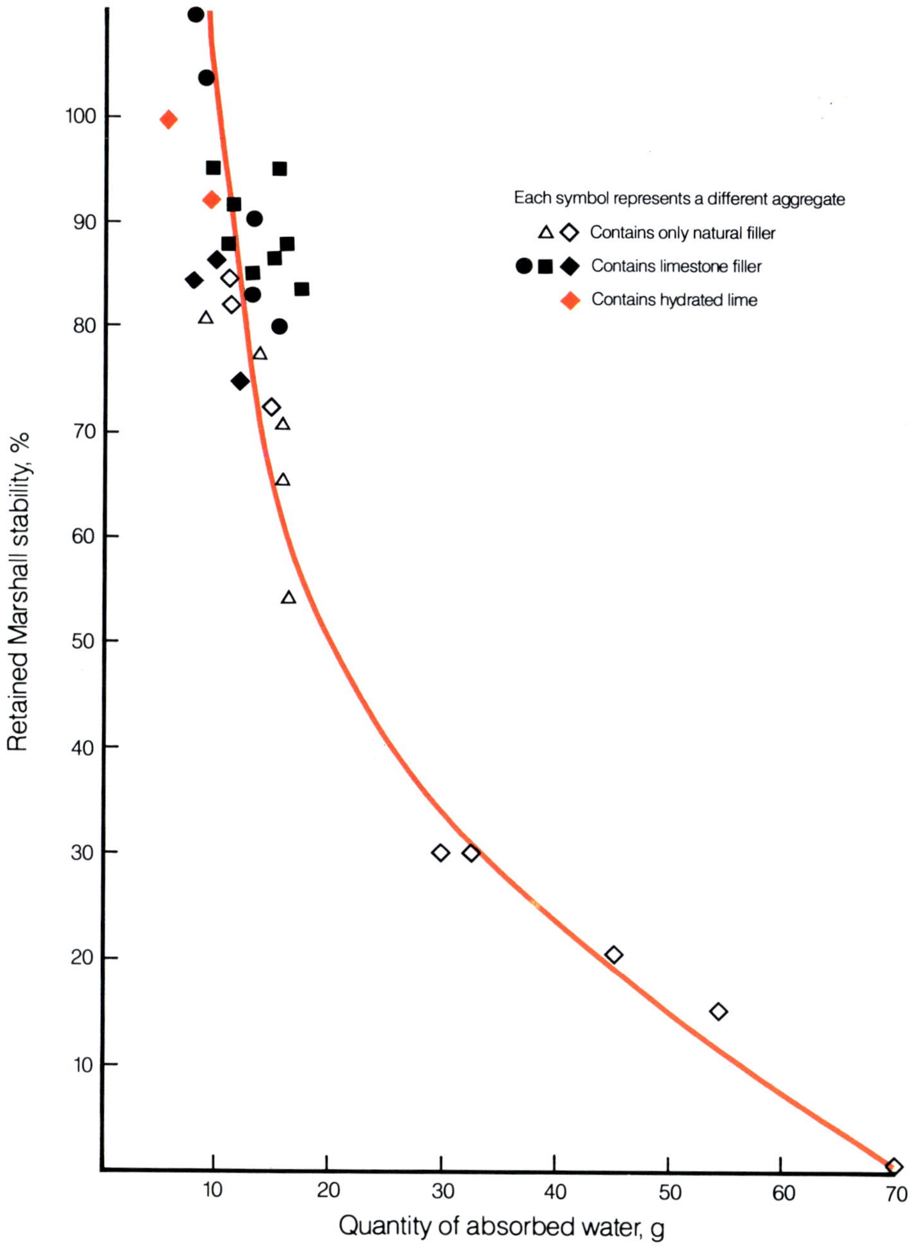

Figure 9.1 – The effect of absorbed water on the retained Marshall stability[4]

which is a common constituent of igneous rocks, possesses a weak negative surface charge resulting from the presence at the surface of oxygen atoms which are not fully saturated electrically. The chemical bond between bitumen and aggregate is primarily due to relatively weak dispersion forces, whereas water, which is very polar, is strongly attracted to the charged aggregate surface by orientation forces.

Aggregate surfaces are therefore to a greater or lesser degree hydrophilic (water loving) or oleophobic (oil hating). Acidic aggregates are more hydrophilic than basic aggregates. If the surface of an aggregate is wet, it is almost impossible to disperse the water with bitumen and for the bitumen to adhere to the aggregate. However, it is possible for water to penetrate a film of bitumen and detach the bitumen from the aggregate.

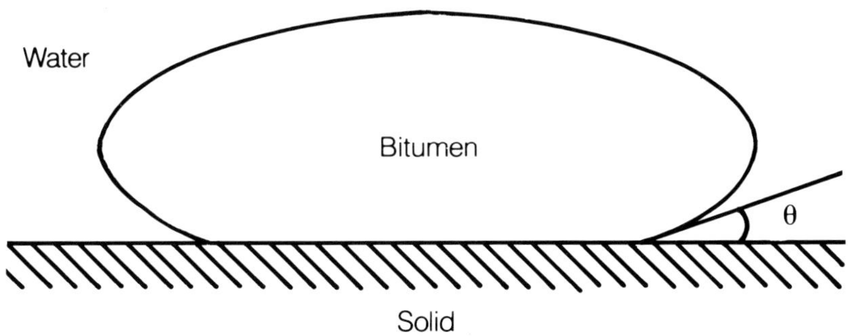

Figure 9.2 – A diagrammatic illustration of bitumen droplet applied to a solid surface under water

The conditions existing when a solid is already wetted by water and a bitumen is introduced into the system are shown in figure 9.2. If the contact angle between the three phases is θ and the energies of the solid/bitumen, solid/water and bitumen/water interfaces are respectively γ_{sb} γ_{sw} and γ_{bw} then the work of displacing water from a unit area of stone is given by:

$$W = \gamma_{sb} + \gamma_{bw} - \gamma_{sw}$$

Now for equilibirum to be reached, Young and Dupre's equation must be satisfied:

$$\gamma_{sb} = \gamma_{sw} + \gamma_{bw} \cos \theta$$

It follows that:

$$W = \gamma_{bw} (1 + \cos \theta)$$

Thus $W \alpha \gamma_{bw}$ (bitumen/water interfacial energy)

and $W = f(\theta)$ (contact angle).

So, a reduction in γ_{bw} and an increase in θ will improve adhesion.

In the base of a bitumen, θ is always $< 90°$ and so the factor $(1 + \cos \theta)$ is always > 1, ie an appreciable amount of work is required to make the bitumen displace the water and as a result, the bitumen tends to collect in drops on the surface. (However, it is possible to produce wetting if a very high pressure is applied).

The values of γ_{bw} and θ can both be altered by adding certain chemicals to the system. When a cationic surface-active agent is dissolved in the bitumen, it is strongly absorbed at the aggregate surface and the contact angle is increased (as well as the interfacial energy between bitumen and water being reduced), so that spreading of the bitumen against the water is able to take place. The one danger implicit in the use of a chemical additive to reduce the energy of the water/bitumen interface, is that the tendency for the bitumen to form an emulsion in water is increased.

9.4 The factors affecting bitumen/aggregate adhesion

The factors which influence bitumen/aggregate adhesion are numerous and complex. Table 9.1 identifies some of these influences, of which approximately 80% are considered to be controllable during production and construction. The following highlights some of the factors which are particularly important for attaining and retaining adhesion.

Aggregate properties	Bitumen properties	Mix properties	External influences
Mineralogy Surface texture Porosity Dust Durability Surface area Absorption Moisture content Shape Weathering	Rheology Electrical polarity Constitution	Void content Permeability Bitumen content Bitumen film thickness Filler type Aggregate grading Type of mix	Rainfall Humidity Water pH Presence of salts Temperature Temperature cycling Traffic Design Workmanship Drainage

Table 9.1 – Material properties and external influences which affect the bitumen/aggregate bond

The importance of the mineralogical composition of the aggregate has been recognised for many years [5,6]. The physio-chemical character of the aggregate is important, with the adhesive capacity being a function of its chemical composition, shape and structure, residual valency and surface area. Generalisations about the influence of mineralogy are difficult because of the effects of grain size, shape and texture which are also important. However, the majority of adhesive failures are associated with siliceous aggregates such as granites, rhyolites, quartzites, cherts, etc. The fact that satisfactory performance is achieved with these same aggregates and that failures occur using aggregates which have good resistance to stripping, eg limestone, emphasises the complexity of bitumen/aggregate adhesion.

The residual valence or surface charge on an aggregate has been shown to be important[7,8]. Aggregates with unbalanced surface charges possess a surface energy. If the aggregate surface is coated with a liquid of opposite polarity, the surface energy demands can be satisfied and an adhesive bond will result. Where two liquid phases are present, eg bitumen and water, the liquid which can best satisfy the energy requirement will most tenaciously adhere to the aggregate. The phenomenon of stripping of the bitumen in the presence of water can therefore be related to the surface charges. Where water equilibrates the surface charge better than bitumen, the bitumen may separate from the aggregate surface in favour of the water.

Physio-mechanical adsorption of bitumen into the aggregate depends on several factors including the total volume of permeable pore space, the size of the pore openings[9], and the viscosity and surface tension of the bitumen. The presence of a fine micro-structure of pores, voids and microcracks can bring about an enormous increase in the adsorptive surface available to the bitumen. It has also been shown[10,11] that fractions of the bitumen are strongly adsorbed in the aggregate surface to a depth of approximately 180 angstroms.

Other factors affecting the initial adhesion and subsequent bond are the surface texture of the aggregate, the presence of dust on the aggregate and to a lesser extent the pH of the water in contact with the interface. It is generally agreed that rougher aggregate surfaces have better adhesion characteristics. However, a balance is required between wetting of the aggregate (smooth surfaces being more easily wetted) and rougher surfaces which hold the bitumen more tenaciously once wetting has been achieved. It has been suggested that good mechanical bond achieved on a rough aggregate can be more important than the aggregate mineralogy in maintaining bitumen/aggregate adhesion[7].

As indicated in table 9.1, the properties of the bitumen are also important in the acquisition and subsequent retention of the bitumen/aggregate bond. In particular the viscosity of the bitumen during coating and in service, polarity and constitution all influence adhesion characteristics. However, it is the nature of the aggregate which is by far the most dominant factor influencing bitumen/aggregate adhesion.

9.5 Disbonding mechanisms[12]

A large number of studies have been carried out in the past to determine the mechanism of bitumen disbonding in bituminous mixes[13,14]. There are two modes of failure in a bitumen/aggregate system; adhesive and cohesive. If the aggregate is clean and dry and the mix is effectively impermeable, then the mode of failure will be cohesive. However, in the presence of water, the failure mode will almost certainly be a loss of adhesion caused by stripping of the bitumen off the aggregate. Several mechanisms of disbonding can be identified, viz:

9.5.1 Displacement[15]

The displacement theory arises from consideration of the thermodynamic equilibrium of the three-phase bitumen/aggregate/water system. If water is introduced at a bitumen/aggregate interface, then consideration of the surface energies involved shows that the bitumen will retract along the surface of the aggregate. Figure 9.3 shows an aggregate particle embedded in a bitumen film with point A representing the equilibrium contact position when the system is dry. When in contact with water the equilibrium point shifts and the new interface moves or retracts over the surface to point B. This new equilibrium position has a contact angle which will depend on the type and viscosity of the bitumen used.

9.5.2 Detachment

Detachment occurs when the bitumen and aggregate are separated by a thin film of water or dust although no obvious break in the surface of the bitumen film is apparent. Although the bitumen film completely encapsulates the aggregate particle no adhesive bond exists and the bitumen can easily be peeled from the aggregate surface. This process is reversible; if the water is removed the bitumen may re-adhere to the aggregate. A second mechanism of disbonding must be occurring to allow the ingress of water between the bitumen and the aggregate.

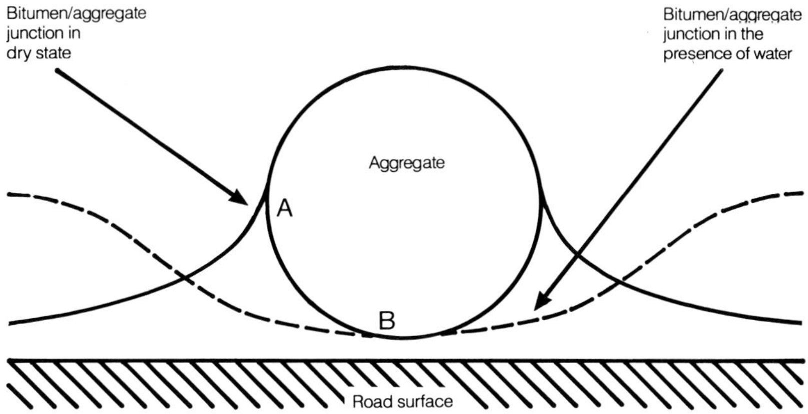

Figure 9.3 – Retraction of the bitumen/water interface in the presence of water[12]

9.5.3 Film rupture

Film rupture occurs where the bitumen fully coats the aggregate, but at sharp edges or asperities, where the bitumen film is thinnest, it has been shown[13] that water can penetrate through the film to reach the aggregate surface. This movement of water to the aggregate surface may occur with the water in vapour or liquid form. Once such a process has started it is possible for the water to spread between the bitumen and the aggregate to produce a detached film of bitumen. The speed with which the water can penetrate and detach the bitumen film will depend on the viscosity of the bitumen, the nature of the aggregate surface, the thickness of the bitumen film and the presence of filler and other components such as surface-active agents. Once significant detachment of the bitumen film from the aggregate has occurred, stresses imposed by traffic will readily rupture the film and the bitumen will retract, exposing water-covered aggregate.

9.5.4 Blistering and pitting[9]

If the temperature of the bitumen in a pavement increases, the viscosity of the bitumen will be reduced. If this is associated with a recent rainfall, the bitumen may creep up the edges of water droplets to form a blister, as shown in figure 9.4. If the temperature is increased, the blister will expand, leaving a pit which may allow access for water to the aggregate surface.

9.5.5 Hydraulic scouring

Hydraulic scouring occurs in the wearing course and is caused by the action of vehicle tyres on a saturated pavement surface. Water can be forced into surface voids in front of the vehicle tyre. On passing, the action of the tyre sucks up this water, thereby inducing a compression-tension cycle in these surface voids, which may result in debonding of the bitumen from the aggregate. Suspended dust and silt in the water can act as an abrasive and accelerate disbonding.

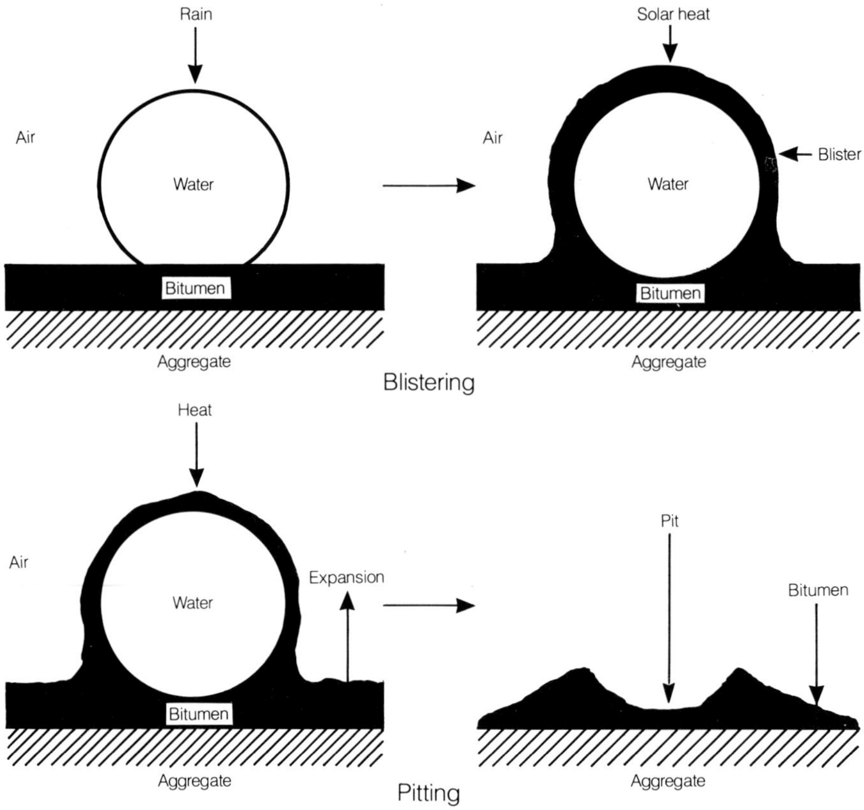

Figure 9.4 – Formation of blisters and pits in a bituminous material[9]

9.5.6 Pore pressure

This disbonding mechanism is most important in open or poorly compacted mixes. In these types of mixes water can be trapped as the material is compacted by traffic. Once the material becomes effectively impermeable subsequent trafficking induces a pore water pressure[7]. This creates channels around the bitumen/aggregate interface leading to loss of bond. Higher temperatures acting on the entrapped water result in expansive stresses accelerating water migration and disbonding. Low temperatures may lead to the formation of ice which is equally destructive.

9.5.7 Chemical disbonding

Diffusion of water through the bitumen films can lead to double layers of water at the aggregate surface. The presence of the water causes the aggregate surface to adopt a negative surface charge against a slightly negatively charged bitumen. This results in two negatively charged surfaces in contact and a net repulsion results. As more water is attracted to the aggregate surface disbonding of the bitumen film will result[16].

9.6 Adhesion testing

It is clearly desirable that there should be a reliable method of determining, quantitatively in the laboratory, the quality of adhesion between bitumen and an aggregate in the presence of water. Numerous arbitrary tests have been developed which can be used to compare different combinations of aggregate and bitumen. However, for the majority of these tests little information is available to correlate test data with performance in practice. In most adhesion tests aggregate coated with bitumen is immersed in water under controlled conditions. The degree of stripping after a period of time is determined. The various techniques differ in the type of specimen used, the conditions under which the sample is immersed in water and the method by which the degree of stripping is assessed. In all these tests the coated aggregate is immersed in water. Tests are also available in which an attempt is made to coat an aggregate in the presence of water; in this case the degree of coating obtained is used as an index of adhesivity.

Adhesion tests fall into the following six categories:

- static immersion tests;
- dynamic immersion tests;
- chemical immersion tests;
- immersion mechanical tests;
- immersion trafficking tests;
- coating tests.

There are several examples of each of these types of test, but in most cases the individual tests of one type differ in detail rather than in principle.

9.6.1 Static immersion tests

In this type of test, aggregate coated with bitumen is immersed in water and the degree of stripping estimated by a visual inspection after a period under water. An example of this type of test is the total water immersion test, often colloquially known as the 'twit' test. In essence this test involves coating 14 mm single size chippings with a known quantity of bitumen. The coated aggregate is then immersed in distilled water at 25°C for 48 hours. The percentage of bitumen stripped off the aggregate is estimated visually (see chaper 18, section 18.3.1.2). Although the reproducibility of this test is poor it has been shown that the ranking order of aggregates produced by different laboratories was very similar. Thus it is possible to compare the test results with the performance on the road. However, there are unfortunately cases where aggregates with poor static immersion test results have performed satisfactorily in practice and, conversely, aggregates which do not strip in the test which perform poorly on the road.

9.6.2 Dynamic immersion tests

This type of test is very similar to the static immersion test but the sample is agitated mechanically by shaking or kneading. Again the degree of stripping is estimated visually together with a subjective judgement whether the mixture remains cohesive or separates into individual particles of aggregates. The reproducibility of this type of test is also very poor.

9.6.3 Chemical immersion tests

In this type of test aggregate coated in bitumen is boiled in solutions containing various concentrations of sodium carbonate. The strength of the solution of sodium carbonate in which stripping is first observed is used as a measure of adhesivity. However, the artificial conditions of the test make it of doubtful value in predicting road performance.

9.6.4 Immersion mechanical tests

Immersion mechanical tests involve measurement of a change in a mechanical property of a compacted bituminous mix after immersion in water. Thus the ratio of the property after immersion divided by the initial property is an indirect measure of stripping.

A number of mechanical properties can be measured including – shear strength, flexural strength and compressive strength. Probably the most popular is the Marshall test; the ratio of Marshall stability after immersion and before immersion is known as retained Marshall stability and is usually quoted as a percentage.

As a part of the QUALAGON testing of bitumens (see chapter 3, section 3.5.2) Shell carry out a retained Marshall test[(17)]. In the Shell version of this test at least eight Marshall specimens are manufactured using a prescribed aggregate type, aggregate gradation, bitumen content and void content. Four specimens are then tested according to the standard Marshall method, yielding a standard stability value. The four remaining specimens are vacuum-treated under water at a temperature of between 0°C and 1°C to saturate the pore volume of the mix with water. Subsequently the specimens are stored in a water bath at 60°C for 48 hours. Finally, their Marshall stabilities are determined and the ratio of the Marshall stability of the stored specimens divided by the standard Marshall stability is termed the retained Marshall stability.

A relatively wide scatter in absolute Marshall stability values and poor reproducibility are inherent to the Marshall test. However, it has been found that consideration of the percentage relative to the result from the standard procedure reduced the differences between results from different laboratories.

9.6.5 Immersion trafficking tests

Almost all the tests described above tacitly assume that traffic passing over a surfacing does not influence stripping, whereas traffic can play an important role in stripping. A test which simulates the effect of traffic is the immersion-wheel tracking test[(18)]. This test consists of three solid-tyred wheels which traverse three specimens of an open-textured bitumen macadam. The wheels travel with a reciprocating motion with a frequency of 25 cycles per minute. Each wheel is loaded to give a total weight of approximately 20 kg bearing on the specimen. The specimens are maintained horizontally in a water bath so that the water level is just above their top (see figure 9.5). The bitumen macadam is compacted in the moulds under standardised conditions, and cured for a short time before immersion. The temperature of the water bath is normally 40°C. The criterion which is adopted to measure stripping is that of the time necessary to produce failure; incipient and ultimate failure times are obtained.

It has been shown that good correlation exists between stripping failures on heavily trafficked roads and the behaviour of similar materials in the immersion-wheel tracking test[(19,20)]. Studies on seventeen different aggregates[(21)] showed a wide variation of failure times with a repeatability of about 30% of mean median failure times. A study of the effect of bitumen viscosity[(22)] on failure time

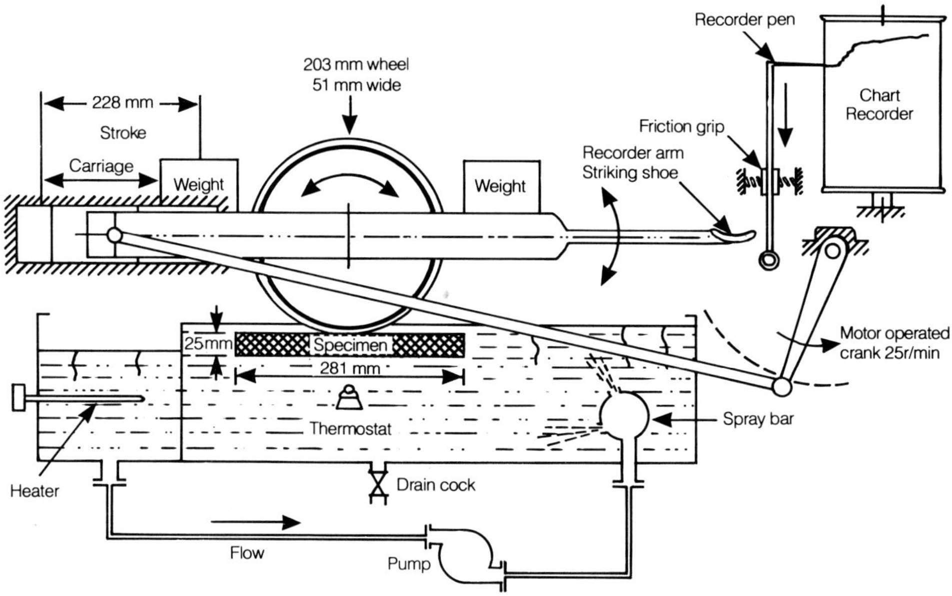

Figure 9.5 – The immersion wheel tracking test[18]

showed a linear relationship between the logarithm of viscosity and median failure time. Factors affecting failure times include:

aggregate shape;
aggregate interlock;
bitumen viscosity;
sample preparation.

9.6.6 Coating tests

Coating tests are an attempt to obtain adhesion between an aggregate and a bitumen when water is also present. For example, the immersion tray test involves the immersion of a tray of bitumen in water and then the application of chippings to the surface of the bitumen through the water. This test is extremely convenient for examining how surface-active agents improve adhesion between bitumen and aggregates for use in surface dressing under wet conditions (see chapter 18, section 18.3.1.2).

9.7 Improving bitumen/aggregate adhesion

Hydrated lime has traditionally been used as an antistripping additive in hot bituminous mixes. It is usually incorporated as part of the mineral filler at between 1% to 3% w. It is suggested[23] that the hydrated lime reacts with the carboxylic acids which are present in the bitumen and allows other carbonyl groups such as ketones to attach themselves to the aggregate surface. These compounds are not as easily removed by water as the acids would have been and so the material is less susceptible to stripping. Alternatively, it has been suggested[24] that in mixes containing hydrated lime, if water is present at the bitumen/aggregate interface, a hydrated lime solution will result. The calcium ions in the solution cause the surface of the aggregate to become basic and the electro-chemical balance forces the water away from the interface and into an emulsion in the bitumen. The balance will then force the attachment to the hydrophobic surface of the mineral. The hydrated lime in this case remains dispersed in the bitumen or in solution in the water. The benefit of adding 2% w of hydrated lime to a mix is clearly shown in figure 9.1.

For some types of aggregate adhesion can be improved by the addition of certain chemical additives. Generally these act by either:

- changing the interfacial conditions between the bitumen and the aggregate so that the bitumen preferentially wets the aggregate;
- reinforcing the adhesive bond between the aggregate and the bitumen thereby improving long-term resistance to bitumen detachment by ingress of water.

Long chain primary amines form the basis of a number of commercially available adhesion promoters for bituminous mixes. The amines are cationic in nature and are attracted to the surface of mineral aggregates. The other end of the molecule is a long hydrocarbon chain which is compatible with the bitumen. The net result is an ionically bonded cross-link between the aggregate and the bitumen.

The quantity of adhesion agent added to the bitumen is usually 0.1% to 1.0% (by weight of the bitumen). It can be added to the bitumen prior to mixing with the aggregate or by pre-coating the aggregate. Although the initial wetting of the aggregate by the bitumen is improved, the effectiveness of some adhesion additives to improve adhesion long-term when exposed to prolonged continuous exposure to water is questionable(11,25).

As the function of the additive is to interact with polar groups on the aggregate surface, they will also interact with polar groups in the bitumen; as a result a threshold concentration must be exceeded before the adhesion agent is likely to become effective. In addition both laboratory studies and field experience suggest that different bitumen/aggregate combinations may require different types of additives or different concentrations to achieve optimal performance.

The use of these additives does not imply any deficiency on the part of the bitumen but rather the opportunity to enhance properties for a demanding site or to enable aggregate types to be used that might otherwise be rejected.

9.8 References

(1) Young, C P: *The prediction of time for which a road is wet.* Road Research Laboratory, Technical Note TN 402, 1969.

(2) Hargreaves, A E, Moores, T A, Bonner, D G, and Cawsey, D C: *Interim report on the investigation of the premature failure of bituminous macadam wearing courses.* The Hatfield Polytechnic. Interim Report Number 1, 1985.

(3) Cawsey, D C and Massey, S W: *Monitoring the in service performance of highway aggregates in bituminous macadam wearing courses.* Bulletin of the International Association of Engineering Geology, vol 30, pp 35-40, 1984.

(4) Dormon, G M: *Some observations on the properties of bitumen and their relation to performance in practice and specifications.* Paper presented to the first conference on asphalt pavements for South Africa, 1969.

(5) Saal, R N J: *Adhesion of bitumen and tar to solid road building materials.* Bitumen, vol 3, p 101, 1933.

(6) Winterkorn, H F: *Surface chemical aspects of the bond formation between bituminous materials and mineral surfaces.* Proceedings of the Association of Asphalt Paving Technologists, vol 17, pp 79-85, 1936.

(7) Lee, A R and Nicholas, J H: *The properties of asphaltic bitumen in relation to its use in road construction.* Journal of the Institute of Petroleum, vol 43, pp 235-246, 1957.

(8) Orchard, D F: *Properties and testing of aggregates.* Concrete Technology, vol 3, third edition. Applied Science Publications Ltd, p 281, 1976.

(9) Thelan, E: *Surface energy and adhesion properties in asphalt/aggregate systems.* Highway Research Board Bulletin, vol 192, pp 63-74, 1958.

(10) Plancher, H, Dorrene, S M and Petersen, J C: *Identification of chemical types strongly absorbed at the asphalt-aggregate interface and their relative displacement by water.* Proceedings of the Association of Asphalt Paving Technologists, vol 46, pp 151-175, 1977.

(11) Scott, J A N: *Adhesion and disbonding mechanisms of asphalt used in highway construction and maintenance.* Proceedings of the Association of Asphalt Paving Technologists, vol 47, pp 19-48, 1978.

(12) The Asphalt Institute: *Cause and prevention of stripping in asphaltic pavements.* Educational series No 10, 1981.

(13) Hughes, R I, Lamb, D R and Pordes, O: *Adhesion in bitumen macadam.* Journal of Applied Chemistry, vol 180, pp 433-440, 1960.

(14) Taylor, M A and Khosla, N P: *Stripping of asphalt pavements: State of the art.* Transportation Research Record 911, pp 150-157, 1983.

(15) Blott, J F T, Lamb, D R and Pordes, O: *Weathering and adhesion in relation to the surface dressing of roads with bituminous binder. Adhesion and adhesives: Fundamentals and practice.* Society of Chemical Industry, 1954.

(16) Güsfeldt, K H and Scott, J A N: *Factors governing adhesion in bituminous mixes.* Proceedings of the second international symposium devoted to tests on bitumens and bituminous materials, vol 1, pp 45-65, Budapest, 1975.

(17) van Gooswilligen, G, de Bats, F Th and Harrison, T: *Quality of paving grade bitumen – A practical approach in terms of functional tests.* Proceedings of the fourth Eurobitume symposium, pp 290-297, Madrid, October 1989.

(18) Mathews, D H and Colwill, D M: *The immersion wheel-tracking test.* Journal of Applied Chemistry, pp 505-509, 1962.

(19) Nicholas, J H and Mathews, D H: *The immersion wheel tracking test: Correlation with road performance.* Road Research Laboratory, Research Note RN/2182/JHN, 1954

(20) Lee, A R and Nicholas, J H: *Adhesion in the construction and maintenance of roads.* Society of Chemical Industry, 1954.

(21) Mathews, D H, Colwill, D M and Yuce, R: *Adhesion tests for bituminous materials.* Road Research Laboratory, Laboratory Note LN 698, 1964.

(22) Colwill, D M and Edler, A C: *Immersion wheel-tracking test: Effect of binder viscosity on failure time.* Road Research Laboratory, Laboratory Note LN 584, 1964.

(continued)

(23) Plancher, H, Green, E and Paterson, J C: *Reduction of oxidative hardening of asphalts by treatment with hydrated lime – a mechanistic study*. Proceedings of the Association of Asphalt Paving Technologists, vol 45, pp 11-24, 1976.

(24) Ishai, I and Craus, J: *Effect of the filler on aggregate-bitumen adhesion properties in bituminous mixtures*. Proceedings of the Association of Asphalt Paving Technologists, vol 46, pp 228-257, 1977.

(25) Fromm, H J: *The mechanisms of asphalt stripping from aggregate surfaces*. Proceedings of the Association of Asphalt Paving Technologists, vol 43, pp 191-223, 1974.

Chapter 10 – Modified bitumens

On the majority of our roads conventional bituminous materials perform satisfactorily. However, demands made upon roads increase year by year. Ever increasing numbers of commercial vehicles with increased axle loads take their toll and as we look forward to the year 2000, it is clear that this trend will continue. To enable the highway engineer to meet this growing challenge there now exists a wide range of road-proven modified bitumens.

Good basic design, stage construction and timely maintenance enable asphalt roads not only to withstand all the demands made, but also to provide the high standards of safety and comfort which have become synonymous with bitumen-bound surfacings. However, certain areas of road are, predictably, much more highly stressed than others and therefore, if the entire length is to perform satisfactorily during the design life, these areas may require special attention. We are all aware of the need for special materials on bridge decks for example, but the needs of other critical areas can often be overlooked during construction or maintenance. It is clearly advantageous that maintenance or strengthening work be required along the entire length at a given point in time, rather than repeated work being required at certain 'black spots' at shorter intervals. Disruption to traffic flow during maintenance work is a considerable cost to society. The closure of one lane of a U.K. motorway in 1987 was estimated to cost the community £20,000 per day and this is probably a conservative estimate(1).

The use of modified bitumens offers one solution to reduce the frequency of maintenance required at 'black spots' and provides much longer service life for maintenance treatments at difficult sites. To optimise the cost effectiveness of modified bitumens, they are best used in combination with high quality aggregates, in unusual mixes such as pervious macadam, or in areas requiring exceptional resistance to fatigue and deformation.

The degree of improvement required, and hence the cost, will depend upon the needs of the site to be treated. Modest improvements in deformation resistance can be achieved using reclaimed tyre rubber. Moving up-market, sulphur, organo-manganese compounds and thermoplastic polymers can improve both the workability of the asphalt during compaction and its deformation resistance in service. Moving further up-market, thermoplastic rubbers improve both deformation resistance and fatigue characteristics. Thermosetting systems offer the ultimate performance of exceptional deformation resistance coupled with outstanding flexibility.

10.1 The role of bitumen modifiers in bituminous mixes

As the bitumen is responsible for the visco-elastic behaviour characteristic of all bituminous materials, it plays a large part in determining many aspects of road performance. Resistance to permanent deformation and cracking are typical examples. In general the proportion of any induced strain in a bituminous material that is attributable to viscous flow, ie non-recoverable, increases with both loading time and temperature. The effect of this is illustrated in figures 10.1a and 10.1b(2). Figure 10.1a shows the response of a bituminous material to a simple creep test. The strain resulting from the applied loading shows an instantaneous elastic response followed by a gradual increase in strain with time until the load is removed. The change in strain with time is caused by the viscous behaviour of the material. On removal of the load, the elastic strain is recovered instantaneously and some additional recovery occurs with time. This is known as 'delayed elasticity'. Ultimately, a permanent residual strain remains which is irrecoverable and is directly caused by viscous behaviour.

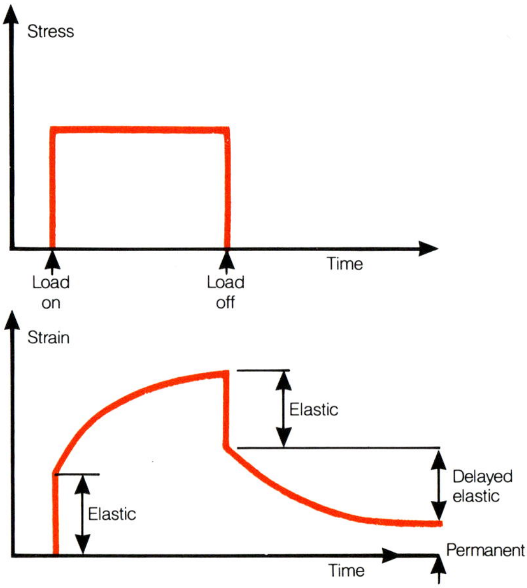

Figure 10.1a – Visco-elastic response to static loading[(2)]

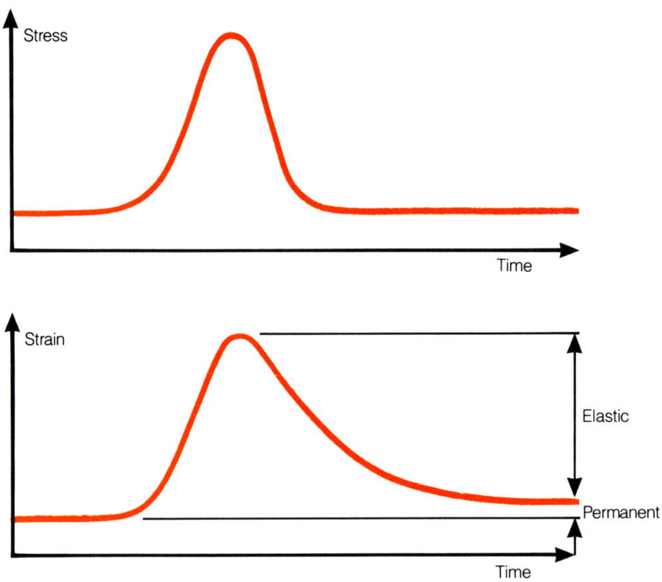

Figure 10.1b – Visco-elastic response to a moving wheel load[(2)]

The response to a load pulse induced in an element of a bituminous material due to moving traffic loads is shown in figure 10.1b. Here it is not possible to distinguish between the two components of elastic response, but the small permanent strain and larger elastic strain are shown. Although the permanent strain illustrated in figure 10.1b is small for a single pulse load, when many millions of load applications are applied to a pavement, a large accumulation will develop and this is the component which results in surface deformation[2]. From the above it is clear why more deformation occurs at high ambient temperatures and where traffic is slow moving or stationary.

One of the prime roles of a bitumen modifier is to increase the resistance of the asphalt to permanent deformation at high road temperatures, without adversely affecting the properties of the bitumen or asphalt at other temperatures. This is achieved by either stiffening the bitumen so that the total visco-elastic response of the asphalt is reduced, with a corresponding reduction in permanent strain, or by increasing the elastic component of the bitumen, thereby reducing the viscous component, which again results in a reduction in permanent strain.

Increasing the stiffness of the bitumen is also likely to increase the dynamic stiffness of the asphalt, thus improving the load spreading ability of the material and increasing the structural strength and expected design life of the pavement; alternatively it may be possible to achieve the same structural strength but with a thinner layer. Increasing the elastic component of the bitumen will improve the flexibility of the asphalt which is important where high tensile strains are present.

10.2 The modification of bitumen

For many years researchers and development chemists have experimented with modified bitumens, mainly for industrial uses, adding asbestos, special fillers, mineral fibres and rubbers. In the last twenty years many researchers have looked at a wide spectrum of modifying materials for bitumens used in road construction.

For the modifier to be effective and for its use to be both practicable and economic, it must[3]:

be readily available;

resist degradation at asphalt mixing temperatures;

blend with bitumen;

improve resistance to flow at high road temperatures without making the bitumen too viscous at mixing and laying temperatures or too stiff or brittle at low road temperatures;

be cost effective.

The modifier when blended with bitumen should have the following characteristics:

maintain its premium properties during storage, application and in service;

be capable of being processed by conventional equipment;

be physically and chemically stable during storage, application and in service;

achieve a coating or spraying viscosity at normal application temperatures.

10.2.1 The modification of bitumen by the addition of sulphur

Sulphur is used in two types of asphalt paving processes and a number of proprietary products. Sulphur-extended asphalt utilises a relatively small amount of sulphur as a diluent to the bitumen. The second process, known as THERMOPAVE, uses high sulphur contents where the excess sulphur behaves as a mouldable filler, producing a very workable mix which can be laid by machine without roller compaction and when cool is very resistant to deformation[4,5].

Shell have also developed THERMOPATCH, which is a high sulphur content asphalt for repairing potholes, utility reinstatements, etc, in both bituminous and concrete pavements. It is free-flowing and is simply poured into the hole, the surface is levelled and, when cool, it is ready for trafficking.

The amount of sulphur that will react with bitumen depends on the temperature and composition of the bitumen. It has been shown[6,7] that sulphur reacts predominantly with the naphthene-aromatic fraction of bitumen, either by adding to the molecule, or by oxidising it with the extraction of hydrogen as hydrogen sulphide. Between 119.3°C, the melting point of pure monoclinic sulphur, and 150°C, the reaction is mainly one of addition, producing an increase in the polar aromatic fraction and a relatively small change in the rheological properties of the bitumen. Above 150°C the competitive oxidation reaction increases sharply, producing an increase in asphaltenes and an effect on the properties of the bitumen similar to air blowing[8].

The evolution of hydrogen sulphide at temperatures in excess of 150°C is clearly unacceptable; however, it has been shown that by careful control of temperature and by the use of purpose-built equipment the problem of fume emissions can be overcome[4].

Depending on the chemical constitution of the bitumen, approximately 15% to 18%m of sulphur can be dispersed in bitumen and remain stable for a prolonged period[9]. Increasing the sulphur content above this level results in a temporary suspension of the excess sulphur during mixing. When the asphalt is hot the excess sulphur makes the asphalt very workable because sulphur is a low viscosity liquid at temperatures between its melting point and about 160°C. As the asphalt cools, the excess sulphur partially fills and conforms to the shape of the voids in the compacted material, keying together aggregate particles, thereby increasing interparticular friction within the asphalt and imparting high mechanical strength to the mix.

In practice, the stiffening mechanism of the sulphur is not quite as simple as this, and changes in the behaviour and properties of the asphalt modified by sulphur are also time dependent. Initially the Marshall stability of asphalt mixes made with sulphur is usually slightly lower than when unmodified bitumen is used. However, after 7 to 21 days the Marshall stability increases to approximately twice that of the unmodified material. The most likely explanation for this behaviour is that crystallisation of the sulphur is retarded by some interaction with the bitumen.

The influence of sulphur content on the Marshall stability of a sand asphalt manufactured using a 200 pen bitumen is shown in figure 10.2[4,5]. This shows the substantial increases in Marshall stability that can be achieved by the addition of sulphur.

10.2.2 The modification of bitumen by the addition of rubbers

Polybutadiene, polyisoprene, natural rubber, butyl rubber, chloroprene, random styrene-butadiene-rubber, etc, have all been used with bitumen but their effect is mainly to increase viscosity. In some instances the rubbers have been used in a vulcanised (cross-linked) state, eg reclaimed tyre crumb, but this is difficult to disperse in bitumen, requiring high temperatures and long digestion times and can result in a heterogeneous binder, with the rubber acting mainly as a flexible filler.

10.2.3 The modification of bitumen by the addition of organo-manganese compounds

The use of organo-manganese compounds in bitumen is the subject of both a U.S. patent (No 4234346) and a U.K. patent (No 1600897) which indicate that paving materials of improved strength are obtained by adding to the bitumen a soluble organo-manganese compound, either alone, or in combination with organo-cobalt or organo-copper compounds. This product is better known under

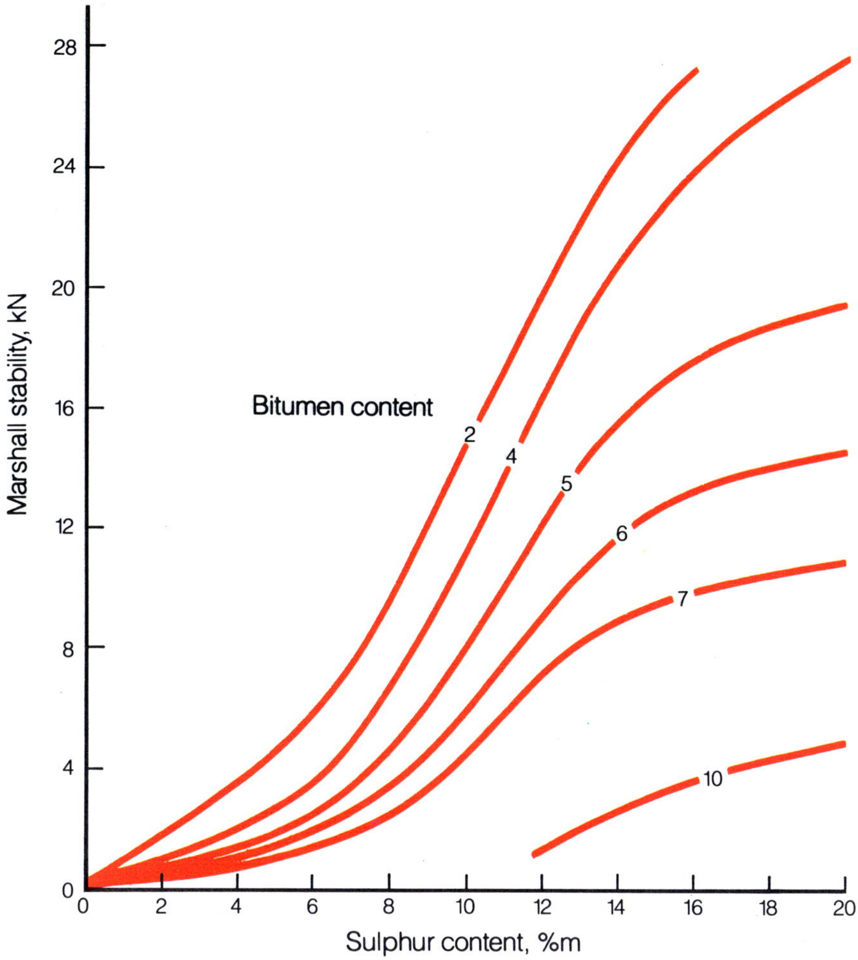

Figure 10.2 – Effect of sulphur content on Marshall properties[4,5]

its brand name – CHEMCRETE. The use of manganese modified bitumens in asphalt and macadam is claimed to improve the temperature susceptibility of the mixes, thereby improving their physical properties, ie Marshall stability, resistance to permanent deformation and dynamic stiffness[10,11,12].

To enable the organo-manganese compound to be dispersed rapidly in bitumen the material is blended with a carrier. The modifiers in the range currently differ from each other with respect to the type and concentration of the organo-metallic compounds and the viscosity of the carrier oil used. Both aspects need to be considered to ensure the appropriate modifier is used for a particular application.

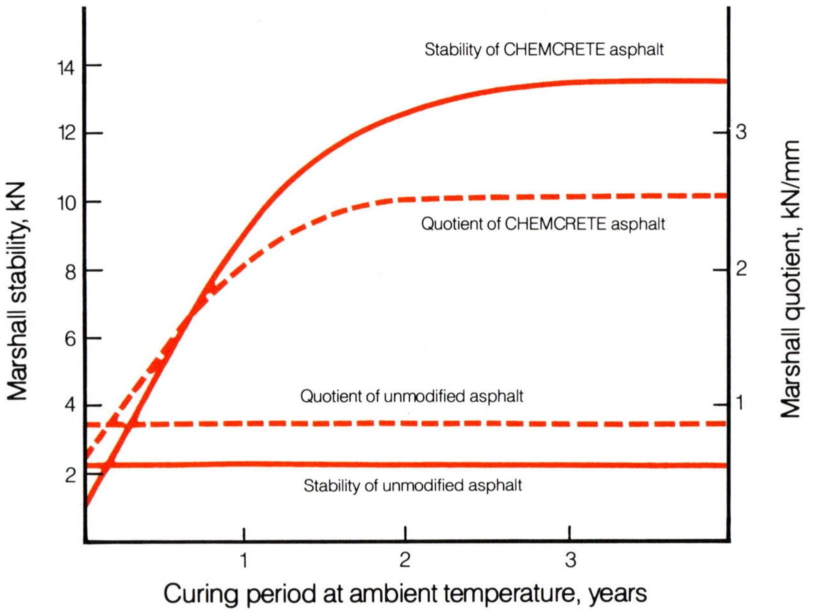

Figure 10.3 – Marshall properties of CHEMCRETE modified asphalt[14]

The mechanism put forward in the CHEMCRETE patent is that certain ring structure compounds in bitumen, for example tetrahydronaphthalenes and similar structures, are readily auto-oxidised. It is believed that manganese promotes the formation of a diketone compound which then forms stable, chemically resistant complexes with the metal. These complexes tie together bitumen molecules which ultimately improves the temperature susceptibility of the bitumen and increases the strength of the bituminous mixture[13].

The curing reaction is dependent on oxygen availability, bitumen film thickness and temperature. Storage experiments carried out by CHEMCRETE have shown that a 100 pen bitumen containing 2%w of manganese modifier (U1-49) stored at 160°C for 28 days hardened from 127 pen to 100 pen.

Figure 10.3[14] shows the Marshall stability and quotient measured on two sand asphalts, one manufactured using 100 pen bitumen containing CHEMCRETE, the other unmodified 50 pen bitumen. The initial Marshall properties of these two mixes are very similar; however, after 12 months the Marshall stability of the modified sand asphalt had increased by over 400%. During this 12 month period the Marshall flow increased by about 75%, resulting in an overall increase in Marshall quotient of over 200%. The increase in Marshall flow is believed to show that the gain in strength during curing is not due to embrittlement of the bitumen. Similar improvements were determined, using a laboratory wheel-tracking test, on cores cut from a road trial. Deformation rates at 45°C, immediately after laying, on modified and unmodified asphalt were very similar – 5 mm/h. However, after six months the deformation rate on the modified asphalt had dropped to less than 1 mm/h.

A large number of road trials[15] covering a wide range of highway maintenance applications have been undertaken in the last five years. The use of organo-manganese modifiers in HRA is currently being evaluated by the TRRL[10] and it is understood that it is to be included in the Department of Transport specification in 1990.

10.2.4 The modification of bitumen by the addition of thermoplastic polymers

Polyethylene, polypropylene, polyvinyl chloride, polystyrene and ethylene vinyl acetate (EVA) are the principal thermoplastic polymers which have been examined in modified road binders. As thermoplastics, they are characterised by softening on heating and hardening on cooling.

Thermoplastic polymers, when mixed with bitumen, associate at ambient temperature, increasing the viscosity of the bitumen. Unfortunately they do not significantly increase the elasticity of the bitumen and when heated they can separate, which may give rise to a coarse dispersion on cooling(1). However, accepting these limitations, EVA at a concentration of 5% in 70 pen bitumen has become very popular.

EVA copolymers are thermoplastic materials with a random structure produced by the copolymerisation of ethylene and vinyl acetate. Copolymers with low vinyl acetate content possess properties similar to low density polyethylene. As the level of vinyl acetate increases so the copolymer properties change(16,17). The properties of EVA copolymers are controlled by molecular weight and vinyl acetate content:

- *molecular weight;* the molecular weights of many polymers, are defined in terms of an alternative property; standard practice for EVAs is to measure melt flow index (MFI), a viscosity test that is inversely related to molecular weight; the higher the MFI, the lower the molecular weight and viscosity; this is analogous to the penetration test for bitumen – the higher the penetration the lower the average molecular weight and viscosity of the bitumen;

- *vinyl acetate content;* in order to appreciate the main effects of vinyl acetate on the properties of the modified binder it is useful to consider a simple illustration of its structure, as shown in figure 10.4(16,17); this shows how the regular polyethylene segments of the chain can pack closely together and form what are known as crystalline regions; it also shows how the bulky vinyl acetate groups disrupt this closely packed arrangement to give non-crystalline or amorphous rubbery regions; the crystalline regions are relatively stiff and give considerable reinforcing effect, whereas the amorphous regions are rubbery; the more vinyl acetate groups (or the higher the vinyl acetate content), the higher the proportion of rubbery regions, and conversely, the lower the proportion of crystalline regions.

There is a wide range of EVA copolymers available (18) which are specified by MFI and vinyl acetate content. For example EVA with an MFI of 150 and a vinyl acetate content of 19 is the 150/19 grade. From the wide range available the 150/19 and to a lesser extent the 45/33 grades are the most popular, usually blended in 70 pen bitumen. EVA copolymers are easily dispersed in, and have good compatibility with bitumen. They are thermally stable at normal asphalt mixing temperatures. However, during static storage some separation may occur and it is therefore recommended that the blended product should be thoroughly circulated before use.

Table 10.1(18) on page 158 shows typical changes in penetration and softening point when 5% of these two grades of EVA are added to 70 pen bitumen, and the Marshall and wheel-tracking properties of hot rolled asphalt manufactured using the modified binders. The wheel-tracking test results show a four fold reduction in tracking rate when either EVA grade is added to 70 pen bitumen. The resistance to deformation is also markedly less sensitive to changes in temperature, as illustrated in figure 10.5(18). The increase in Marshall stability is not as large as one would predict based on the wheel-tracking test results. This is true of the majority of polymer modified mixes examined in the Marshall test and therefore Marshall tests results on polymer modified materials should be interpreted with care.

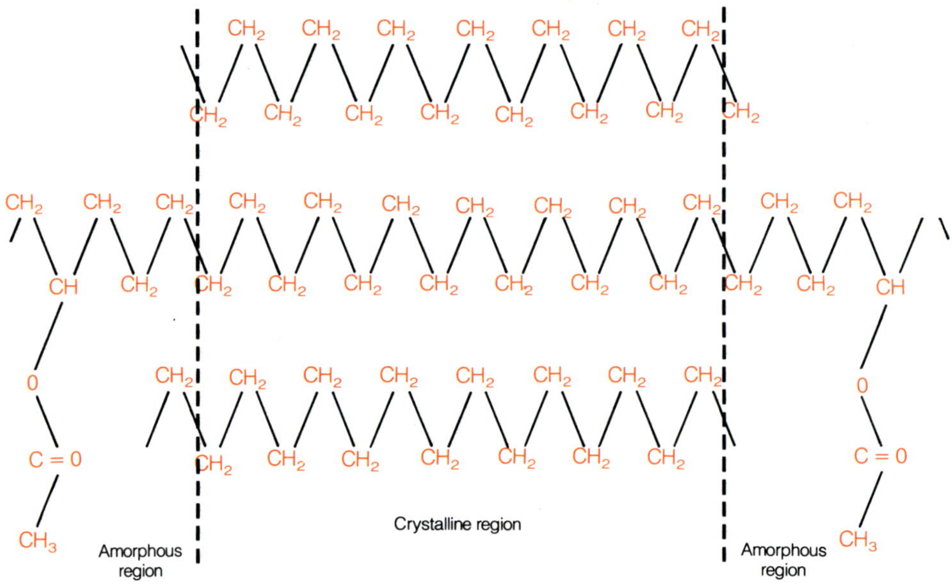

Figure 10.4 – Packing of the polyethylene in EVA(16,17)

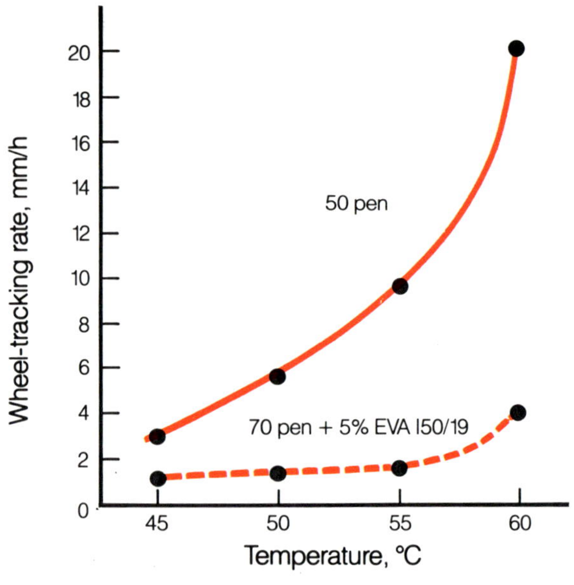

Figure 10.5 – Effect of temperature on wheel-tracking rate on hot rolled asphalt wearing course(18)

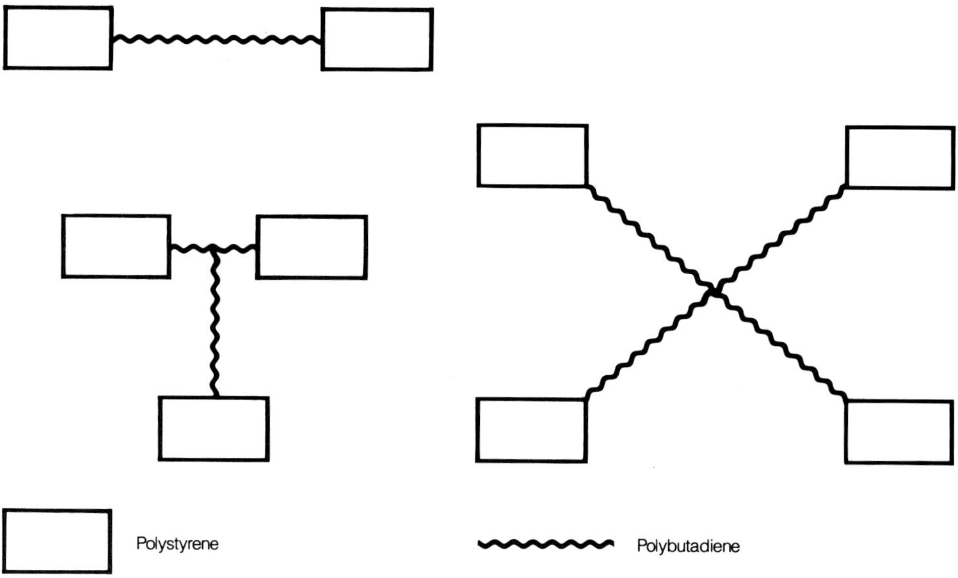

Figure 10.6 – Linear and branched thermoplastic rubber(19)

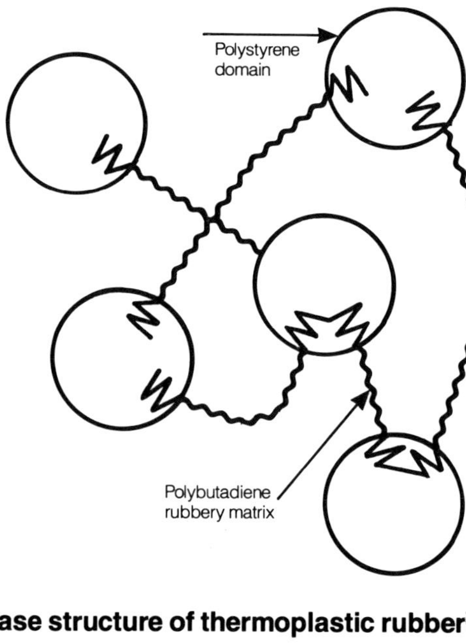

Figure 10.7 – Phase structure of thermoplastic rubber(19)

Binder	Binder properties		Marshall properties			Wheel-tracking rate at 45°C, mm/h
	Penetration at 25°C, dmm	Softening point (IP), °C	Stability, kN	Flow, mm	Quotient, kN/mm	
70 pen	68	49.0	6.3	3.3	1.9	4.4
70 pen + 5% EVA 150/19	50	65.5	7.6	3.2	2.4	0.8
70 pen + 5% EVA 45/33	57	58.0	8.0	2.7	3.0	1.0

Table 10.1 – Effect of EVA grades 150/19 and 45/33 on binder and hot rolled asphalt wearing course properties[18]

Despite the improvements in performance that are achievable using EVA, currently a significant quantity of EVA is used to enable 'cold weather working'. EVA tends to improve asphalt workability because of its susceptibility to shear[17] and because it uses a slightly softer base bitumen. However, care should be exercised when using EVA to lay hot rolled asphalt wearing course in cold ambient conditions, particularly on exposed sites open to cross-winds. EVA modified bitumen stiffens rapidly as the EVA polymer crystallises. (This process occurs below the minimum rolling temperature). The relevance of this is that asphalt workability drops equally rapidly. On exposed sites surface chilling may occur, resulting in a stiff surface crust. If this occurs before the chippings have been rolled into the asphalt they will puncture the surface crust and sit in the puncture hole rather than be embedded into the asphalt with surface mortar moulded around the chipping.

If the optimal performance of EVA modified bitumens is to be achieved in practice it is essential that they are specified and costed on a rational basis and applied accepting that they have limitations. If these are acknowledged, then EVA modified bitumen will make a significant contribution to improving the cost-effectiveness of Britain's roads.

10.2.5 The modification of bitumen by the addition of thermoplastic rubbers

Of the four main groups of thermoplastic elastomers – polyurethane, polyether-polyester copolymers, olefinic copolymers and styrenic block copolymers, it is the styrenic block copolymers which have proved to present the greatest potential when blended with bitumen[19].

Styrenic block copolymers, commonly called thermoplastic rubbers (TR), may be produced by a sequential operation of successive polymerisation of styrene-butadiene-styrene (SBS) or styrene-isoprene-styrene (SIS). Alternatively, a di-block precursor can be produced by successive polymerisation of styrene and mid-block monomer, followed by a reaction with a coupling agent. Thus, not only linear copolymers but also multi-armed copolymers can be produced; these are often referred to as star-shaped, radial or branched copolymers (see figure 10.6)[19].

Thermoplastic rubbers derive their strength and elasticity from a physical cross-linking of the molecules into a three-dimensional network. This is achieved by the agglomeration of the polystyrene end-blocks into separate domains, as shown in figure 10.7, providing the physical cross-links for a three-dimensional polybutadiene or polyisoprene rubbery matrix. It is the polystyrene end-blocks which impart strength to the polymer and the mid-block which gives the material its exceptional elasticity[20].

At temperatures above the glass transition point of polystyrene (100°C) the polystyrene softens as the domains weaken and will even dissociate under stress, thus allowing easy processing. Upon cooling, the domains reassociate and the strength and elasticity is restored, ie the material is thermoplastic.

10.2.5.1 The influence of bitumen constitution on thermoplastic rubber/bitumen blends

As discussed in chapter 6, bitumens are complex mixtures which can, to a limited extent, be subdivided into groups of molecules which have common structures:

saturates;
aromatics;
resins;
asphaltenes.

The saturates and aromatics can be viewed as carriers for the polar aromatics, ie the resins and asphaltenes. The polar aromatics are responsible for the visco-elastic properties of the bitumen at ambient temperatures. This is due to association of the polar molecules which leads to large structures, in some cases even to three-dimensional networks, ie 'gel' type bitumen. The degree to which this association takes place depends on the temperature, molecular weight distribution, the concentration of the polar aromatics and on the solvency power of the saturates and aromatics in the maltene phase. If the concentration and molecular weight of the asphaltenes is relatively low, the result will be a 'sol' type bitumen.

The addition of thermoplastic rubbers, with a molecular weight similar to or higher than that of the asphaltenes disturbs the phase equilibrium. The polymer and the asphaltene 'compete' for the solvency power of the maltene phase and if insufficient maltenes are available, phase separation may occur. Photograph 10.1 shows the microscopic structures of 'compatible' and 'incompatible' bitumen/polymer systems. The compatible system has a homogeneous 'sponge-like' structure whereas the incompatible system has a coarse discontinuous structure.

Phase separation or incompatibility can be demonstrated by a simple hot storage test. Photograph 10.2 shows the storage vessel used for the test; it has three discharge arms along its length to enable the top, middle, and bottom thirds to be obtained separately. Tests are carried out on the top and bottom third and the middle third is discarded. Figure 10.8 shows the numerical difference in softening point of the top and bottom samples of two bitumens containing 7% of SBS thermoplastic rubber after 1, 3, 5 and 7 days storage at 140°C. The results clearly show that the compatible system is extremely stable, whereas the incompatible system has separated dramatically with virtually no polymer remaining in the bottom sample after 7 days storage. The implications of this in full-scale storage are obvious.

The main factors influencing storage stability are:

the amount and molecular weight of the asphaltenes;
the aromaticity of the maltene phase;
the amount of polymer present;
the molecular weight and structure of the polymer;
the storage temperature.

The complex inter-relationship between these factors is shown in figure 10.9[21].

Photograph 10.1 – The microscopic structures of compatible and incompatible bitumen/polymer systems

10.2.5.2 The manufacture of thermoplastic rubber/bitumen blends

The quality of the polymer dispersion achieved is influenced by a number of factors, but is primarily dependent on the shear rate applied by the mixer. When the polymer is added to the hot bitumen, the bitumen immediately starts to penetrate into the polymer particles, causing the styrene domains of the polymer to become solvated and swollen. Once this has occurred the level of shear exerted on the swollen particles is critical if a satisfactory dispersion is to be achieved within a realistic blending time. Thus, medium or preferably high shear mixers are required to adequately disperse thermoplastic rubbers in bitumen.

10.2.5.3 Properties of thermoplastic rubber/bitumen blends

Shell Bitumen U.K. have developed two SBS polymer-modified binders:

- CARIPHALTE DM;
- CARIPHALTE DA.

Photograph 10.2 – The storage stability test

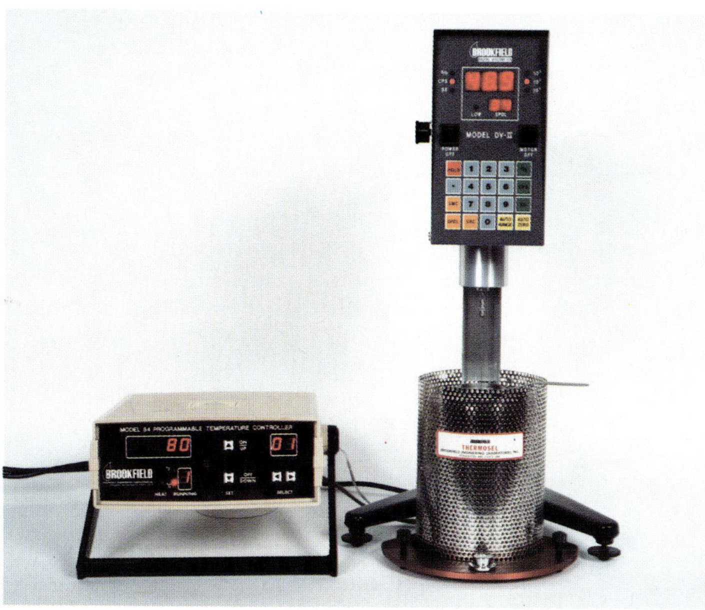

Photograph 10.3 – The Brookfield viscometer, used to measure the viscosity of bitumen/polymer systems at high temperature

Test \ Binder		CARIPHALTE DM	CARIPHALTE DA
Penetration at 25°C,	dmm	90±20	130±20
Softening point (IP),	°C	85±10	80±10
Brookfield viscosity, 150°C, shear rate 3.36 sec^{-1},	poise	10±4	9±4
Ductility at 5°C,	cm	>50	–
Minimum stiffness modulus 40°C, loading time 1000 secs, shear rate 2.5 x 10^{-4} sec^{-1},	kN/m^2	> 3	–
Storage stability. Difference in softening point top-minus-bottom after storage in a cylinder at 160°C for 7 days,	°C	< 5	< 5
Polymer content,	%	7.0±1.0	6.0±1.0

Table 10.2 – Specifications for CARIPHALTE DM and CARIPHALTE DA(23,24)

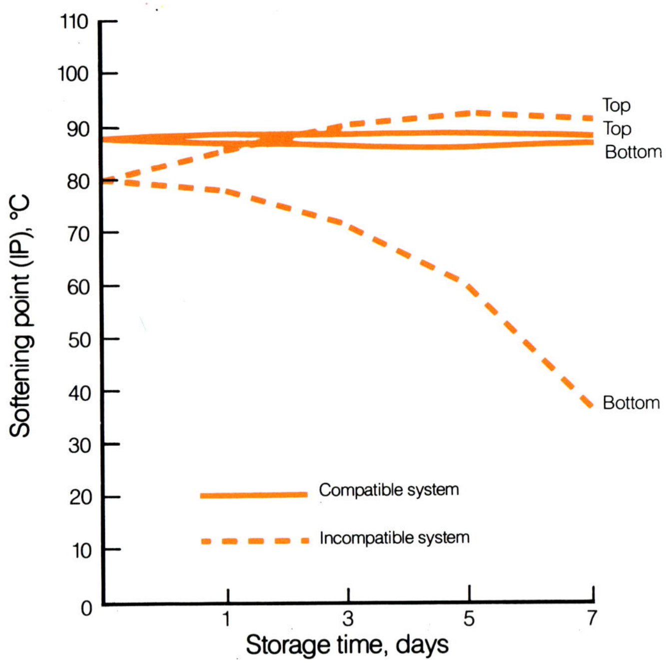

Figure 10.8 – Effect of the compatibility of the system on the storage stability of bitumen/thermoplastic rubber blends

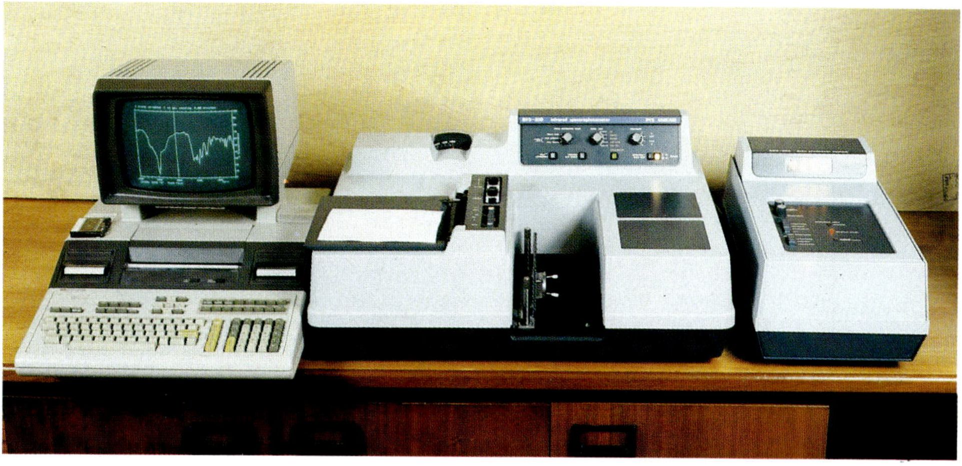

Photograph 10.4 – An infrared spectrophotometer, used to determine the polymer content of bitumen/polymer systems

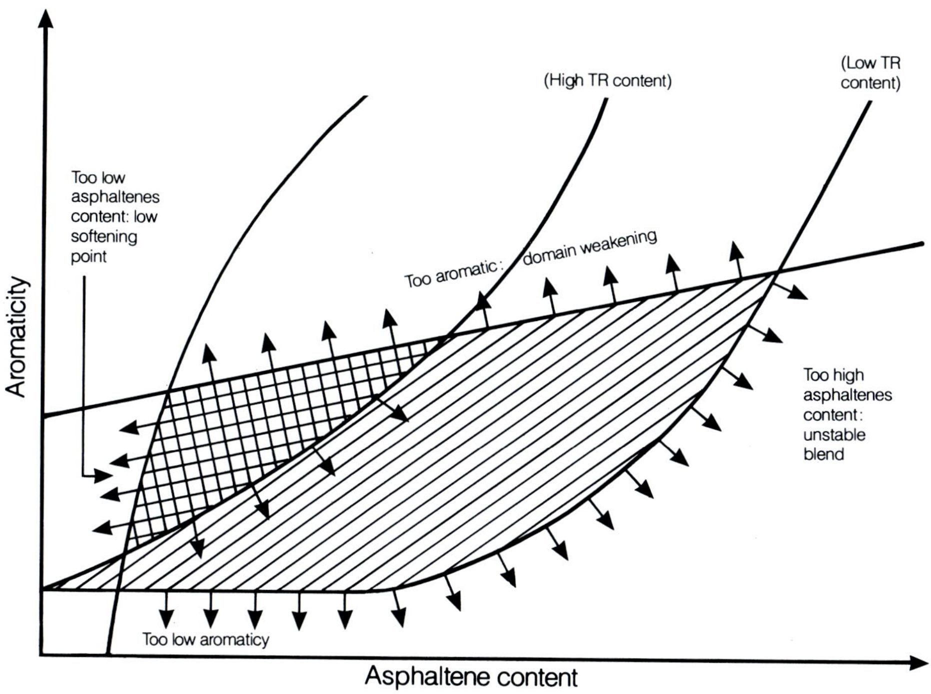

Figure 10.9 – Influence of bitumen constitution on bitumen/thermoplastic rubber blends(21)

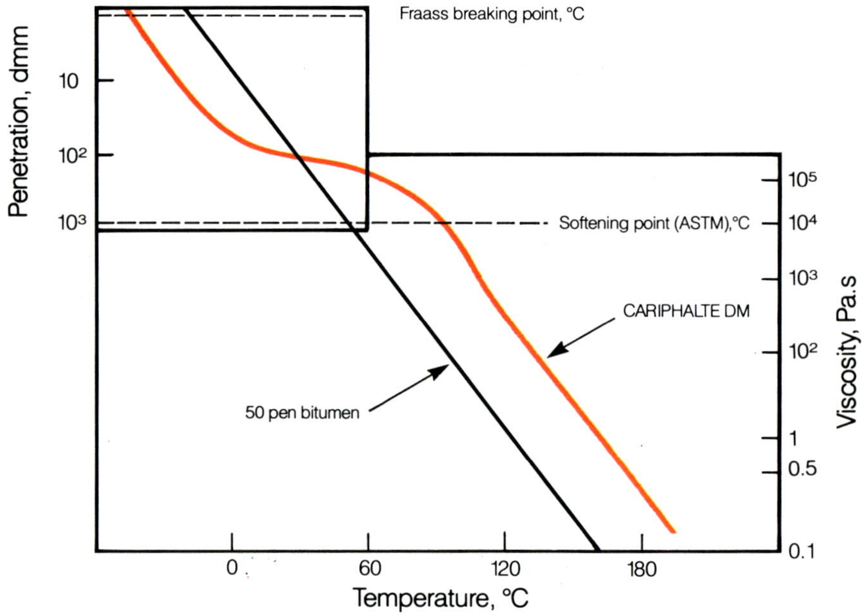

Figure 10.10 – Relationship between viscosity and temperature

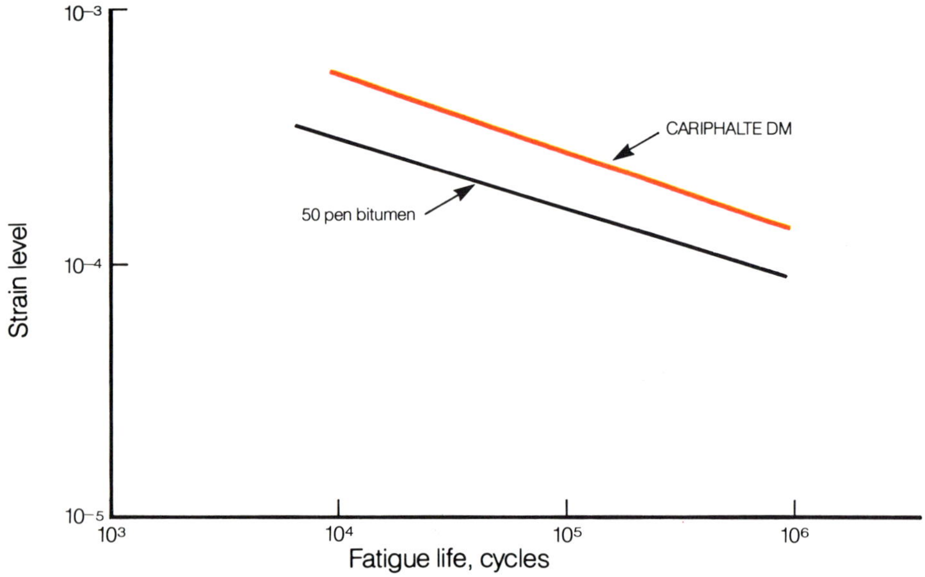

Figure 10.11 – Relationship between strain level and fatigue life

CARIPHALTE DM was developed for use in dense hot rolled asphalt and asphaltic concrete mixes to overlay lean concrete roadbases and old concrete carriageways, where temperature induced movements in the cementitious layer were resulting in cracks being 'reflected' up to the surface.

CARIPHALTE DA was developed for use in pervious macadam and friction course mixes.

The specifications of these two binders are detailed in table 10.2 on page 162.

The viscosity/temperature characteristics of CARIPHALTE DM and a 50 pen bitumen are shown on a Heukelom test data chart in figure 10.10. This clearly shows that at high service temperature, say 60°C, CARIPHALTE DM is significantly stiffer than 50 pen bitumen and is therefore more resistant to deformation. At low service temperatures, <0°C, CARIPHALTE DM is more flexible than 50 pen bitumen and is therefore more resistant to cracking.

The degree of improvement in deformation resistance has been demonstrated by wheel-tracking tests carried out by both the Transport and Road Research Laboratory (TRRL)(3) and Shell Research Limited(22) as shown in table 10.3. Clearly there is a substantial increase in deformation resistance, similar to that of heavy duty (HD) bitumen which was specifically developed to resist deformation. These improvements have been confirmed by full-scale trials.

The degree of improvement in the flexibility of bituminous mixes has been quantified by constant strain fatigue tests again carried out by both the TRRL(3) and Shell Research Limited(22). Fatigue tests carried out by Shell Research Limited on hot rolled asphalt at 5°C and at a frequency of 50 Hz show that over a wide range of applied strain levels CARIPHALTE DM improves fatigue life by a factor of at least 3 as shown in figure 10.11(23,24). Thus this material is eminently suitable for locations where repeated high tensile strains are present.

Binder	Binder properties		Wheel-tracking rate at 45°C, mm/h
	Penetration at 25°C, dmm	Softening point (IP), °C	
50 pen	56	52.0	3.2
HD 40	42	68.0	0.7
CARIPHALTE DM	84	90.0	0.7

Table 10.3 – Comparison of the wheel-tracking rates of hot rolled asphalt wearing course made with 50 pen, HD 40 and CARIPHALTE DM(3)

10.2.6 Thermosetting binders(25)

Thermosetting polymers are produced by blending two liquid components, one containing a resin and the other a hardener, which react chemically to form a strong three dimensional structure. Two-component epoxy resins blended with bitumen display the properties of modified thermosetting resins rather than those of bitumen. The remarkable family of two-component thermosetting resins was developed around thirty years ago and found wide application as surface coatings and adhesives.

The principal differences between bitumen (a thermoplastic) and thermosetting binders are as follows:

- when the two components in a thermosetting binder are mixed the usable life of the binder is limited; the amount of time available is strongly dependent on temperature; the higher the temperature the shorter the usable life;
- after a thermoset product is applied it continues to cure and increase in strength, cf organo-manganese compounds; the rate of curing on the road is dependent on the ambient temperature;
- as temperature increases, bitumen softens and flows; thermosetting binders are much less temperature susceptible and are virtually unaffected by the temperature changes experienced on a road;
- the cured thermosetting binder is an elastic material and does not exhibit viscous flow;
- when cured, thermosetting binders are very resistant to attack by chemicals, including solvents, fuels and oils.

There are three fully road-proven thermosetting systems available:

- SHELLGRIP/SPRAYGRIP;
- EROPHALT;
- SHELL EPOXY ASPHALT.

The first two are discussed in chapter 18, but in particular it is SHELL EPOXY ASPHALT(25) which most clearly demonstrates the combination of strength and flexibility which can be achieved using thermosetting binders. Table 10.4(26) gives a comparison of the properties of two hot rolled asphalt mixes, one containing bitumen and the other containing the epoxy asphalt binder. The Marshall stability of the epoxy asphalt can be up to ten times greater than the conventional asphalt and the wheel-tracking results show no measurable rutting on the epoxy material, demonstrating its superb resistance to permanent deformation. The dynamic stiffness of the epoxy asphalt is significantly higher than that of the conventional material, particularly at high ambient temperatures. However,

Test		SHELL EPOXY ASPHALT	CONVENTIONAL ASPHALT
Marshall stability,	kN	45	7.5
Marshall flow,	mm	4.0	4.0
Marshall quotient,	kN/mm	11.2	1.9
Wheel-tracking rate at 45°C,	mm/h	0	3.2
Stiffness modulus, (3 point blending, frequency 10 Hz, strain 1.8×10^{-4})	N/m^2		
0°C		2.0×10^{10}	1.5×10^{10}
20°C		1.2×10^{10}	3.0×10^{9}
40°C		3.3×10^{9}	4.0×10^{8}
60°C		9.5×10^{8}	–
Flexural fatigue resistance, cycles to failure. (Composite specimen: 50mm asphalt on 10mm steel plate, constant deflection 1.0 mm, 25°C, 5 Hz haversine loading)			
6 % binder		1.0×10^{6}	3.0×10^{4}
7 % binder		$>2.0 \times 10^{6}$	2.0×10^{5}

Table 10.4 – Comparison of typical properties of SHELL EPOXY ASPHALT and conventional asphalt(26)

flexural fatigue tests using asphalt bonded to a steel plate (simulating flexing of an orthotropic bridge deck) showed that in this test the fatigue life of the epoxy asphalt was at least an order of magnitude greater than that of the conventional asphalt.

Since its introduction in the sixties, SHELL EPOXY ASPHALT has been used in a variety of situations all over the world[26,27]. In 1986, a full-scale road trial using the epoxy binder in a hot rolled asphalt mix was laid on the M6 at Keele, Staffordshire. The evaluation of this trial is continuing and to date the material is performing well.

10.3 Cost-performance relationships for modified binders[22]

For any premium product to gain credence with highway engineers it must be shown to be cost-effective. Total-life or life-cycle costing is an economic evaluation technique which is increasingly being used to justify new construction or maintenance expenditure. In this technique all the costs are considered in terms of discounted costs and benefits over the period of time that the road is required (net present cost concept).

Downes[22] et al considered it particularly appropriate to apply the principles of life-cycle costing to the consideration of the cost effectiveness of polymer modified bitumens vis-a-vis conventional bitumens. CARIPHALTE DM was one of the modified materials evaluated and it was shown that at a discount rate of 7% the net saving using CARIPHALTE DM was marginally above 20%. If user costs and accidents were taken into account then the savings increased to 45%.

Although much has been done already there is perhaps still more to be done. That the majority of polymer-modified bitumens provide technically sound solutions to a number of road construction problems is proven. That most of them achieve this with only minimal changes to established working practices is equally sure. The next step must be for all sectors of the industry to work together to develop a greater understanding of the potential of these materials in order to provide better, more cost-effective roads for the benefit of both the industry and the community as a whole.

10.4 References

(1) Hoban, T W S: *Modified binders: an underexploited road maintenance tool.* Highways, vol 55, No 1926, pp 13-19, June 1987.

(2) Brown, S F: *Bituminous pavements: materials, design and evalation.* Residential course at The University of Nottingham, section D, *Basic concepts*, pp D6-D8, April 1988.

(3) Denning, J H and Carswell, J: *Improvements in rolled asphalt surfacings by the addition of organic polymers.* Transport and Road Research Laboratory, Report LR 989, 1981.

(4) Deme, I: *Shell sulphur-asphalt products and processes.* Paper presented at the International Road Federation symposium on sulphur-asphalt in road construction, Bordeaux, France, February 1981.

(5) Deme, I: *Shell sulphur-asphalt products and processes.* Shell Bitumen Review 59, pp 1-7, June 1981.

(6) Petrossi, U, Bocca, P L and Pacor, P: *Reactions and technological properties of sulphur-treated asphalt.* Ind. Eng. Chem. Prod. Res. Develop., vol 11, pp 214-219, 1972.

(7) Lee, D Y: *Modification of asphalt and asphalt paving mixtures by sulphur additives.* Ind. Eng. Chem. Prod. Res. Develop., vol 14, pp 171-177, 1975.

(8) Denning, J H and Carswell, J: *Improvements in rolled asphalt surfacings by the addition of sulphur.* Transport and Road Research Laboratory, Report LR 963, 1981.

(9) Deme, I: *Sulphur as an asphalt diluent and mix filler.* Advances in Chemistry Series, No 165, 1977.

(10) Daines, M E, Carswell, J and Colwill, D M: *Assessment of 'Chemcrete' as an additive for binders for wearing courses and roadbases.* Transport and Road Research Laboratory, Report RR 54, 1985.

(11) Cooper, K E and Brown, S F: *Comparative testing to establish the effect of 'Chemcrete' on basecourse and roadbase materials used on a full scale trial.* The University of Nottingham, December 1987.

(12) Cooper, K E and Brown, S F: *Early life tests on rolled asphalt wearing course: the effect of 'Chemcrete'.* The University of Nottingham, November 1987.

(13) Kennedy, T W, Cummings, L O and White, T D: *Changing asphalt through creation of metal complexes.* Proceedings of the Association of Asphalt Paving Technologists, vol 50, 1981.

(14) Loveday, C A: *Improving the performance of bituminous materials without sacrificing workability.* Paper presented at the Institution of Highways and Transportation National Conference, Keele University, September 1985.

(15) Private communication.

(16) Gilby, G W: *Ethylene-vinyl acetate (EVA) copolymers as modifiers for bitumen binders.* Asphalt Technology, No. 36, pp 37-41, June 1985.

(17) Woolley, K G: *Polymer modified bitumen for extra value asphalt.* Asphalt Technology, No. 38, pp 45-51, September 1986.

(18) Choyce, P W and Woolley, K G: *EVA modified binders.* Highways, vol 56, No. 1933, pp 18-34, January 1988.

(19) Bull, A L and Vonk, W C: *Thermoplastic rubber/bitumen blends for roof and road.* Shell Chemicals Technical Manual TR 8.15, 1984.

(20) Vonk, W C and van Gooswilligen, G: *Improvement of paving grade bitumens with SBS polymers.* Proceedings of the fourth Eurobitume symposium, Madrid, pp 298-303, October 1989.

(21) van Gooswilligen, G and Vonk, W C: *The role of bitumen in blends with thermoplastic rubbers for roofing applications.* Paper presented at 'Roofing and Waterproofing Worldwide', London, May 1986.

(22) Downes, M J W, Koole, R C, Mulder, E A and Graham, W E: *Some proven new binders and their cost effectiveness.* Paper presented at the seventh Australian Asphalt Pavement Association conference, Brisbane, Australia, August 1988.

(23) Whiteoak, C D: *Broadening the boundaries of binder technology.* Highways, vol 56, No. 1942, p 35, October 1988.

(24) Whiteoak, C D: *Shell Cariphalte DM. An SBS modified bitumen.* Shell Bitumen Review 64, pp 2-5, September 1989.

(25) Dinnen, A: *Shell epoxy bitumens for high performance surfaces.* Shell Bitumen Review 59, pp 11-12, June 1981.

(26) Dinnen, A: *Epoxy bitumen binders for critical road conditions.* Proceedings of the second Eurobitume symposium, pp 294-296, Cannes October 1981. Reprinted in a special publication of Asphalt Technology, pp 294-296, January 1982.

(27) Rebbechi, J J: *Epoxy asphalt surfacing of the Westgate Bridge.* Proceedings of the Australian Road Research Board, vol 10, pp 136-146, 1980.

Chapter 11 – Types and application of bituminous mixes

The range of possible mix compositions is almost infinite, from single size coated stone at one extreme to mastic asphalt at the other. However, the types of bituminous mixes used in the U.K. can be broadly classified as either asphalts or macadams.

The principal differences between asphalts and macadams are as follows:

- the grading of the aggregate in an asphalt is usually a relatively single-sized coarse aggregate with a large proportion of fine aggregate and very little intermediate sized material; as a result of this 'gap' in the grading this type of mix is often referred to as 'gap-graded'; in macadams the aggregate grading is continuous (see figure 11.1);

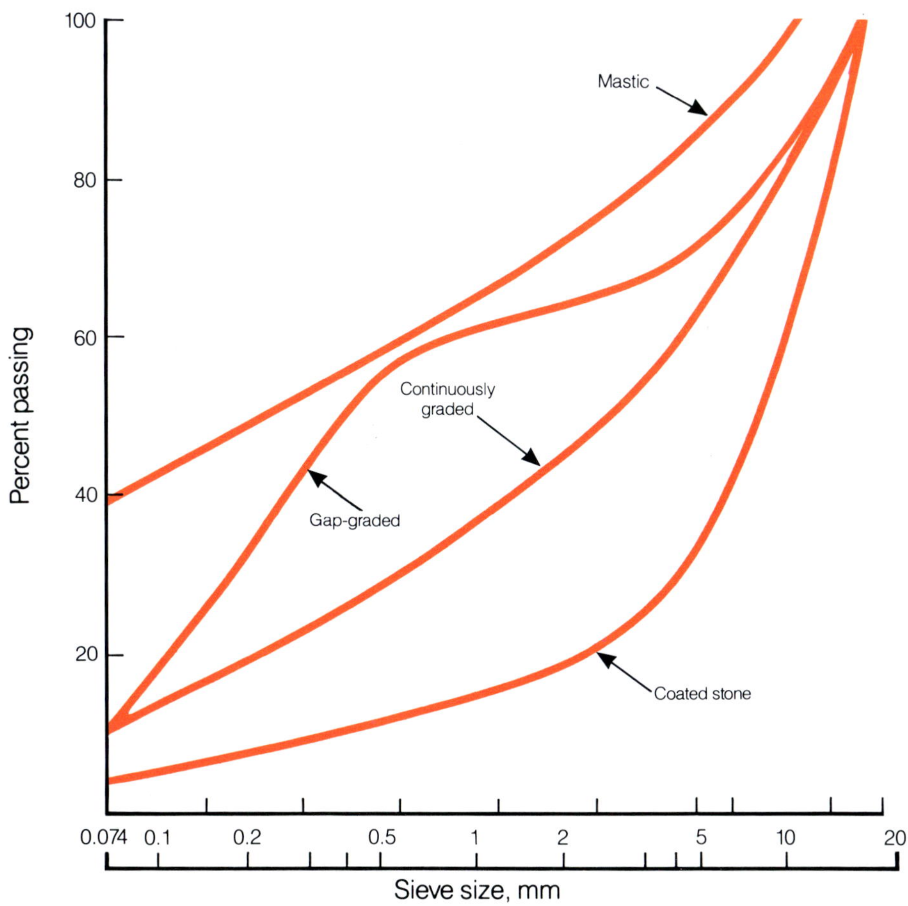

Figure 11.1 – Typical grading curves for various mixes

- the strength of an asphalt is dependent on the stiffness of the sand/filler/bitumen fraction, ie the mortar; to achieve a mortar of adequate stiffness, asphalts are manufactured using relatively hard bitumens and high filler contents; the strength of a macadam is primarily achieved through friction and mechanical interlock of the aggregate particles, enabling softer grades of bitumen to be used;
- as asphalts contain a high proportion of filler, ie a large surface area of aggregate to coat, they have a relatively high bitumen content; on the other hand macadams have less fine aggregate, thus less bitumen is required to coat the aggregate satisfactorily; in macadams the primary function of the bitumen is to act as a lubricant during compaction and to bind the aggregate particles in service;
- asphalts have a low void content, are relatively impermeable and are sufficiently durable for the heaviest traffic categories; macadams have a relatively high void content, are permeable and are not as durable as asphalts.

Typical compositions are given in table 11.1 with the corresponding aggregate gradings in figure 11.1. From table 11.1 it can be seen that as the mix changes from coated stone towards mastic asphalt, harder bitumen grades are used, the bitumen, filler and sand content increases and the coarse aggregate content decreases.

		Coated stone	Continuously graded	Gap-graded	Mastic
Coarse aggregate,	%w	86.0	52.0	30.0	30.0
Fine aggregate,	%w	7.0	38.0	53.0	26.0
Filler,	%w	3.0	5.0	9.0	32.0
Bitumen,	%w	4.0	5.0	8.0	12.0
Coarse aggregate,	%v	64.5	44.1	25.7	27.5
Fine aggregate,	%v	5.1	32.2	46.0	18.9
Filler,	%v	2.1	4.2	7.8	27.0
Bitumen,	%v	8.3	11.5	17.5	26.6
Void content	%v	20.0	8.0	3.0	< 1.0
Grade of bitumen,	pen	100-300	100-200	35-100	15-25

Table 11.1 – Typical compositions of various mixes

11.1 Asphalt

Asphalts or hot process asphalts are mixtures of mineral aggregate, filler and bitumen which are processed at temperatures as high as 230°C. *In-situ*, asphalts distribute traffic stresses primarily through the fine aggregate/sand/filler/bitumen mortar. Therefore, to resist deformation under the imposed stresses, the mortar must have high stiffness; this is achieved by using a relatively hard bitumen and a high filler content. In service, asphalts are impermeable and very durable. Two types of asphalts are used in the U.K.: hot rolled asphalt (HRA) and mastic asphalt. HRA is widely used in the surfacing of major roads, whereas the use of mastic asphalt in road construciton is limited to specialist applications, eg to provide resistance to abnormally heavy loadings, to provide a waterproof membrane, etc.

11.1.1 Mastic asphalt

Hand-laid mastic asphalt was first applied successfully as a covering to footways in Paris in 1835. This was followed by its introduction into London the following year. There are records of a number of important streets in the City of London laid with mastic asphalt between 1870 and 1873, including Lombard Street, Moorgate Street and Carter Lane. Following the success which was achieved using mastic asphalt, its use grew and in the years following the 1914-18 war it was applied extensively on town and country roads as well as city streets.

Today, mastic asphalt consists of a mortar of bitumen and fine aggregate and a porportion of coarse aggregate. Full details of the basic raw materials are given in British Standard (BS) 1446: 1973[(1)] and BS 1447: 1988[(2)]. The majority of the mastic asphalt laid in the U.K. is to BS 1447; material to BS 1446 is seldom used because of the need to import natural rock asphalt. To comply with BS 1447 the fine aggregate has to be naturally occurring limestone rock with between 40% to 55% passing the 75 micron sieve and no more than 3% retained on the 2.36 mm sieve. The percentage of coarse aggregate added varies depending upon the application. Normally the aggregate is mixed with either a 25 pen or 15 pen bitumen.

The manufacture of mastic asphalt is somewhat complex. It can be carried out in a single process only when the material is to be used immediately; in such a case all the aggregate is combined with the bitumen in a large mixer equipped with slow-moving blades. Otherwise only the fine aggregate is mixed with the bitumen, producing a mortar which is cast into 25 kg blocks and left to cool. These blocks can be stored and supplied to sites when required. They are then remelted in a special mastic mixer, together with the desired proportion of coarse aggregate.

Mastic asphalt is normally hand-laid by skilled and experienced 'asphalters' who use wooden floats to 'work' the material at temperatures of between 175°C and 230°C until it has stopped flowing. Thicknesses may vary between 20 mm and 50 mm, according to the intended end use. Its impermeability is assured by the fact that the void content of the laid material is generally below 1%. These stringent requirements – and the relatively small areas normally laid at any one time – have in the past ruled out machine-laying. Large areas such as on the Humber Bridge, have been recent exceptions; the routine machine-laying of mastic asphalt on German roads is not strictly comparable since the German 'gussasphalt' mix differs from mastic asphalt to BS 1446 and 1447.

The high percentage of fines in mastic asphalt gives a smooth surface with poor skid-resistance properties. To increase the latter, a variety of treatments may be applied. Rolled-in pre-coated chippings are used for busy road surfaces in the same way as with rolled asphalt: the chippings are applied while the material is still plastic enough to allow partial but secure embedment and are rolled in using either hand or light power-rollers. On footways and other lightly-trafficked surfaces, the desired finish may be obtained by applying a special sand, using the floats. Another method, particularly used for multi-storey car park surfaces, is to crimp the surface with an indentation roller while it is still hot.

11.1.2 Hot rolled asphalt

HRA was first used in the U.K. in 1895 on the Kings Road, Chelsea and Pelham Street, Kensington. Unfortunately the importance of sand grading and filler content was not fully appreciated and as a result the material failed within a matter of months.

One of the pioneers of HRA was the American engineer Clifford Richardson. As a part of a survey of asphalt pavements in the U.S.A. and Europe, Richardson visited the U.K. in 1896. Based on his experience he recommended the use of an asphalt mix, very similar to sand carpet, which performed well for many years. This material formed the basis of the first British Standards for asphalt[3,4]. In 1935 these two standards were combined into one single standard, BS 594[5]. Since this date BS 594 has been revised six times[6,7,8,9,10]. In the latest revision (1985) it was again divided into two parts[11,12], each revision reflecting the views of diferent sectors of the road construction industry, including research data.

As stated earlier, one of the most important characteristics of HRA is that the aggregate is 'gap-graded' that is to say, it contains very little medium-sized aggregate, ie 2.36 mm to 10 mm, but consists of a matrix of sand, fine mineral filler and bitumen into which coarse aggregate, usually nominal 14 mm, is incorporated. Although the inclusion of the coarse aggregate fraction increases the stiffness of the material, its primary role is to extend the mortar, thereby making the material more economical. It is the gap-grading which gives HRA wearing courses their characteristic weather-resistant, durable surface which can accommodate heavy loads without cracking.

BS 594 : 1985[11] contains specifications for HRA roadbase, basecourse and wearing course mixtures with various coarse aggregate contents. Roadbase and basecourse mixtures generally contain 60% coarse aggregate and at this level stresses are distributed by a combination of aggregate contact and through the mortar. The bitumen contents of roadbase and basecourse mixtures are generally lower than wearing course materials because of their lower fines and filler contents. The nominal size of the coarse aggregate in these two mixes is usually larger than in wearing course mixtures and the gradings are coarser, in keeping with the laid thickness of the material, which, in the case of roadbases, may be up to 150 mm.

Photograph 11.1 – Lightly chipped hot rolled asphalt

Photograph 11.2 – Heavily chipped hot rolled asphalt

Wearing course mixtures can contain 0%, 15%, 30%, 40% or 55% coarse aggregate. Wearing course mixes generally contain 30% coarse aggregate and are laid 40 mm thick and this has been found suitable for the vast majority of applications, although 40% coarse aggregate content material laid 50 mm thick is becoming increasingly popular. Increasing the thickness significantly increases the time available to satisfactorily compact the material and it is therefore a means of improving winter workability (see chapter 13, section 13.3.1.1). Up to 40% coarse aggregate the surface of the laid material is smooth and a rugous surface is achieved by rolling into the asphalt surface 14 mm or 20 mm precoated chippings (see photographs 11.1 and 11.2). BS 594[11] details the required shape, size and rate of spread of the chippings. The required properties of the chippings such as polished stone value and abrasion value, are usually specified in individual specification documents taking into account the requirements of the individual site. Wearing course mixtures containing 55% coarse aggregate do not require surface chippings to achieve a rugous texture. Indeed at coarse aggregate contents greater than 45% it is difficult satisfactorily to embed the chippings into the asphalt.

The composition of HRA is shown diagrammatically in figure 11.2. The function of the four constituents can be summarised as follows:

- coarse aggregate; extends the mortar making the mix more economical and increases the stablity of the mix;
- fine aggregate; this forms the major proportion of the mortar and is probably the most important component influencing the performance of the material both during application and in service;
- filler; this fraction can be considered to be acting in two ways; firstly, it can be considered that it modifies the grading of the fine aggregate, thereby producing a denser mix with more points of contact between aggregate particles; secondly, and probably a better

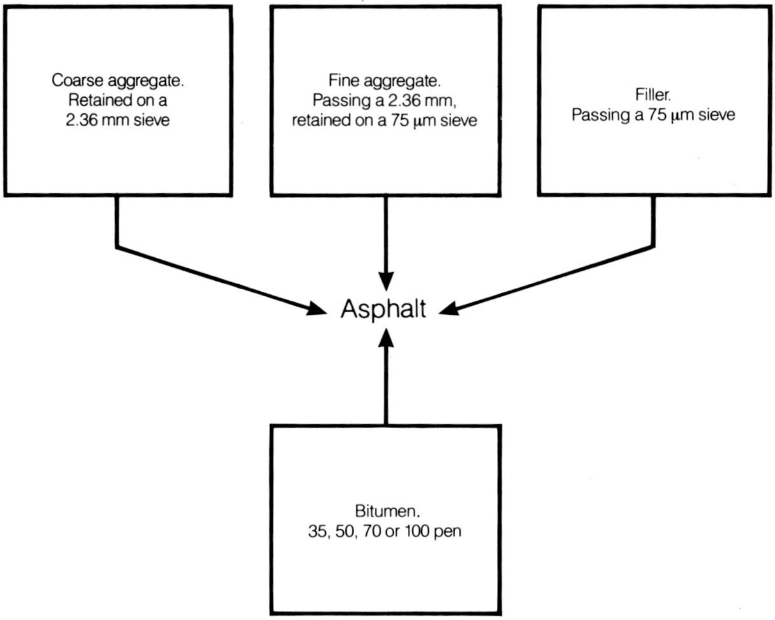

Figure 11.2 – Composition of rolled asphalt

way of regarding the filler is to consider that the filler and bitumen together form a binder which both lubricates and binds the fine aggregate to form the mortar; the properties of the mortar will depend on the nature of the fine aggregate and on the amount and viscosity of the binder;

- bitumen; acts as a lubricant during compaction and as a visco-elastic binder of high viscosity in service; five grades of bitumen can be used, 35 pen, 50 pen, 70 pen, 100 pen and HD 40; 35 pen and HD 40 pen are used only for the most heavily trafficked sites, and 70 pen and 100 pen are generally used on more lightly trafficked sites; 50 pen is suitable for most purposes.

HRA is a high-quality material which is used principally on heavily trafficked roads, ie motorways, trunk roads and to a lesser extent city streets. Traditionally the specifications for composition are in the form of recipes specifying the quality and proportions of aggregate and bitumen to be used, depending upon climate and traffic loading. Alternatively, HRA wearing course mixes can be 'designed' using the Marshall test(13); this is discussed in some depth in chapter 16.

In 1989 more than 60 million square metres of HRA wearing course were laid in the U.K. If properly specified, designed, manufactured and compacted, the vast majority of this material should be performing satisfactorily well into the next century.

11.2 Coated macadam

Coated macadam made with tar is the oldest form of coated material used in the U.K. The earliest record of its use was in Gloucestershire in 1832 and in Nottinghamshire in 1884. Bitumen macadam began to be used when bitumen became available at the turn of the century. The term macadam perpetuates the memory of the famous Scottish road engineer, John Loudon McAdam. McAdam understood that to optimise the strength of a layer of crushed aggregate it was necessary to use aggregate of a mixture of sizes, ie a graded aggregate. Using this technique McAdam appreciated that the spaces between the larger aggregate would be filled with smaller aggregate producing a layer of relatively open-textured interlocking aggregate. It was the manufacture of open-textured bitumen and tar macadam mixes, a development of McAdam's principle, that was the foundation of the growth of the coated materials industry in the middle of the twentieth century.

The composition of coated macadam is shown diagrammatically in figure 11.3. The mixture can be divided into three aggregate fractions and the bitumen. The functions of the constituents can be summarised as follows[14]:

- coarse aggregate; providing the main skeleton of interlocking aggregate to distribute the traffic loads;
- fine aggregate; fills or partially fills the voids in the coarse aggregate skeleton;
- filler; increases the viscosity of the binder thereby reducing the risk of bitumen draining from the aggregate; combined with the bitumen the filler assists to fill small voids;
- bitumen; acts as a lubricant during compaction and also as a waterproofing and bonding agent in service; in dense macadams it also assists to fill the voids and provide strength to the mix.

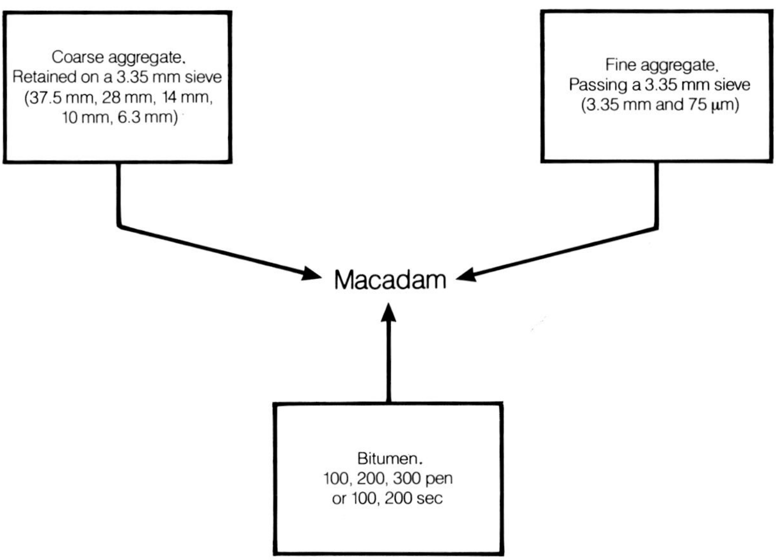

Figure 11.3 – Composition of coated macadam

Recipe specifications for the composition of a wide range of coated macadams, ie bitumen content and aggregate grading, are given in BS 4987 : 1988[15], viz:

dense macadam roadbases;
dense macadam basecourses;
open graded basecourse;
single course macadam;
open-graded wearing courses;
close-graded wearing courses;
dense wearing course;
medium-graded wearing course;
fine-graded wearing course;
pervious wearing courses.

11.2.1 Open and medium-graded macadams

These are used for both basecourse and wearing course mixes. The main feature of these mixes is that they have a low fines content which results in good workability during laying, with void contents after compaction of 15% to 25%. To prevent ingress of water into the road structure a waterproof membrane such as a surface dressing is necessary.

The strength of the material is provided in the main by aggregate interlock and only to a limited extent by the bitumen. One of the main functions of the bitumen is to provide tensile strength at the top surface, to prevent loss of the fine aggregate from the surface, when heavily stressed by traffic. The mix is usually bound by 200 pen or 300 pen bitumen or a cutback bitumen, the choice of bitumen depending on the intensity of traffic and the time of year the material is laid.

The texture of a wearing course plays a major role in its skidding resistance. Open and medium-textured macadams provide satisfactory texture for light and medium traffic conditions, eg private drives, car parks, playgrounds, etc. This type of material does not contribute significantly to the strength of the overall road structure.

The permeable nature of this type of mix has been used to advantage on airfield runways and on heavily trafficked roads to prevent aquaplaning and reduce spray. These are known as friction course when used on airfields and pervious macadam when used on highways. Pervious macadams were introduced into BS 4987 for the first time in the 1988 revision of this standard. The development, advantages and disadvantages of pervious macadam/friction course are discussed in detail in section 11.4.

11.2.2 Dense and close-graded macadams

Dense macadam roadbases and basecourses have been used in substantial quantities in the last 25 years in major road construction and are suitable for the heaviest traffic conditions. The high fines content results in a dense material, typically 5% to 10% void content, which has good load spreading properties, high resistance to deformation and is sufficiently flexible to resist cracking under repeated loading.

Photograph 11.3 – 10 mm dense bitumen macadam wearing course

In BS 4987 : 1973[16] 14 mm and 10 mm 'dense' macadam wearing courses were specified. However, these two materials (see photograph 11.3) have a relatively high void content and are certainly not impermeable. In the 1988 edition of BS 4987 these two materials have been renamed 'close'-graded macadam and a new 6 mm dense wearing course mix of lower void content was introduced. Dense and close-graded wearing course mixes are only suitable for light and medium traffic categories as they do not possess the long-term durability nor the high-speed skid resistance required for the heaviest traffic categories.

11.2.3 Heavy duty macadam

Dense bitumen macadam has served well as a high strength roadbase and basecourse on most heavily trafficked roads for a quarter of a century. However, to cope with the predicted increase in traffic loads, even stronger roadbases will be required, with greater resistance to permanent deformation than traditional roadbases. One such material is heavy duty macadam (HDM) which is based on dense bitumen macadam roadbase but incorporates a harder grade of bitumen (50 pen) and a higher filler content (8%).

These two changes in the composition increase the dynamic stiffness by a factor of up to 3. This enables a 10% to 15% reduction in roadbase thickness to be made to that used for conventional dense macadam roadbase, for equivalent performance. Reduced thickness with equivalent performance is attractive for reconstruction work where existing levels, shallow service ducts or limited bridge headroom may dictate the use of thinner than normal constructions. Alternatively, retaining the same thickness will result in a correspondingly longer life (see chapter 17, section 17.4).

11.2.4 Fine-graded macadams

Prior to the introduction of BS 4987: 1988 this material was called fine cold asphalt. However, despite this name this material falls within the macadam category. The majority of fine-graded macadams are manufactured using 200 pen or 300 pen bitumen or 100 second cutback bitumen and therefore they require to be laid and compacted at 80°C to 100°C. However, heavily cutback grades are produced for cold laying and depot storage.

When newly laid, fine-graded macadam has a high void content and it is thus susceptible to the effects of water. Therefore, care should be taken in the selection of the aggregate. The void content gradually reduces under traffic and the mix ultimately becomes impervious. The material is usually laid 15 mm to 25 mm thick and will thus remove small irregularities in the road profile when used as an overlay. The material provides little additional strength to the road structure. In view of these points a sound and preferably impervious substrate is required. In addition, a base of good regularity is required beneath the material as it is prone to problems if laid at variable thickness.

11.3 Asphaltic concrete

Like macadams, asphaltic concrete mixes are continuously graded but contain less coarse aggregate, more fine aggregate and filler, coated with a greater quantity of a slightly harder bitumen. Asphaltic concrete was originally developed in the U.S.A. to meet the need for stiff and strong pavements to carry the heavy loads and high tyre pressures of aircraft. In the U.K. asphaltic concrete is used for airfields and is often termed Marshall asphalt. It is not normally used on roads in the U.K., firstly, because it requires a very strong stiff structure below it: if the structure is not sufficiently stiff it is liable to crack under the pavement deflections caused by heavy vehicles; secondly it provides very little surface texture and would require a surface dressing treatment to provide an adequate level of skid-resistance on high-speed roads.

Asphaltic concrete derives its strength and stability primarily through aggregate interlock and to a lesser extent through the sand/filler/bitumen mortar. The composition of asphaltic concrete is determined by the U.S.A. Asphalt Institute Marshall mix design procedure(17) the object of which is to arrive at an optimum bitumen content for maximum stability and density. This results in an economical blend of aggregate and bitumen having high stability under traffic and good durability in service, while at the same time being sufficiently workable to lay and compact to a typical void content of 3% to 5%. Bitumen contents are similar to those of dense and close-graded macadams but harder bitumens are generally used, ie 70 pen and 100 pen. During laying, the compaction of the material has to be carefully controlled to ensure that the maximum compacted density is achieved.

11.4 Pervious macadam/friction course

In any one year roads and airfields in the U.K. are wet for about the number of hours equal to 100 times the rainfall in inches(18). Effectively this means that roads in some parts of the U.K. can be wet for up to 50% of the year. To overcome the problems of aquaplaning and skidding on aircraft runways in the mid fifties, the then Air Ministry Works Organisation carried out extensive trials of a 10 mm open-textured bitumen macadam (now commonly known as airfield friction course). This material was designed to allow water to drain through to an impermeable barrier and, following the crossfall of the runway, drain into an aggregate filled trench. The trials of friction course proved to be very successful and since 1962 a large number of both civil and military airfield runways have been surfaced with friction course(19).

On roads in the U.K. it has been estimated that in up to 10% of all wet road accidents, water spray from fast moving vehicles (particularly lorries) is a major contributory factor[20]. The cost of accidents caused by spray could be as much as one-third that of accidents caused by skidding on wet roads[21]. Thus the adaptation by the Transport and Road Research Laboratory (TRRL) of the highly successful airfield 10 mm friction course to a 20 mm pervious macadam for roads was a logical development.

11.4.1 The development of pervious macadam

Since 1967 a number of full-scale trials have been carried out to determine the suitability of pervious macadam for heavily trafficked roads[22,23,24,25,26,27]. Szatkowski and Brown[28] reported that in these trials the macadams made with the larger aggregate generally possessed the highest void content, permeability and texture depth but with both aggregate sizes the values fell considerabaly during the first few years after laying. However, the spray reduction remained effective for longer periods on the material manufactured using 20 mm aggregate, lasting at least three years with heavy traffic (>7000 cvd/lane) and up to six years with medium traffic (2,500 cvd/lane). However, the surfacing is likely to remain sound for many years after its effective spray-elimination properties have reduced, and the ultimate life may be as long as 15 years. It was further concluded that resistance to skidding at low speeds with the 20 mm aggregate was approximately the same as with other surfacings made with the same aggregate. The change in skid-resistance with speed on the 20 mm aggregate correlated well with texture depth, similar to the behaviour observed with HRA and surface dressings.

Structurally, 40 mm of pervious macadam manufactured using a 100 or 200 penetration bitumen is equivalent to either 16 mm of HRA or 20 mm of a dense bitumen macadam[29]. However, if epoxy bitumen is used instead of a penetration grade bitumen, the dynamic stiffness of the pervious macadam is substantially increased and its structural contribution is then equal to or greater than HRA. Bitumens containing polymers such as ethylene vinyl acetate and styrene-butadiene-styrene lie between these two extremes.

11.4.2 The benefits of pervious macadam

11.4.2.1 Reduction in spray

The hazard created by the spray thrown up by the tyres of vehicles on wet roads, especially at high speeds, has long been recognised[20]. The spray restricts visibility with the result that overtaking, in particular, becomes very dangerous. Extensive road trials have demonstrated conclusively that spray from traffic can be considerably reduced by pervious macadam (see photograph 11.4). Pervious macadam has a high void content, ca 25%, and consequently will absorb rainwater like a sponge. A 40 mm thick layer of pervious macadam can absorb up to 8 mm of rainfall before becoming saturated. It also acts as a drainage layer, enabling rainwater to percolate through the material rather than ponding on the surface.

11.4.2.2 Reduction in noise

Road surfaces are laid with a coarse macrotexture to enhance contact of the tyre tread with the surface at high speeds in wet weather and thus help to maintain good skid resistance. The relationship between noise and the frictional properties of the road surface have been studied by the TRRL[30]. The frictional properties were assessed from the texture depth, measured using the sand patch technique, and from the estimated percentage change in the braking-force coefficient for a test trailer braking at 130 km/h and the same trailer braking at 50 km/h. This work concluded that the

Photograph 11.4 – Comparison of the spray generated from hot rolled asphalt (top) and pervious macadam (bottom) *(Photograph reproduced by kind permission of the Transport and Road Research Laboratory)*

peak noise from light and heavy vehicles is a function of the texture depth and is a logarithmic function of the percentage change in braking-force coefficient. This means that in practice it is impossible with conventional road surfacing to achieve high standards of skid resistance without an associated high noise level.

Fortunately, however, this relationship is not applicable to pervious macadam where the mechanism of water dispersion from the road surface is quite different. Following experiments on a large number of sites, a further report was prepared on the noise from vehicles running on open textured road surfaces(31). This work proved conclusively that the noise from all vehicles running on pervious macadam surfaces was significantly lower than that generated on conventional bituminous surfacings for similar levels of skid resistance: 3 to 4 dB(A) lower in dry conditions and 7 to 8 dB(A) lower in wet conditions. Recent research in the Netherlands(32) has also shown that the structure of pervious macadam can absorb a considerable proportion of vehicle engine noise.

An additional benefit arising from the surface characteristics of pervious macadam is the reduction in the rolling resistance it offers to vehicle tyres. It has been calculated that a change in texture from that represented by HRA or brushed concrete to that of pervious macadam on the motorways and trunk roads of the U.K. could result in a fuel saving of some £20 million per annum(33).

11.4.3 Disadvantages of pervious macadam

Although pervious macadam surfaces possess many obvious advantages there are a few disadvantages – some minor, some not so minor – that have to be overcome.

The allowable variation of bitumen content in pervious macadam mixes is relatively small. If the bitumen content is too low it may result in inadequate coating of the aggregate, or a bitumen film which is so thin that it oxidises rapidly, resulting in premature failure of the surfacing. On the other hand, if the bitumen content is too high (or if the mix temperature is too high) bitumen may drain off the aggregate during transport, resulting in variation in bitumen content in the laid material. In extreme cases bitumen has been seen running from the tailboards of delivery vehicles.

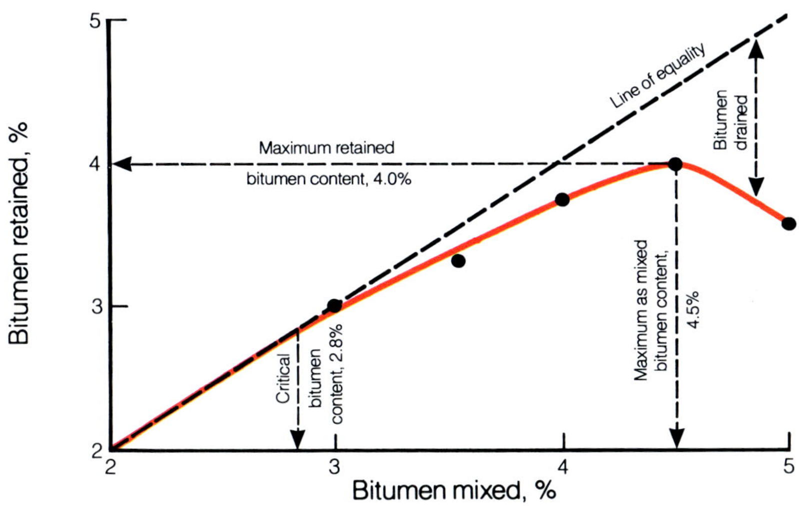

Figure 11.4 – Typical binder drainage test result(33)

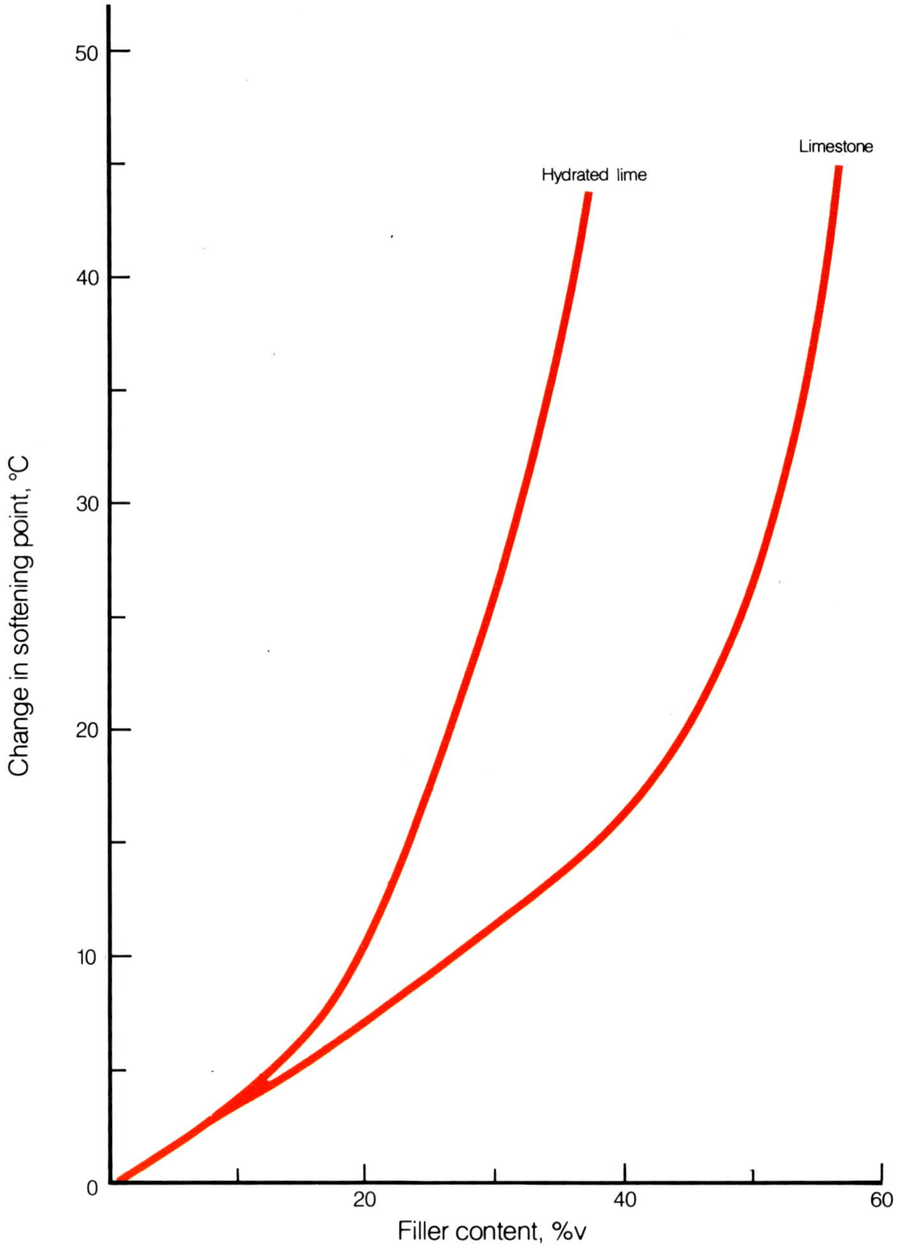

Figure 11.5 – The effect of filler content and filler type on the softening point of bitumen/filler mixtures

Until 1984 the determination of the bitumen content for a pervious macadam in the U.K. has been made largely on a trial and error basis. However, the TRRL have developed a simulative quantitative bitumen drainage test on the mixed material(34). The test involves storing the pervious macadam in a perforated metal basket in an oven over a preweighed tray that collects any drained bitumen. The test is performed in duplicate over a range of bitumen contents. The quantity of bitumen retained is plotted as a function of the initial bitumen content as shown in figure 11.4. Up to a 'critical' bitumen content no drainage occurs. However, above this bitumen content drainage increases markedly until a peak is reached beyond which the bitumen content retained in the mix actually reduces; this is known as the 'avalanche' effect. This drainage test enables bitumen contents to be specified on a rational basis. However, full-scale bitumen drainage trials(35) have shown that the laboratory test tends to underestimate the actual drainage in practice. The reasons for this are being investigated by the TRRL.

The addition of hydrated lime filler significantly increases the viscosity of the bitumen as shown in figure 11.5, thereby reducing bitumen drainage. It also improves adhesion (see chapter 9, section 9.7) which is clearly beneficial for a porous material which is saturated in water for prolonged periods.

As pervious macadam and friction course are, by definition, permeable, the bitumen films within the compacted material are constantly exposed to oxygen. The result is that the bitumen oxidises relatively quickly *in-situ*. Figure 11.6 shows the *in-situ* penetration of bitumen in airfield friction course between 1 and 9 years old. After nine years the bitumen has a penetration of 20 which is said by some to be a 'critical' penetration, ie at penetrations below 20 the material is thought to be vulnerable to fretting.

In the full-scale road trial on the A38(34) initial results suggest that hydrated lime and some polymers may reduce the oxidation rate of the bitumen *in-situ*. Laboratory tests(36) have shown that bitumen oxidation is reduced if 0.5% to 1.0% of hydrated lime is added. It is hypothesised that the hydrated lime absorbs polar oxidation products which then cannot participate in further oxidation reactions.

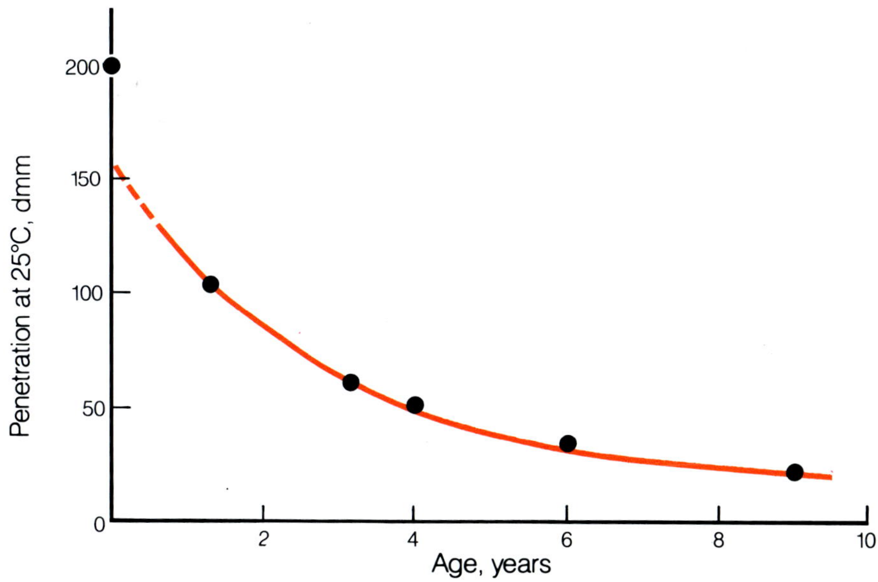

Figure 11.6 – Hardening of bitumen in friction course

The removal of snow and ice from pervious macadam necessitates two to three times as much salt treatment as required for conventional HRA. The reasons for this are twofold. Firstly, if snow is compacted into the voids of the pervious macadam, salt will tend preferentially to remove snow from other areas, first leaving the compacted material in the wheel-tracks. Secondly, as the snow thaws forming a briny slush, it will drain away below the running surface and the salt will become ineffective. However, although the extra salt treatments are a nuisance, and undoubtedly add to the cost of winter maintenance, the substantial benefits that accrue to the travelling public and the substantial reduction in traffic noise for adjacent residents considerably outweigh the need for extra salt(37).

11.5 Comparison of the performance of various surfacings

The road engineer is frequently faced with the problem of choosing different types of wearing course surfacing materials. In broad terms the decision will be dictated by the cost of the treatment, the requirements of the site and the benefits of the different types of surfacing. In 1975 the TRRL laid a full-scale road trial(25) to compare the performance of open-textured macadam, pervious macadam, HRA, dense bitumen macadam, Delugrip (a proprietary material) and a surface dressing. A simplified comparison of various properties of the different surfacings after six years is given in table 11.2(27). Similarly table 11.3(28) compares the 'effective ratios' for pervious macadam, HRA and surface dressings over a wider range of factors. The values from both tables should be regarded as only approximate. However, they do provide the highway engineer with some performance data to assist him to make an engineering judgement as to which material is the most suitable for a particular site.

Photograph 11.5 – Pervious macadam

	Merit rating (1 = poor, 5 = good)					
	Open textured macadam	Pervious macadam	'Delugrip'	Dense bitumen macadam	Rolled asphalt	Surface dressing
Resistance to deformation	3	3	4	4	4	–
Durability	3	2	3	2	4	3
Skid-resistance (low speed)	5	5	5	5	4	5
Skid-resistance (high speed)	4	5	3	4	4	5
Riding quality	4	3	5	5	3	1
Spray suppression	3	5	2	1	2	3
Noise reduction	4	4	4	3	2	1
Ease of application	3	3	3	3	2	4
Area/unit cost	2	2	1	2	1	5

Table 11.2 – Relative performance of different surfacings after six years[27]

	Effective ratio		
	Surface dressing	Pervious macadam	Rolled asphalt
Skid resistance	1	1	1
Texture depth	1	1	1
Average life	1	2	3
Structural strength	1	3	6
Improved ride	1	4	4
Low noise	1	4	2
Low spray	2	6	1
Ease of resurfacing	4	1	1
Area treated/unit cost	6	2	1

Table 11.3 – Relative merits of surface dressing, pervious macadam and rolled asphalt[28]

11.6 References

(1) British Standard 1446:1973: *Specification for mastic asphalt (natural rock asphalt and fine aggregate) for roads and footways.*

(2) British Standard 1447:1988: *Specification for mastic asphalt (limestone fine aggregate) for roads, footways and pavings in buildings.*

(3) British Standard 344 : 1928 : *Single-coat asphalt (sand and stone aggregate).*

(4) British Standard 345 : 1928 : *Single-coat asphalt (clinker aggregate).*

(5) British Standard 594 : 1935 : *British specification for rolled asphalt, fluxed lake asphalt and asphaltic bitumen (hot process).*

(6) British Standard 594 : 1945 : *Rolled asphalt, asphaltic bitumen and fluxed lake asphalt (hot process). Part 1, single course. Part 2, two course.*

(7) British Standard 594 : 1950 : *Rolled asphalt, asphaltic bitumen and fluxed lake asphalt (hot process).*

(8) British Standard 594 : 1958 : *Rolled asphalt, asphaltic bitumen and fluxed lake asphalt (hot process).*

(9) British Standard 594 : 1961 : *Rolled asphalt hot process.*

(10) British Standard 594 : 1973 : *Specification for rolled asphalt (hot process) for roads and other paved areas.*

(11) British Standard 594 : Part 1 : 1985 : *Hot rolled asphalt for roads and other paved areas. Part 1. Specification for constituent materials and asphalt mixtures.*

(12) British Standard 594 : Part 2 : 1985 : *Hot rolled asphalt for roads and other paved areas. Part 2. Specification for the transport, laying and compaction of rolled asphalt.*

(13) British Standard 598 : Part 3 : 1985 : *Sampling and examination of bituminous mixtures for roads and other paved areas. Part 3. Methods for design and physical testing.*

(14) Asphalt and Coated Macadam Association : *The types and scope of coated macadam.* Third edition, 1975.

(15) British Standard 4987 : Part 1 : 1988 : *Coated macadam for roads and other paved areas. Part 1. Specification for constitutent materials and for mixtures.*

(16) British Standard 4987 : 1973 : *Specification for coated macadam for roads and other paved areas.*

(17) The Asphalt Institute : *Mix design methods for asphalt concrete and other hot-mix types.* Manual series No 2, 1988.

(18) Young, C P : *The prediction of the time for which a road is wet.* Road Research Laboratory, Technical Note TN 402, 1969.

(19) Hutson, R M : *Development of open textured bitumen macadam friction course for airports in Great Britain.* Proceedings of the international symposium on porous asphalt, pp 41-47, June 1976.

(20) Maycock, G : *The problem of water thrown up by vehicles on wet roads.* Road Research Laboratory, Report No 4, 1966.

(21) Sabey, B E : *Accidents : their cost and relation to surface characteristics.* Paper presented to the symposium on safety and the concrete road surface and design, specification and construction, 1973.

(22) Please, A, O'Connell, B J and Buglass, B F : *A bituminous surface texture experiment, High Wycombe bypass (M40).* Road Research Laboratory, Report LR 307, 1967.

(23) Brown, J R : *Pervious bitumen-macadam surfacings laid to reduce splash and spray at Stonebridge, Warwickshire.* Transport and Road Research Laboratory, Report LR 563, 1973.

(24) Jacobs, F A : *A full-scale experiment to examine improved bituminous surfacings for heavily trafficked motorways – M1, 1973.* Transport and Road Research Laboratory, Supplementary Report 125 UC, 1974.

(25) Brown, J R: *A full-scale experiment to compare surfacings for maintaining bituminous roads: A1 Buckden 1975*. Transport and Road Research Laboratory, Supplementary Report 323, 1977.

(26) Brown, J R: *Interim report on the performance of surfacings for maintaining bituminous roads: A1 Buckden (1975-1977)*. Transport and Road Research Laboratory, Supplementary Report 476, 1979.

(27) Brown, J R: *The performance of surfacings for maintaining bituminous roads : A1 Buckden (1975-1982)*. Transport and Road Research Laboratory, Report RR 70, 1986.

(28) Szatkowski, W S and Brown, J R: *Design and performance of pervious wearing courses for roads in Britain 1967-1976*. Highways and Road Construction International, January/February 1977.

(29) Potter, J F and Halliday, A R: *The contribution of pervious macadam surfacing to the structural performance of roads*. Transport and Road Research Laboratory, Report LR 1022, 1981.

(30) Franklin, R E, Harland, D G and Nelson, P M: *Road surfaces and traffic noise*. Transport and Road Research Laboratory, Report LR 896, 1979.

(31) Nelson, P M and Ross, N F: *Noise from vehicles running on open textured road surfaces*. Transport and Road Research Laboratory, Report LR 696, 1981.

(32) van der Heide, J P J: *Asphalt friction course in built-up areas*. Asfalt (NL), pp 19-25, March 1989.

(33) Williams, A R: *The influence of tyre and road surface design on tyre rolling resistance*. Institute of Petroleum, Report IP 81-003, 1981.

(34) Daines, M E: *Pervious macadam: trials on trunk road A38 Burton bypass, 1984*. Transport and Road Research Laboratory, Report RR 57, 1986.

(35) Farrington, J J and Roberts, K: *Practical aspects of pervious macadams*. Paper presented at the Institution of Highways and Transportation national conference, September 1985.

(36) Plancher, H, Green, E L and Peterson, J C: *Reduction of oxidative hardening in asphalts by treatment with hydrated lime – a mechanistic study*. Proceedings of the Association of Asphalt Paving Technologists, vol 45, p 1, 1976.

(37) Farrington, J J and Roberts, K: *Experiences with pervious macadam in Staffordshire*. Highways and Transportation, vol 33, pp 8-13, October 1986.

Photograph 12.1 – An asphalt plant at night *(Photograph reproduced by kind permission of Wimpey Asphalt Limited)*

Chapter 12 – The manufacture and testing of bituminous mixes

The manufacture of a bituminous mix involves mixing aggregate in specified proportions with bitumen to produce a homogeneous mix which complies with the specified composition and temperature. Before mixing, the aggregate must be blended to ensure the required grading is achieved and then heated to dry it and to bring it to a temperature suitable for coating with bitumen.

The earliest known method of producing coated aggregate (ca 1850) was by using a shallow steel tray mounted over a closed-grate coal fire. Broken and roughly graded aggregate was heated and dried on the tray before hot tar was poured over it and mixed manually[1]. Mechanical means of drying and mixing were developed by the turn of the century; these machines were based on concrete mixers and cylindrical dryers using coal fires to dry and heat the aggregate. Subsequently the paddle-mixer or pug-mill was developed employing a single-shaft with blades; the first types of these mixers remained in use until the mid twenties.

Three types of aggregate dryer were developed in the U.K.: the long rotating cylinder, a tower dryer where the aggregate is slowly passed downwards whilst hot gases were forced upward through it and the short drum type of batch dryer; this was the forerunner of the batch-heater mixer. In the late thirties and early forties oil started to replace coal as the fuel for heating. Initially hot burner gases were ducted into the drums and further development brought the burner inside the drum, much as we now see in modern plants.

12.1 Types of mixing plants

In general, mixing plants comprise two main sections; the first is the actual mixing unit and the second is the equipment necessary to support the mixer, ie drying and aggregate proportioning. Often the latter equipment is common to different types of mixers. There are basically three types of plant used in the U.K.:

- indirectly heated batch mixing plant;
- directly heated batch mixing plant;
- directly heated continuous mixing plant.

12.1.1 Indirectly heated batch mixing plant

In this plant, as is evident from its designation, a pre-set amount of proportioned aggregate and bitumen is mixed at a time, giving batches which may vary in size from 250 kg to 5000 kg. Both the type of mix and the batch size determine plant capacity: mixing times vary with the type of mix. A 2500 kg batch, for example, requiring a mixing time of 60 seconds, will give a maximum plant capacity of 150 tonnes per hour.

The cold aggregate feed is the first major component of any mixing plant. The aggregate is collected from a stockpile and is loaded into a cold feed holding bin. Between 4 and 9 bins are normally used, each charged with a different aggregate (see photograph 12.2). Feeder units are located under the bin to control the flow of aggregate onto a conveyor belt. The control of the feeders can be individual or combined to provide a proportional arrangement whereby the ratio of one feeder to another is maintained but the total flow rate of all the feeders can be varied. The conveyor belt transports the combined aggregates to the drying and heating unit.

Photograph 12.2 – Cold feed holding bins *(Photograph reproduced by kind permission of Wimpey Asphalt Limited)*

Photograph 12.3 – Aggregate heater/dryer *(Photograph reproduced by kind permission of Wimpey Asphalt Limited)*

A basic part of any mixing plant is the aggregate dryer/heater. Its function is to ensure that as much water as possible is removed from the aggregate (in order to allow intimate and durable bonding of the bitumen to the aggregate see chapter 9), and to heat the aggregate to the required mixing temperature. With aggregate moisture contents in the cold feed as high as 5% of dry weight (47 kg of water per tonne of aggregate!)[2] it is clear that this part of the plant plays a vital role in assuring the ultimate quality of the finished mix. It normally consists of a large rotating metal drum and is mounted at an angle of between 2 and 10 degrees along its axis (see photograph 12.3). A gas or oil burner at the lower end causes hot gases to rise through the drum and emerge at the upper end. The cold aggregate enters the rotating drum from the upper end and is moved through the hot gas and flame by means of steel angles or flights fitted to the inside, aided by the downward inclination of the drum, finally discharging onto an elevator. The length of time during which the aggregate remains in the drum is known as the 'retention time'. The drying and heating process is very fuel- (and cost-) intensive: it takes 10 litres of fuel per tonne of aggregate to remove 5% moisture and increase aggregate temperature by 160°C[2]. Aggregate temperature at the lower end of the drum is measured by a thermocouple or a radiation thermometer, compared to the specified temperature, and corrected as required.

To eliminate dust from the dryer exhaust, dust collectors are fitted. In general there are three types of dust collector: cyclone, fabric filters or wet collectors. Two or more of these devices may be fitted – a primary collector and one or more final or secondary collectors. Provision is usually made for this dust to be returned to the hot aggregate. However, if the dust is unsuitable or is produced in excessive quantities, it can be removed from the collector and disposed of.

When the aggregate is discharged from the dryer/heater it is transported vertically to the screening unit housed at the top of of the tower as shown in figure 12.1[3]. Here the aggregate is separated into fractions of specified size and stored temporarily in hot storage bins. Material is withdrawn from each bin in pre-determined amounts and charged into a weigh hopper, sequentially commencing with the

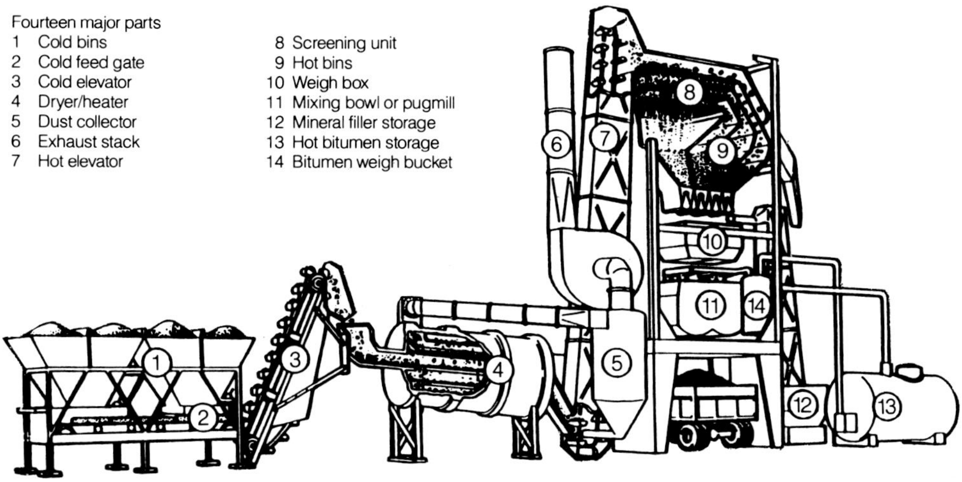

Figure 12.1 – Schematic diagram of an indirectly heated batch mixing plant[3]

largest size aggregate and progressing down to the finest size. Each aggregate is individually weighed to a quantity determined by the mix formulation as a proportion of the batch size.

The bitumen binder is kept in a storage tank and fed into a recirculating system of pipes which returns it to the tank. When bitumen is needed for the mixer, a valve in the system is opened to let the required quantity of bitumen flow out. A weight bucket is one method of measuring bitumen quantity; another is a flow meter which measures it volumetrically. In the latter case care must be taken to use the specific gravity of bitumen at the pumping temperature when converting to a gravimetric measurement. Table 20.2 shows the effect of temperature on the specific gravity of bitumen.

The dust collected from the dryer exhaust can be re-introduced into the hot aggregate if the gradation of this dust complies with the specified aggregate grading. If not, the dust may be collected in a fines silo which feeds fines into the batch in controlled amounts. Where collected dust is not suitable for adding to the mix, mineral filler will have to be used instead. This is normally limestone filler, supplied in bulk into a silo and added to the mix in the same way as the collected dust.

The weighed aggregate, fines and bitumen are discharged into the pugmill. This is usually a large semi-circular bowl with two shafts which have angled mixing blades. These blend the hot aggregate and when the bitumen is added the movement of the aggregate ensures that all the aggregate surfaces are coated with a thin film of bitumen. Depending on the efficiency of the pugmill and the material being mixed, the duration of the mixing varies from 30 to 60 seconds. To minimise oxidation of the bitumen the mixing time should be as short as possible; however, it is important that it is long enough to ensure that uniform distribution of the aggregate sizes and uniform coating of the aggregate is achieved. The mixed material is then discharged directly into the delivery vehicle or transported by skip into a hot-storage bin.

Photograph 12.4 – An indirectly heated batch plant *(Photograph reproduced by kind permission of Wimpey Asphalt Limited)*

Photograph 12.5 – A temporary asphalt plant on an airfield *(Photograph reproduced by kind permission of Redland Aggregates Limited)*

The indirectly heated batch mixing plant is so called because the aggregates are pre-heated before they are batched. This type of plant is the most common in the U.K. and is very flexible, changing from one mix type to another without difficulty. Thus this type of plant is ideal for handling the multiplicity of mix specifications deemed necessary in the U.K.[4]. These plants are usually permanent, with capacities ranging from 100 to over 400 tonnes per hour.

12.1.2 Directly heated batch mixing plant

In this type of plant a batch of cold proportioned aggregate from the aggregate cold feed is charged into the dryer. This is dried and heated as a batch and is discharged into a conventional pugmill for mixing with the bitumen. The cold aggregate feed is almost identical to that of the indirectly heated batch plant; the proportional cold feed is charged into a hopper for check weighing before drying. The proportions of aggregate cannot be changed in the process and therefore feed control of the cold aggregate must be very accurate. From the weigh hopper the aggregates are charged into the batch heater – in effect a shortened version of the aggregate dryer – and the batch is dried and heated. The heated aggregate is discharged into a conventional pugmill and is mixed with the bitumen.

The reason why this type of plant is called a directly heated batch plant or a batch heater plant is because the aggregates are dried and heated in batches before mixing. These plants are very common in the U.K.: they allow rapid changes in mix specification and are not limited by hot aggregate storage capacity. Plant outputs vary from some 50 to more than 200 tonnes per hour.

However, aggregates having moisture contents >2% cannot be dried economically in these plants. They are thus better suited to low- and medium-temperature materials, such as the coated macadams, than to mixes with high sand contents, such as rolled asphalt.

12.1.3 Directly heated continuous mixing plant

These plants produce bituminous materials without the use of screens, hot aggregate storage bins, weigh hoppers or pugmill mixers. In principle, the cold aggregate feed system is identical to that described in section 12.1.1. However, in the directly heated continuous or drum mix plant the aggregate flows continuously from the aggregate cold feed bins without any interruption to the flow. The aggregate proportioning must be precise, as any error in feed ratio will be detected only when samples of the mixed material are analysed in the laboratory.

The cold feed bins discharge the aggregate onto a collecting conveyor which in turn feeds them into a charging conveyor. This is provided with a weighbridge enabling the aggregate flow rate to be read off (making an appropriate deduction for the amount of moisture in the aggregate) and thus allowing the calculation of the fine aggregate and bitumen addition.

The drum used here differs from that used in the indirectly heated batch plant dryer in two respects:

- the burner is at the top end of the drum and the aggregate thus flows away from the flame;
- the configuration of the flights inside the drum is more complex.

Photograph 12.6 – A drum mix plant *(Photograph reproduced by kind permission of Bardon Quarries Limited)*

Photograph 12.7 – Hot mix storage silos *(Photograph reproduced by kind permission of Bardon London Limited)*

Photograph 12.8 – A modern drum mix plant control room *(Photograph reproduced by kind permission of Redland Aggregates Limited)*

The drum interior has two zones. The first zone dries and heats the proportioned aggregate and the second zone mixes the hot aggregate with the bitumen. The mixing zone is protected by either a flame shield or a device to increase the density of the aggregate curtain to protect the coated aggregate from the radiant heat of the heating unit.

The bitumen is injected into the drum at the start of the second zone. The flow rate of the bitumen is proportional to the aggregate flow rate and is controlled by either variable-speed fixed displacement pumps or variable orifice control valves.

Compared with other types of mixing plant, this plant's exhaust gases contain only 10% of the dust carried by the former(2). This is because most of the airborne dust generated in the heating and drying zone is absorbed by the bitumen-coated aggregate in the second zone. U.K. environmental legislation, however, still requires dust collection. This dust, once it is collected, may be recirculated to the plant in two ways: either through the aggregate charging conveyor or by blowing it into the drum pneumatically. Either of these means may also be used to introduce controlled amounts of mineral filler into the drum. The completed mix flows continuously from the bottom of the drum; bulk storage is normally in the form of well-insulated high-capacity hot storage silos(5), (see chapter 8, section 8.2.3).

Directly heated continuous mixing plants are so called because the aggregates are dried and heated on a continuous basis just prior to mixing with the bitumen. This type of plant is not very flexible and requires considerable effort to change mix type. However, the increased use of computerised control is beginning to overcome this problem. These plants are relatively simple and the smaller models are ideal for contracts where a portable plant is required. Their simplicity also reduces maintenance costs and improves reliability. However, the overriding advantage of drum mixers is their throughput. Capacities range from 100 tonnes per hour to an astounding 500 tonnes per hour which makes this type of equipment ideal for large contracts where large quantities of the same material are required for a long period.

12.2 Production control testing of bituminous mixes

Production control testing is used to ensure that the manufactured material complies with the specification, ie aggregate grading and bitumen content. As it is impractical to test every batch of material produced, representative samples of the material are taken at prescribed intervals. It is *essential* that the sample is representative, otherwise the data generated will be meaningless and could be totally misleading. It is equally important that the tests are carried out by trained and experienced operators using well maintained and properly calibrated equipment.

There is a number of procedures for the correct sampling of bituminous mixes; these are detailed in British Standard (BS) 598: Part 100(6). This standard details procedures for sampling material at a number of locations: from the mixer during discharge, delivery vehicles, paver, and from the laid material before or after compaction. The advantages and disadvantages of sampling at these different locations are described in the above BS.

Sampling at the mixing plant is obviously necessary for production control testing. Once the sample has been obtained it requires preparatory treatment, eg reducing it in size, before testing. The necessary preparatory procedures are detailed in BS 598: Part 101(7). The test methods for determination of the composition of the mix in terms of aggregate grading and bitumen content are detailed in BS 598: Part 102(8). These are required for production control and laboratory investigations in cases of failure. Four methods are detailed in the above BS, all involving the separation of the bitumen from the aggregate using chlorinated hydrocarbon solvents (see table 12.1) followed by wet grading of the aggregate into its individual fractions. The majority of bituminous

Method	Solvent	Bitumen content	Filler content
Funnel	Trichloroethylene or Methylene chloride	By difference	Directly
Extraction bottle (bitumen by difference)	Trichloroethylene or Methylene chloride	By difference or directly	Directly
Extraction bottle (bitumen directly determined)	Methylene chloride	Directly	By difference or directly
Sieving extractor	Methylene chloride	By difference or directly	By difference or directly
Hot extractor	Trichloroethylene	By difference	Directly

Table 12.1 – Test methods for analysis of bituminous materials[8]

mixes can be analysed by any of the methods; however the funnel and hot extractor methods take longer to carry out. Consequently the extraction bottle and sieving extractor methods have gained in popularity. Table 12.1 shows that in some of the tests the filler fraction (<75 micron) is totally recovered and weighed directly, with the bitumen content being calculated by difference, whereas in other methods the bitumen content is directly determined by measuring its concentration in a known volume of solvent; in these cases the filler content is calculated by difference. Collection and direct determination of filler content and direct determination of bitumen content provides a check on the results but it increases the duration of the test.

Determining the bitumen content directly involves evaporation of the solvent from a known quantity of the bitumen/solvent solution (known as an aliquot) and weighing the recovered bitumen. The percentage of bitumen can then be calulated from the weight of the sample and the volume of solvent originally used. However the procedure is relatively complicated and prone to error. The most common sources of error are:

- loss of solvent by evaporation;
- incomplete separation of the insolvable matter from the bitumen/solvent when centrifuge methods are used;
- incomplete removal of the solvent during recovery;
- dust entering the bitumen/solvent solution.

All of the above will increase the calculated bitumen content.

When determining the filler fraction directly a pressure filter is used. The disadvantage with this procedure is that it is slow and great care must be taken to avoid leaks from the pressure filter seals.

All four methods involve sieving the fine aggregate in the presence of the solvent. This 'wet' sieving method washes more material through the 75 micron sieve than would be the case for dry sieving. Prior to 1985 all gradings in BSs related to dry sieving and a correction factor had to be agreed between consumer and producer to allow for the extra material washed through the 75 micron sieve. However, now both BS 594: Part 1: 1985[9] and BS 4987: Part 1: 1988[10] specify aggregate grading on a wet sieving basis.

12.3 Determination of recovered bitumen properties

In addition to routine production control tests there is a number of additional tests available to highway engineers to investigate unsatisfactory performance. One very useful test is to recover the bitumen from the mix and determine its penetration and softening point.

In this test a sample of the mix is soaked in methylene chloride to remove the bitumen from the aggregate into solution. The bitumen/solvent solution is separated from the fine mineral matter by filtration and centrifuging. The solvent is then evaporated under controlled conditions, using the apparatus shown in figure 12.2, ensuring that all of the solvent is removed without removing any of the lighter components of the bitumen. When the recovery procedure is complete, the penetration and softening point of the recovered bitumen can be determined in the normal way. The method is described in detail in BS 598: Part 103[11] which is identical to the Institute of Petroleum (IP) method IP 105/85. Alternatively the solvent can be removed using a rotary evaporator[12] (see photograph 12.9) which has the advantage that it removes the solvent very rapidly from the solution, but there is at present no BS or IP test method for this apparatus.

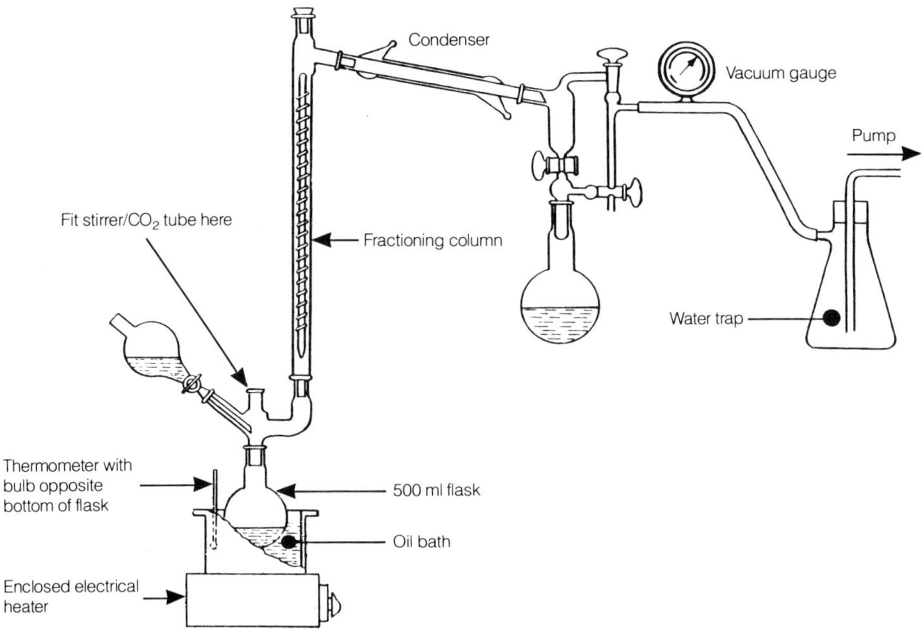

Figure 12.2 – Distillation apparatus used for recovery of bitumen[11]

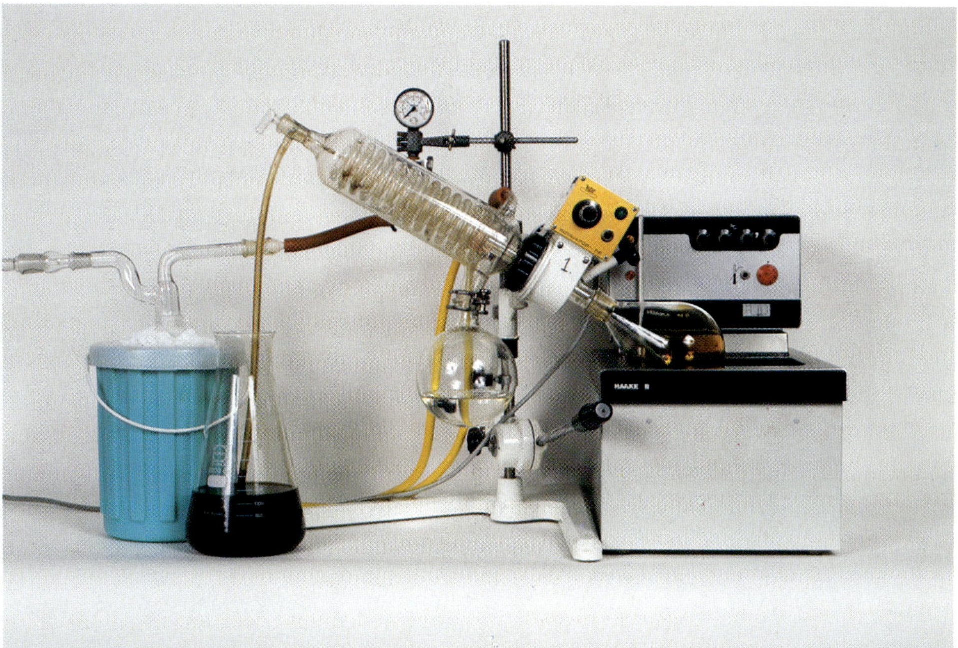

Photograph 12.9 – The rotary evaporator

The precision for the BS 598 : Part 103 test method is as follows:

- *Repeatability*

  Penetration – 5 dmm

  Softening point – 2.5°C

- *Reproducibility*

  Penetration – $1.9\sqrt{x}$ dmm

  Softening point – 3.5°C

  where x is the mean of the two results. These precision values only apply to bitumens with a penetration $\leqslant 120$.

The test is very operator sensitive and deviation from the standard method will almost certainly result in misleading data[12]. For these reasons the test is not suitable for contractual purposes but is a very useful tool for investigating unsatisfactory performance. With sensible engineering interpretation of test results the bitumen recovery method can be used to determine if the bitumen in the mix is abnormally hard or soft.

In addition to penetration and softening point, chromatographic examination of the recovered bitumen can be useful to detect the presence of any contaminants. Figures 12.3a and 12.3b show chromatographic plots (using true boiling point – gas liquid chromatography) of a 50 pen bitumen and the same bitumen containing 2%w diesel. The presence of the contaminant is very clear. Although the technique is only qualitative and not quantitative, it is a useful tool for examining recovered bitumens which are abnormally soft.

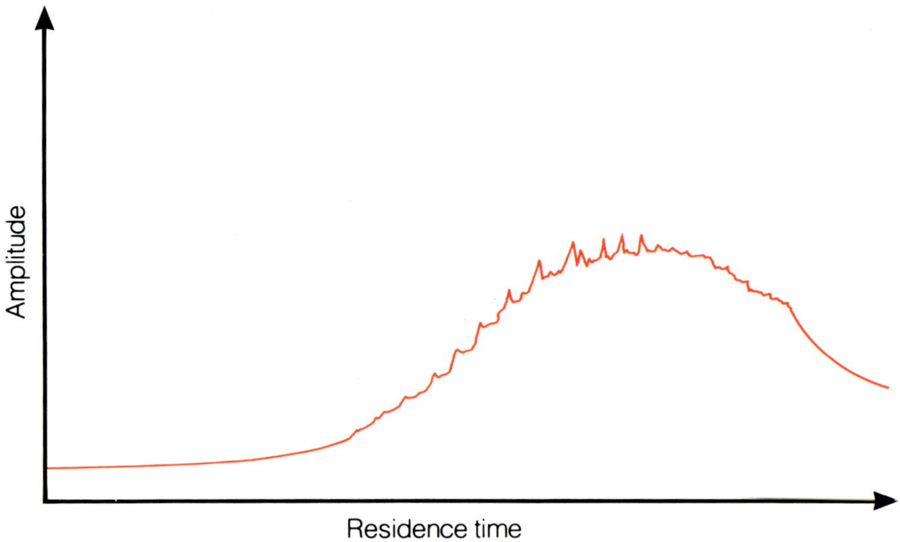

12.3a – 50 pen bitumen

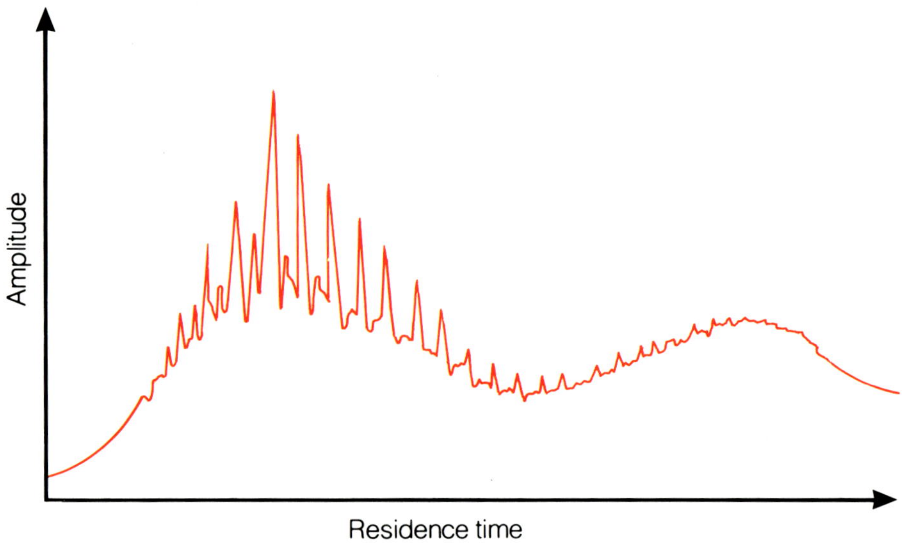

12.3b – 50 pen bitumen + 2%w diesel

Figure 12.3 – True boiling point – gas liquid chromatography traces on two bitumens

12.4 Recycling bituminous materials[13]

A limited amount of re-use or recycling of old bituminous surfacings has been carried out in the U.K. for many years. Material removed from roads prior to resurfacing has been used as a fill material and for regulating layers for paved areas carrying light traffic. A form of *in-situ* recycling involving the scarification of the existing road surface, mixing this material with fresh bitumen, followed by profiling and compaction, was carried out in the U.K. as early as 1937.

Recycling of bituminous materials can be broadly divided into two categories; *in-situ* (or surface) recycling and off-site (or hot mix) recycling.

12.4.1 In-situ recycling

In the U.K. *in-situ* recycling can be further subdivided into three processes known as 'Repaving', 'Remixing' and 'Retread'. Repaving and Remixing are hot *in-situ* recycling processes, whereas the Retread treatment is a cold process. The Repave process, which was introduced into the U.K. in 1975, involves heating and scarifying the existing surface to a depth of about 20 mm (see photograph 12.10). Approximately 20 mm of new asphalt is laid directly on the hot scarified material and then the material is compacted. These operations can be completed by a single pass of a purpose-built machine. As both layers are hot when compacted, a good bond is achieved between

Photograph 12.10 – The Repave process *(Photograph reproduced by kind permission of Associated Asphalt Limited)*

the recycled material and the new material. In the Remixing proces a hot scarified material is mixed with new material in the pugmill mixer of a machine and the blended mix is paved and compacted on the scarified surface. A report on the surface characteristics of roads using the Repave process has been published by the Transport and Road Research Laboratory[14], and the Department of Transport (DTp) issued a Departmental Standard on the Repave process in 1982[15,16]. The DTp requirements for this process are now detailed in clause 926 of the Specification for Highway Works[17].

A cold *in-situ* procedure known as Retread has been established U.K. practice for over 30 years[18,19]. Briefly, this process consists of scarifying the existing road to a depth of approximately 75 mm, breaking down the scarified material to the required size and reshaping the road profile. The material is then oversprayed 2 or 3 times with a bitumen emulsion. After each application of emulsion the material is harrowed to distribute the emulsion. When harrowing is complete the material is compacted with an 8 to 10 tonne dead weight roller and finally the surface is sealed with a surface dressing. This process has been successfully carried out on minor roads, housing estate roads, etc for many years. Currently, approximately 1 million m^2 of this type of minor road are annually recycled using the Retread process.

12.4.2 Off-site or plant recycling

In this process material removed from the surface of an existing road is transported to a hot mix plant where it may be stockpiled for future use or processed immediately. Both batch and continuous plants have been successfully converted to produce recycled mixes[20], utilising a range of methods to heat the reclaimed material prior to mixing; successful, in that recycling has been carried out without excessive fuming and blue smoke emissions during manufacture, and successful in that laboratory performance tests on asphalt containing a proportion of recycled material were identical to mixes using virgin components[21].

Recycling in batch plants is achieved by superheating the virgin aggregate and then adding the material to be recycled either immediately after the drier or directly into the pugmill, heat transfer taking place during the mixing cycle. Although this method successfully overcomes the problem of blue smoke, it entails keeping aggregate in the heating drum for longer and consequently the maximum output from the plant is reduced. As it is not possible to obtain adequate heat transfer with high percentages of recycled material, it is widely acepted that the maximum quantity of recycled material that can be added is between 25% and 40%. Other disadvantages using batch plants for recycling are high heating costs and accelerated wear and tear on the drum and dust collectors due to the higher manufacturing temperatures.

Recycling using continuous mixers involves introduction of the reclaimed material into the drum itself. During early trials using this type of plant, environmentally unacceptable 'blue smoke' was produced. Blue smoke is produced when vapourised bitumen condenses. The condensate takes the form of particles too small to be removed by conventional emission control equipment and so escapes through the stack. Bitumen vapourises at about 450°C and as gas temperatures in the drum of a continuous mixer can reach 2000°C the introduction of reclaimed material has to be carefully controlled. By various modifications continuous mixers have been designed which can take up to 60% recycled material, without exceeding statutory pollution standards. The entry point for the recycled material is approximately half way down the drum and the flights in the drum have been modified to produce a homogeneous mix of virgin and recycled material prior to the addition of the bitumen and filler. In addition the redesigned flights shield the recycled material from the intense radiant heat from the heating unit.

Recycling in the U.K. is still in its infancy and although plant developments have enabled significant quantities of reclaimed material to be successfully recycled and full-scale road trials and laboratory evaluations of these materials are encouraging[22,23], the long-term performance of recycled surfacings has yet to be established.

12.5 References

(1) Jones, L: *Recent developments in coating plant technology*. Quarry Management, pp 25-30, October 1986.

(2) Grant, R M: *Flexible pavements and bituminous materials*. Residential course at the University of Newcastle upon Tyne, section J, *Manufacture of coated materials*, pp J1-J12, September 1989.

(3) The Asphalt Institute: *Asphalt plant manual*. Manual series No 3, 1987.

(4) Curtis, C R: *Blacktop specifications – should we rationalise?* A paper given at the ACMA seminar, The road ahead for blacktop specifications, November 1988.

(5) Baxter, J: *Hot storage of bituminous mixes. Problems and advantages*. Quarry Management and Products, pp 213-218, August 1977.

(6) British Standard 598: Part 100: 1989: *Sampling and examination of bituminous mixtures for roads and other paved areas. Part 100. Methods for sampling for analysis.*

(7) British Standard 598: Part 101: 1989: *Sampling and examination of bituminous mixtures for roads and other paved areas. Part 101. Methods for preparatory treatment of samples for analysis.*

(8) British Standard 598: Part 102: 1989: *Sampling and examination of bituminous mixtures for roads and other paved areas. Part 102. Analytical test methods.*

(9) British Standard 594: Part 1: 1985: *Hot rolled asphalt for roads and other paved areas. Part 1. Specification for constituent materials and for mixtures.*

(10) British Standard 4987: Part 1: 1988: *Coated macadam for roads and other paved areas. Part 1. Specification for constituent materials and for mixtures.*

(11) British Standard 598: Part 103: 1990: *Sampling and examination of bituminous mixtures for roads and other paved areas. Part 103. Method of recovery of soluble bitumen for examination.*

(12) Wadelin, F A: *Workshop on binder recovery from asphalt and coated macadam*. Asphalt Technology, No 32, pp 28-42, August 1982.

(13) Dinnen, A: *Recycling – A summary of current methods*. Shell Bitumen Review 59, pp 15-18, June 1981.

(14) Cooper, D R C and Young, J C: *Surface characteristics of roads resurfaced using the Repave process*. Transport and Road Research Laboratory, Supplementary Report 744, 1982.

(15) Department of Transport: *In-situ recycling – The Repave process*. Departmental Standard HD/7/82, 1982.

(16) Department of Transport: *In-situ recycling – The Repave process*. Advice Note HA/14/82, 1982.

(17) Department of Transport: *Specification for Highway Works*. Part 3, series 900. Sixth edition, August 1986.

(18) Road Emulsion Association: *The Retread process*. Technical data sheet No 10.

(19) Dinnen, A: *Recycling – The Retread process*. Shell Bitumen Review 60, p 12, March 1983.

(20) Servas, V: *Hot mix recycling of bituminous pavement materials*. The Highway Engineer, December 1980.

(21) Stock, A F: *Structural properties of recycled mixes*. Highways and Transportation, vol 32, pp 8-12, March 1985.

(22) Mayhew, H C and Edwards, A C: *Performance of pavements containing recycled bituminous material*. Institution of Highways and Transportation tenth national workshop, pp 75-98, Royal Leamington Spa, April 1989.

(23) Edwards, A C and Mayhew, H C: *Recycled asphalt wearing courses*. Transport and Road Research Laboratory, Report RR 225, 1989.

Photograph 13.1 – A modern delivery vehicle *(Photograph reproduced by kind permission of ECC Quarries Limited)*

Chapter 13 – Transport, laying and compaction of bituminous mixes

13.1 Transport

Bituminous mixes are transported to site in well insulated, sheeted vehicles (see photograph 13.1) to prevent an excessive drop in temperature or crusting on the top of the load between the time of mixing and discharge into the paver hopper. It is usual to double-sheet vehicles during cold weather conditions.

To facilitate the discharge of the hot material, sand or dust can be applied sparingly to the floor of the vehicle before loading. Diesel oil must not be used because it tends to pond on the floor of the vehicle and is subsequently absorbed into the hot material. This softens the bitumen and if this is not fully dispersed during laying it could result in a soft patch of laid material. This would be particularly detrimental if the material was a wearing course.

13.2 Laying bituminous materials

Figure 13.1 shows a schematic sketch of a floating screed paver. The paver consists of two units, the tractor and the screed (see photograph 13.2). The tractor unit has three main functions:

- to provide mechanical, electrical and hydraulic power;
- to collect material from the delivery vehicle and convey and distribute material to the face of the screed;
- to pull the screed.

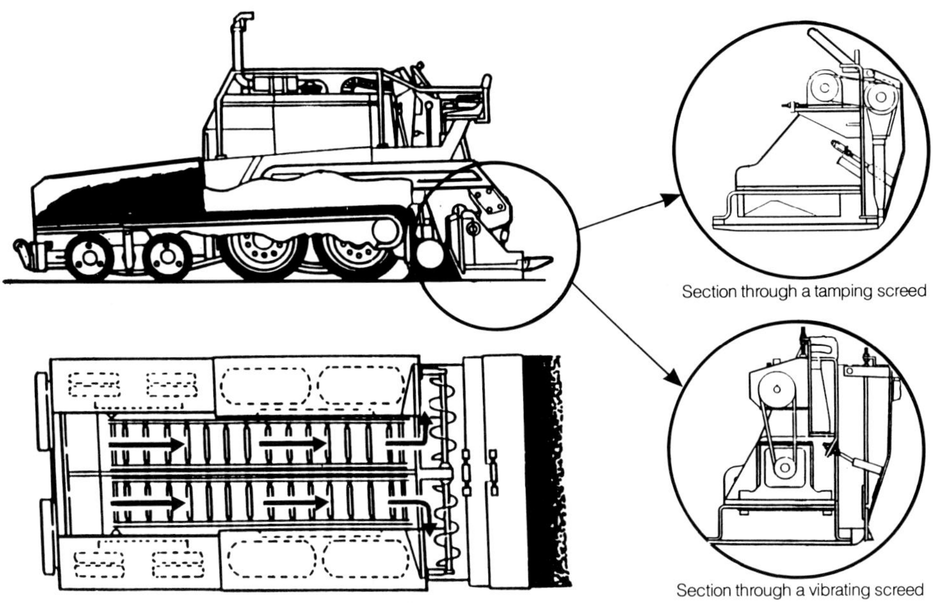

Figure 13.1 – Schematic sketch of a floating screed paver

Photograph 13.2 – A modern paver *(Photograph reproduced by kind permission of Alfred McAlpine Quarry Products Limited)*

The screed also has three main functions:

- to lay material accurately to a specified level and shape;
- partially to compact the laid mat;
- to smooth the mat to give a good surface finish.

The material to be laid is deposited from the delivery vehicle into the hopper of the tractor unit. Two independent feed systems convey the material through the machine to its rear by means of bar conveyors. Material flow from the hopper is controlled by the speed of the conveyors and/or adjustment of flow gates situated at the rear of the hopper. When material is discharged from the rear of the tractor unit, it is distributed outwards across the face of the screed by left and right augers which are synchronised to the speed of the left and right conveyors.

There is a number of different types of screed. To enable screeds to handle roads of varying width they can be extended. This can be achieved by either bolting suitable braced extensions onto a rigid screed or by hydraulically extending the screed. Rigid screeds can be built up to a width of 12.5 m. However, hydraulic screeds have the advantage that they permit variation in paving width at the touch of a switch whilst the paver is on the move. Typically, working widths can be varied from 2.5 m to 4.75 m and bolt-on sections similar to those of the rigid screed can provide a working width of up to 6.5 m. Heating for both types of screed is provided by forced draught, propane-fuelled burners usually on both the main and extending sections of the screed.

Partial compaction of the laid material is achieved by tamping, vibration or both. Tamping screeds are fitted with hydraulically driven shafts mounted in eccentric bearings on the screed frame. The tamper frame is attached to the shaft, as shown in figure 13.1. These eccentric bearings change the rotary motion of the shaft to a reciprocating vertical motion of the tamper frame, which partially compacts the material. A vibration screed is also fitted with hydraulically driven shafts mounted on the screed frame. However, instead of eccentric bearings, the vibrating screed has eccentric weights attached to the shaft; when the shaft rotates the screed vibrates. The vibrating screed enables material to be laid extremely quickly, whereas the tamping screed gives a better surface finish and generally higher partial compaction, but this is at the expense of a slower laying speed. The combined tamping/vibrating screed is generally used for its greater compaction capability but again at a cost of slower laying speed.

13.3 The compaction of bituminous materials

Compaction is probably the most crucial stage in the construction of any layer within a road pavement. Compaction brings the constituents of a material into intimate contact, and where this does not occur air voids will remain. The importance of compaction as a decisive influence on the performance of bituminous materials has been recognised for some time and the move from the traditional 'method' specification towards 'end result' specifications reflects this appreciation.

In the Department of Transport (DTp) Specification for Road and Bridge Works (1976)[1], the compaction of bituminous materials was specified by defining:

layer thickness;

minimum delivery temperatures;

minimum rolling temperatures;

the type and weight of roller.

The normal procedure was to commence rolling as soon as possible without causing undue displacement of the material, and to continue until all the roller marks had been removed. The compaction achieved was also influenced by the need to meet level and surface regularity specifications.

Although the performance of roadbases and basecourses produced in this way had generally been considered satisfactory, an investigation by the Transport and Road Research Laboratory (TRRL)[2] showed that there was scope for improving compaction in the critical wheel-path by at least 3%. Since the structural performance of the whole pavement is largely governed by the level of compaction at this critical location there was clearly scope for improvement.

The load spreading ability of bituminous basecourses and roadbases is defined by their dynamic stiffness. Figure 13.2 shows the results of three-point bending and uniaxial loading tests on macadams of varying void content[3]. The results show that a 3% reduction in void content increases the dynamic stiffness by 30%. Linear elastic analysis of a typical flexible pavement shows that with this increase in dynamic stiffness the thickness of the roadbase could be reduced by 8% and the vertical stress at the surface of the subgrade would be the same. Alternatively, the additional compaction could be roughly equated to a 30% extension of pavement life[2,4]. Similar results have been obtained over a wide range of test temperatures and frequencies and for different aggregate types[5]. The increase in dynamic stiffness also reduces the tensile strain produced by traffic at the bottom of the roadbase, resulting in a substantially improved resistance to fatigue cracking. Deformation in the bituminous layers within a road pavement makes a substantial contribution to the overall deformation measured at the surface of the pavement. To determine the deformation resistance of macadam roadbase, the TRRL carried out simulative wheel-tracking tests on material at 30°C loaded to 20 kN using a pneumatic tyre[6]. Figure 13.3 shows that the deformation measured after 1000 load cycles decreased rapidly with improved compaction, a 3% decrease in the voids in mineral aggregate resulting in a reduction in deformation of up to 50%.

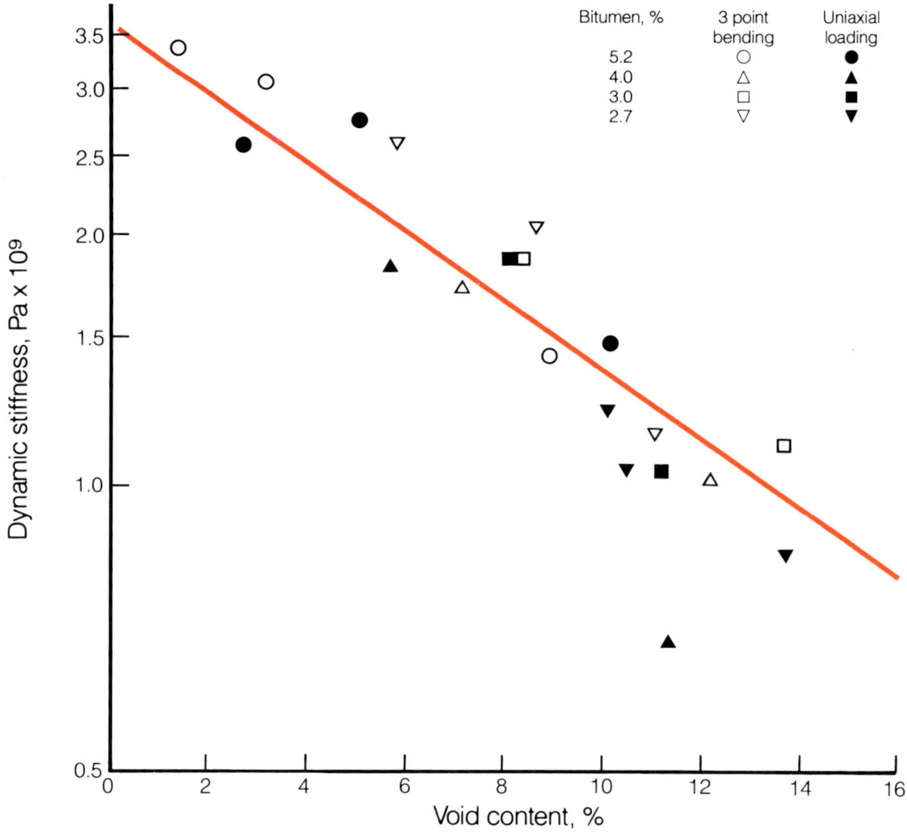

Figure 13.2 –The effect of void content on dynamic stiffness[3]

Improving compaction can, therefore, result in a significant improvement in load spreading, in resistance to fatigue cracking and in the resistance to internal deformation of a bituminous material. These improvements would undoubtedly result in extended pavement life.

13.3.1 Methods of improving compaction

Improved compaction can be achieved by:

increasing the layer thickness;

increasing the compactive effort;

modified rolling techniques.

13.3.1.1 Increasing the layer thickness

The ability of thicker layers to retain heat longer increases the time available after laying for effective compaction to be carried out. It has been shown[7,8,9] that the time it takes for a bituminous layer to cool from an (initial) laying temperature to a final (ambient) temperature is a function of its thickness, viz:

$$t = k\,d^{1.8}$$

where t = is the cooling time from the initial temperature to the final temperature

d = laid thickness

k = a constant, the value of which depends upon environmental conditions, initial laying and final temperatures.

Thus a 50% increase in cooling time can be achieved by increasing the laid thickness by 25%.

The relative importance of laid thickness and a number of other controllable and uncontrollable factors are given in table 13.1[8,9]. For the typical ranges given, the effect of wind speed is greater than that of the ambient air temperature, but both of these uncontrollable factors are smaller than the controllable factors of laid thickness and initial laying temperature.

Factor	Range	Change in compaction time %
Wind speed	0 to 20 km/h	−26
Air temperature	0 to 20° C	+ 19
Solar radiation	+ 10°C	+ 15
Laid thickness	35 mm to 45 mm	+ 57
Initial laying temperature	140°C to 160°C	+ 40
Minimum compaction temperature	100°C to 90°C	+ 30

Notes

1) The value of +10°C for solar radiation is the differential between the road surface temperature over the air temperature

2) The change in compaction time by increasing the initial compaction temperature is calculated based on cooling to 100°C

3) The change in compaction time by reducing the minimum compaction temperature is calculated using an initial temperature of 150°C

Table 13.1 – Effect of various factors on compaction time[8,9]

However, increasing the layer thickness is not a compaction panacea. In thick layers laid hot to counteract adverse weather conditions, a skin of relatively cold, stiff material rests on a hot mobile mass. Under these circumstances movement of the relatively fluid hot material may result in cracking of the stiff surface crust during rolling. In an experiment designed to investigate the possibility of laying material in thick lifts[10], instability was encountered under precisely these circumstances and surface regularity and level specification could not be achieved with layers greater than 200 mm thick.

13.3.1.2 Increasing the compactive effort

The principal requirement of any roller is that it will compact the material being laid in an effective manner without introducing unacceptable surface irregularities. Historically this has been achieved using 8 to 10 tonne three-wheeled deadweight rollers. However, since the late seventies the use of vibratory rollers has become commonplace.

Deadweight rollers are available as either three-wheeled (see photograph 13.3.) or tandem models ranging from 1.5 to 12 tonnes in weight. Compaction is obtained by the application of a fixed mass or

Photograph 13.3 – A traditional deadweight roller

Photograph 13.4 – A modern vibratory roller *(Photograph reproduced by kind permission of Tilcon Limited)*

deadweight variable by changing the number of passes or by increasing the ballast on the roller. The available compaction force may be assessed in terms of a static linear pressure, ie the deadweight of the roller divided by the area in contact with the material. Thus, for rollers of a given deadweight, those having the smallest area of contact, ie those having the smallest diameter wheels, with the material will exert the greatest pressure and therefore give the greatest compaction.

Tandem and three-wheeled deadweight rollers are approximately equivalent on a pass for pass basis(3); the greater load per unit width of the rear wheels of a three-wheeled roller appears to be counterbalanced in tandems by the greater roll width and the smaller diameter of the rear wheel. Three-wheeled rollers tend to be favoured because their tripod configuration gives them greater stability, particularly when rolling close to edges or crowns where tandems may rock and introduce surface irregularities. They are also considered to be more effective for removing roller marks in the final stage of rolling.

Many trials examining the performance of vibratory rollers have been carried out throughout the world in the last fifteen years(11,12,13,14,15). The main conclusion from these trials is that, with appropriate deadweight and vibration characteristics, vibrating rollers can:

- achieve similar levels of compaction to deadweight rollers but with fewer passes;
- attain greater densities than deadweight rollers;
- achieve a surface profile that compares favourably with that produced by deadweight rollers.

Vibratory rollers are almost invariably tandems (see photographs 13.4 and 13.5).

Photograph 13.5 – A modern vibratory roller *(Photograph reproduced by kind permission of Associated Asphalt Limited)*

In addition to the effect of the deadweight of the machine, there is a superimposed force produced by a vibratory mechanism on which the amplitude and frequency of vibration can be adjusted to a limited extent to suit different materials. The vibration generators are normally situated within the drums of the roller, which are isolated from the main frame of the machine by a flexible suspension so that as little of the vibration as possible is transmitted to the frame of the roller. Usually the vibration generator is a rotating eccentric mass whose speed of rotation determines the frequency of vibration. The amplitude of vibration is governed by the mass of the eccentric body and its distance from the centre of rotation.

Vibratory rollers apply dynamic compaction forces and the performane of such machines is dependent on a complex interaction of machine and material variables. The most important factors influencing machine performance are vibrating mass; frequency of vibration; amplitude of vibration and roller speed[3,16].

Table 13.2 compares the results of compaction trials of eight vibratory rollers using different amplitude and frequency settings with densities achieved using an 8 tonne deadweight roller[17] on the same material. The results show that the vibratory rollers usually achieved higher densities than the deadweight machine – up to 4% higher. This work confirmed the potential of vibratory rollers but the results did not show any consistent relationship between the vibrational characteristics, deadweight and dimensions of the roller with the density achieved in the compacted macadam. Therefore, to ensure that a particular vibratory roller is capable of achieving the required level of compaction, it is necessary to carry out comparative trials[18].

13.3.1.3 Modified rolling techniques

The temperature at which a bituminous material is laid and the rate at which roller passes are applied are important parameters affecting the level of compaction achieved. It has been shown that the near-side wheel-path can have void contents up to 5% higher than the centre of the mat. Examination by the TRRL of the factors which influence the level of compaction achieved[19] led to a new method specification[20] designed to increase the number of roller passes in the wheel-path. This resulted in an increase in density in practice of between 1% to 2% at this critical location.

Vibratory roller	Mass, tonne	Average mass per unit width, tonne/m	Density achieved by vibratory roller / Density achieved by 8 tonne deadweight: High amplitude, low frequency	Density achieved by vibratory roller / Density achieved by 8 tonne deadweight: Low amplitude, high frequency
A	8.9	3.23	1.03	1.02
B	9.5	2.81	1.01	1.02
C	8.9	2.71	1.01	1.01
D	7.0	2.52	1.04	1.02
E	6.8	2.45	1.00	1.01
F	6.8	2.44	1.00	1.03
G	6.6	2.37	1.03	1.03
H	7.0	2.33	1.01	1.00
Deadweight	8.0	3.35	1.00	1.00

Table 13.2 – Relative compaction achieved by various vibratory rollers[17]

13.3.2 Compaction specification for bituminous materials

There are two approaches to compaction specification:

- method specification;
- end result specification.

The general view within the bituminous materials industry is that the alternative approach of specifying an end-result requirement for compaction is a more effective way of achieving consistent and high levels of compaction. There are currently three variations of end result specification:

- comparison of the density achieved on site with a standard laboratory compactive effort on uncompacted material, ie percentage Marshall; this method is extensively used in the U.S.A., in many European countries and by the Property Services Agency for airfield pavements in the U.K.;
- comparison of the site density with the maximum theoretical density, ie percentage air voids;
- comparison of the site density with specimens from the road pavement compacted to refusal in the laboratory, ie percentage refusal density (PRD).

13.3.2.1 Air voids specification

In recent years many highway authorities have introduced maximum and minimum void content criteria into their specifications as a means of compaction control, particularly for hot rolled asphalt wearing course. However, it should be recognised that the use of void content as a means of measuring achieved density has a number of disadvantages:

- the calculation of air void content requires accurate determination of both the composition of the individual specimens and the specific gravities of all the constituents of the mix; the specific gravity of the aggregate may vary with the nominal size[21] and therefore it is necessary to examine various aggregate fractions separately; as a result a considerable amount of pre-contract appraisal of the specific gravity is required;
- the allowance made for the absorption characteristics of the aggregate is very simplistic and could result in misleading information;
- the aggregate shape, texture and absorption characteristics will influence the air void content achievable; thus for a fixed aggregate grading envelope the air void content will vary with the type of aggregate used; fixed air void content specifications do not take this into account[22].

Accepting the above limitations, once the maximum theoretical density has been established between the client and supplier, site monitoring of *in-situ* density can be accomplished by the use of a nuclear density gauge[23,24,25] and periodic core extraction.

13.3.2.2 Percentage refusal density

The adoption of modified compaction techniques described in section 13.3.1.3 only resulted in a modest increase in the level of compaction achieved. Thus it was felt that an alternative approach to specifying an end-result requirement for compaction was required to achieve consistently high levels of compaction. However, a satisfactory method of assessing the end result is essential if this method of specification is to have any meaning. The approach adopted was to relate densities achieved on the road with a standard well-defined laboratory test. Initially the Marshall compaction procedure was examined but it was found to be unsuitable because the compacted density achieved depended on composition variables in a very different way to that observed in pilot-scale trials[26]. The ultimate compaction achieved in a laboratory refusal test was found to be far more suitable.

The PRD test is carried out on cored samples taken from the compacted pavement layers, the initial density being expressed as a percentage of the density measured after compaction to refusal. Testing cores from the compacted pavement removes any doubt that the material tested may not be representative of that in the pavement. Tests on uncompacted loose material have been shown to achieve refusal densities 1.5% to 3.0% lower than those obtained on cores of identical composition from compacted material[17], demonstrating that the orientation of the aggregate particles in a compacted material is not reproduced in the laboratory by compacting loose material. A complete description of the test development, method and precision is given elsewhere[27,28,29,30,31,32]. Use of the test on samples using a wide range of aggregrate types and various bitumen contents showed that:

samples of the same composition but having a wide range of density before carrying out the test are compacted to the same refusal density;

samples made with the same aggregate but covering a range of bitumen contents are compacted to the same voids in mineral aggregate (VMA).

These findings confirm that the test brings about a degree of packing of the aggregate that closely approaches the maximum that is possible without degradation of the aggregate, ie it is a true refusal test. As the PRD test is largely insensitive to material variables other than compactive effort, a common target level of compaction can be specified in terms of PRD for all roadbase and basecourse macadams.

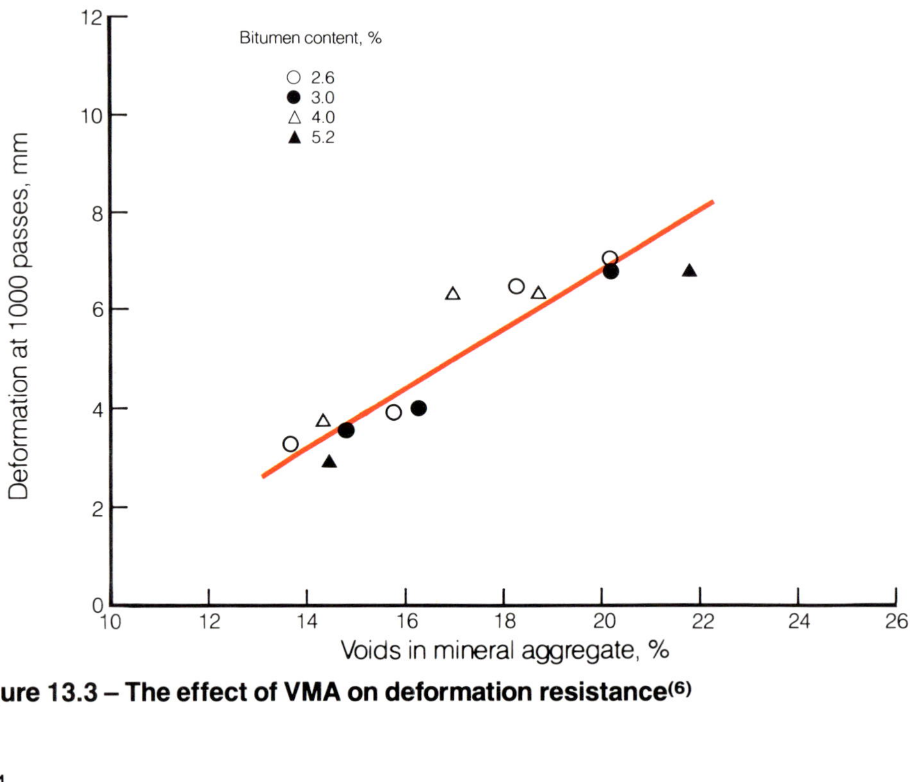

Figure 13.3 – The effect of VMA on deformation resistance[6]

One of the main drawbacks of the PRD test is the time delay for the testing cycle; a full PRD test takes 3 to 4 days and the staff time required in the laboratory is a total of 7 hours for 6 specimens. This may result in delays on site or the risk of overlaying material not yet accepted(33). However, when used in conjunction with a nuclear density meter, the PRD method is less time consuming than the *full* procedure needed for accurate determination of void content(22).

Following a full-scale precision exercise(28,31) the PRD was included in the sixth edition of the DTp Specification for Highway Works(18). The precision tests showed that the standard deviation on single core pairs of macadam roadbase was 1.5%. To achieve a uniform compaction of 95 PRD in practice the specified minimum PRD is 93.

13.4 The workability of bituminous materials(34,35)

The term 'workability' is commonly used to describe the ease, or otherwise, of mixing, laying and most importantly compacting a bituminous material. Less workable mixes are not as mobile during application and compaction. This can affect the height of the screed plate which is particularly critical when replacing sections of existing wearing course. Transverse tearing or dragging of the surface of the mat as it leaves the paver is also characteristic of less workable mixes.

Photograph 13.6 – Paving and compaction of hot rolled asphalt wearing course *(Photograph reproduced by kind permission of Associated Asphalt Limited)*

During compaction less workable mixes will move little under the wheels of the roller. This resistance to internal movement is most noticeable, and critical, when compacting hot rolled asphalt wearing course where the rolling process is embedding the chippings and compacting the layer simultaneously. Satisfactory chipping embedment is more difficult with less workable rolled asphalt mixes, resulting in inadequate compaction of the mortar at the surface. It is likely that this poorly compacted material will fret and ultimately may lose chippings. The perceived need for increased texture depth has resulted in increased application rates of chippings (at ambient temperature) which can result in inadequate amounts of mortar between the chippings, again resulting in possible chipping loss.

In hot rolled asphalt it is primarily the shape and rugosity of the sand that leads to mixes having marked differences in workability. The use of more rugous sands increases the resistance of the mix to deformation, but this has also led to reduced workability during compaction. This has undoubtedly reduced deformation in service but unfortunately it has also resulted in an increased incidence of excessive chipping loss.

The resistance to internal movement of a bituminous mix, which defines its workability, is influenced by the physical properties of the aggregate such as shape, surface area and surface texture, the proportion of aggregate within a mix, the grading of the aggregate and the rheological properties of the bitumen.

There are three components of resistance to internal movement:

- cohesion of the bitumen, which is influenced by the type and amount of added filler;
- internal friction from the mineral aggregate, such as grading, shape, surface area and surface texture;
- mix viscosity, which is influenced by the viscosity of the bitumen and internal friction provided by the included filler and mineral aggregate.

Cohesion and internal friction make up the internal resistance to movement of a mix and therefore it will relate to cessation of compaction. Mix viscosity relates to the initial movement of a mix.

Photograph 13.7 – Paving and compaction of dense bitumen macadam basecourse *(Photograph reproduced by kind permission of Tilcon Limited)*

13.5 Surface texture

With wearing course materials surface texture is important to ensure high-speed skid-resistance in wet conditions. There are two parts to the surface texture of a road that contribute to the overall skid-resistance of a road surface (see figure 18.9). The first of these is microtexture which is the surface roughness of the aggregate exposed at the road surface. Retention of the microtexture of an aggregate under traffic is primarily dependent on the polishing characteristics of the aggregate. This can be quantified by the polished stone value (PSV) test[(36)]. The microtexture is the predominant factor in providing the low-speed skid-resistance, ie up to 50 km/h of a wet road surface, because it governs the tyre penetrating the water film on the fine asperities of the aggregate. It is essential that the surface tension forces are overcome and the water film dispersed in order that the tyre can maintain frictional contact with the surfacing.

The second part of surface texture is the macrotexture. This is provided by the orientation, size and shape of the aggregate particles at the road surface. An indication of the texture depth provided by different bituminous surfacings is given in table 13.3[(37)]. The function of the macrotexture is to provide drainage channels which allow the dispersion of the surface water ahead of and around the rolling tyre. The overall effect of high macrotexture is to enable the low-speed skid-resistance provided by the macrotexture to be maintained at high vehicle speeds. Historically macrotexture has been measured using the sand patch test which is described in chapter 18, section 18.4.2. Although the sand patch test is simple to carry out it is notoriously operator dependent and inaccurate.

A new method based on laser technology has been developed by the TRRL[(38,39)]. A rapidly pulsing semiconductor laser projects infra-red light onto the road surface. The light is reflected from the road surface and is focussed by a receiving lens onto a linear array of 256 photodiodes, as shown schematically in figure 13.4. The position of the diode receiving the most light gives a measure of the distance to the road surface at that instant and the depth of the texture is computed from a series of such measurements. The laser has been mounted in a trailer[(40)] – the high-speed texture meter (HSTM) for operation on public roads at normal traffic speeds, and is also available as a hand

Surfacing material	Texture depth (sand patch), mm
Asphaltic concrete	0.4 – 0.6
Dense bitumen macadam	0.6 – 1.2
Hot rolled asphalt (lightly chipped)	0.5 – 1.2
Hot rolled asphalt (heavily chipped)	1.2 – 2.5
Pervious macadam	1.5 – 3.5
Surface dressing	2.0 – 3.5

Notes
1) For dense bitumen macadam the actual texture depth achieved will depend on the nominal size and grading of the aggregate
2) For rolled asphalt the texture depth achieved will depend on the nominal size, rate of spread and degree of embedment of the chippings

Table 13.3 – Texture depths of newly-laid bituminous surfacings[(37)]

operated manually propelled two-wheeled device – the mini texture meter[41] (MTM). Using these two devices rapid and more reliable measurements of surface texture can be obtained[42].

The TRRL have developed the following correlations between the sand patch method and the high-speed and mini texture meters:

HSTM measured texture = 0.12 + (0.59 x sand patch texture depth);

MTM measured texture= 0.41 + (0.41 x sand patch texture depth).

These two correlations apply only to new hot rolled asphalt with sand patch texture depths between 1.0 mm and 2.0 mm. Different correlations will apply for other levels of texture and for different types of surfacings. However, these two correlations should be interpreted with care as sand patch is a measure of texture over an area, whereas the laser techniques are line measurements of texture. In addition, it is questionable whether sand patch or laser techniques actually measure the ability of the surface to drain surface water, as neither technique measures the continuity of the drainage channels in the surfacing. A more direct measure of surface drainage characteristics would be obtained using an outflow meter[43].

The minimum 1.5 mm sand patch texture depth equates with a minimum 1.03 mm MTM texture depth. The specifications for each method are detailed in the DTp Specification for Highway Works[18].

Since 1978 contract documents relating to the use of hot rolled asphalt wearing course on high speed roads have required a minimum texture depth of 1.5 mm[44,45] measured by the sand patch method. To achieve this, higher rates of spread of chippings were used as well as relatively stiff asphalts to resist excessive embedment. A considerable number of factors influence the degree of embedment of the chippings such as:

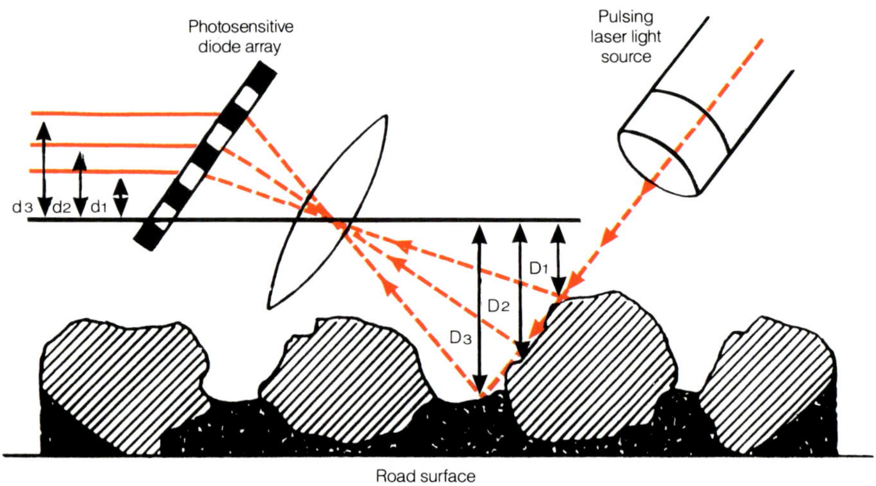

Figure 13. 4 – Principle of the laser contactless sensor for measuring texture depth[40]

rate of spread of chippings;
stiffness of the asphalt;
weather conditions during laying;
rolling temperature;
roller size and technique.

Clearly there is a balance to be struck between:

achieving the required texture depth;
adequate embedment of the chippings;
satisfactory compaction of the mortar around the chippings;
satisfactory compaction of the whole mat.

Achieving a balance between these requires considerable expertise. Even where these four requirements are achieved, some initial chipping loss is likely from hot rolled asphalt where the minimum 1.5 mm texture depth is specified. An overall chipping loss of up to 5% is deemed to be acceptable(44,45). It is important therefore that such texture depths are only specificied for high speed/low stress situations and, bearing in mind that it is the microtexture that is the important factor up to 50 km/h, there is no benefit specifying high texture depths for low speed sites such as junctions and roundabouts. Experience has shown that serious chipping loss will almost certainly occur if high texture depths are applied to these highly stressed sites.

13.6 References

(1) Department of Transport: *Specification for Road and Bridge Works*. Fifth edition, 1976.

(2) Lister, N W and Powell, W D: *The compaction of bituminous base and basecourse materials and its relation to pavement performance*. Transport and Road Research Laboratory, Supplementary Report 260, 1977.

(3) Powell, W D and Leech, D: *Compaction of bituminous road materials using vibratory rollers*. Transport and Road Research Laboratory, Report LR 1102, 1983.

(4) Powell, W D, Lister, N W and Leech, D: *Improved compaction of dense graded bituminous macadams*. Proceedings of the Association of Asphalt Paving Technologists, vol 50, p 394, 1981.

(5) Goddard, R T N, Powell, W D and Applegate, M W: *Fatigue resistance of dense bitumen macadam: the effect of mixture variables and temperature*. Transport and Road Research Laboratory, Supplementary Report 410, 1978.

(6) Leech, D and Selves, N W: *Deformation resistance of dense coated macadams: effect of compaction and binder content*. Transport and Road Research Laboratory, Supplementary Report 626, 1980.

(7) Brown, J R: *The cooling effects of temperature and wind on rolled asphalt surfacings*. Transport and Road Research Laboratory, Supplementary Report 624, 1980.

(8) Daines, M E: *Cooling of bituminous layers and time available for their compaction*. Transport and Road Research Laboratory, Report RR 4, 1985.

(9) Daines, M E: *Cooling of bituminous layers and time available for their compaction*. Third Eurobitume symposium, The Hague, pp 237-242, 1985. Also reprinted in Asphalt Technology, February 1986.

(10) Lister, N W and Powell, W D: *Compaction of bituminous base and basecourse materials and its relation to pavement performance*. Asphalt and Coated Macadam Association seminar, October 1974.

(11) Tunnicliff, D G: *Symposium on vibratory compaction of asphalt pavements*. Proceedings of the Association of Asphalt Paving Technologists, vol 46, pp 259-375, 1977.

(12) Edwards, C D: *Coated macadam compaction tests on a Dynapac CC30 vibrating roller*. Ministry of Defence, Military Vehicles and Engineering Establishment, Report T230, 1978.

(13) Smith, M J M: *Coated macadam compaction tests on a Bomag BW 220A tandem vibrating roller*. Ministry of Defence, Military Vehicles and Engineering Establishment, Report T256, 1980.

(14) Smith, M J M: *Coated macadam compaction tests on an Ingersoll-Rand DA50 tandem vibrating roller*. Ministry of Defence, Military Vehicles Engineering Establishment, Report T296, 1980.

(15) Nunn, M E: *Compaction of coated macadam by a Bomag BW 75ADL vibratory roller*. Transport and Road Research Laboratory, Report RR 56, 1985.

(16) White, M J: *Vibrating rollers for the compaction of bituminous materials*. Asphalt and Coated Macadam Association seminar, October 1981.

(17) Powell, W D and Leech, D E: *Compaction and performance of dense roadbase macadam*. Proceedings of the Compaction Technology conference, organised by New Civil Engineer, pp 15-32, October 1987.

(18) Department of Transport: *Specification for Highway Works*. Part 3, series 900. Sixth edition, August 1986.

(19) Leech, D E and Powell, W D: *Levels of compaction of dense coated macadam achieved during pavement construction*. Transport and Road Research Laboratory, Report LR 619, 1974.

(20) Powell, W D and Leech, D E: *Rolling requirements to improve compaction of dense roadbase and basecourse macadam*. Transport and Road Research Laboratory, Report LR 727, 1976.

(21) McNicol, A and Bell, G: *Air voids versus PRD*. Highways and Transportation, vol 33, pp 3-8, February 1986.

(22) British Aggregate Construction Materials Industries: *A commentary on recent specification clauses for blacktop*. Technical Note, October 1988.

(23) Report of a working party: *Nuclear gauges for measuring the density of roadbase macadam*. Transport and Road Research Laboratory, Supplementary Report 754, 1982.

(24) Reid, J S: *Compaction measurement using a nuclear test gauge*. The Highway Engineer, vol 30, pp 2-7, April 1983.

(25) Elliott, R C: *Using nuclear gauges to monitor asphalt compaction*. Asphalt Technology, No 37, pp 16-19, December 1985.

(26) Powell, W D, Chipperfield, E H and Jenner, L: *Dense graded bituminous macadams: effect of mixture variables on compactability*. Proceedings of the international conference on compaction, Ecole Nationale des Ponts et Chaussées, pp 457-462, 1980.

(27) Powell, W D and Leech, D: *Standards for compaction of dense roadbase macadam*. Asphalt and Coated Macadam Association seminar, October 1981.

(28) Powell, W D and Leech, D: *Standards for compaction of dense roadbase macadam.* Transport and Road Research Laboratory, Supplementary Report 717, 1982.

(29) Ramsey, S: *The percentage refusal density test.* Highways and Transportation, vol 32, pp 12-16, February 1985.

(30) Howe, J H: *The percentage refusal density test.* Highways and Transportation, vol 33, pp 15-18, March 1986.

(31) Wimpey Laboratories: *Precision experiment: percentage refusal density test.* Transport and Road Research Laboratory, Contractor Report 1, 1985.

(32) British Standard 598: Part 104: 1989: *Sampling and examination of bituminous mixtures for roads and other paved areas. Part 104. Methods of test for the determination of density and compaction.*

(33) Sutcliffe, M A: *Bituminous surfacing density and responsibility.* Highways and Transportation, vol 33, pp 13-18, July 1986.

(34) Whiteoak, C D and Fordyce, D: *Asphalt workability. Its measurement and how it can be modified.* Shell Bitumen Review 64, pp 14-17, September 1989.

(35) Kweir, K, Fordyce, D and Whiteoak, C D: *Asphalt workability.* Highways, vol 58, No 1957, Asphalt 90 supplement, pp 16-20, January 1990.

(36) British Standard 812: Part 3: 1975: *Methods for sampling and testing of mineral aggregates, sands and fillers. Part 3. Mechanical properties.*

(37) White, M J: *Flexible pavements and bituminous materials.* Residential course at the University of Newcastle upon Tyne, section G, *Bituminous materials – special requirements, problems and processes*, pp G1-G16, September 1989.

(38) Cooper, D R C: *Measurement of road surface texture by a contactless sensor.* Transport and Road Research Laboratory, Report LR 639, 1974.

(39) Still, P B and Winnett, M A: *Development of a contactless displacement transducer.* Transport and Road Research Laboratory, Report LR 690, 1975.

(40) Roe, P G, Tubey, L W and West, G: *Surface texture measurements on some British roads.* Transport and Road Research Laboratory, Report RR 143, 1988.

(41) Hosking, J R, Roe, P G and Tubey, L W: *Measurement of the macrotexture of roads. Part 2: a study of the TRRL mini texture meter.* Transport and Road Research Laboratory, Report RR 120, 1987.

(42) Daines, M E: *An analysis of texture depth measurements on Motorway M58 (Contracts 1 and 3).* Transport and Road Research Laboratory, Supplementary Report 738, 1982.

(43) Lees, G and Katekhda, I E D: *Prediction of medium and high speed skid resistance values by means of a newly developed outflow meter.* Proceedings of the Association of Asphalt Paving Technologists, vol 43, pp 436-464, 1974.

(44) Department of Transport: *Texture depth of bituminous surfacings for high speed roads.* Departmental Standard HD/3/80.

(45) Department of Transport: *Texture depth of bituminous surfacings for high speed roads.* Advice Note HA/8/80.

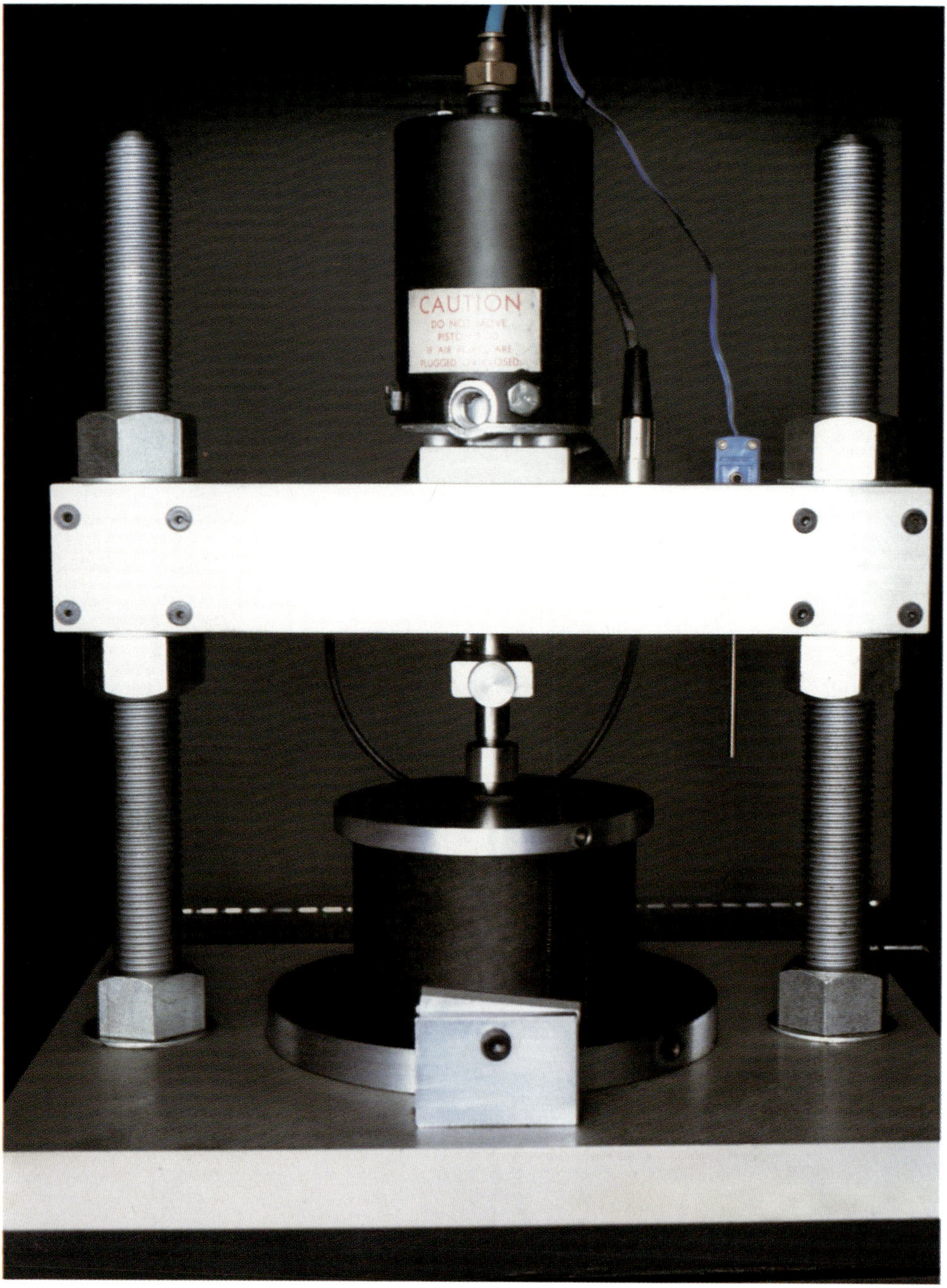

Photograph 14.1 – Creep testing

Chapter 14 – Mechanical testing of bituminous mixes

An analytical approach to pavement design is based on a knowledge of the various materials used in the layers of a pavement structure. Laboratory testing has been used extensively in research and this has resulted in a fundamental understanding of the mechanical behaviour of bituminous mixes.

There are two aspects of material properties relevant to pavement design: firstly, the load-deformation or stress-strain characteristics required for the analysis of the structure; secondly, the performance characteristics which determine the mode of failure. The main performance criteria are fatigue cracking or permanent deformation.

Ideally, the mechanical properties of the material *in-situ* are required for design purposes. However, *in-situ* testing of full-scale trial sections is impractical in most cases, and therefore engineers have to rely on laboratory testing. These tests should reproduce the anticipated *in-situ* conditions as closely as possible, ie temperature, loading time, stress conditions, degree of compaction, etc. However, the *in-situ* conditions are continually changing and therefore the selection of appropriate testing conditions is extremely difficult.

The stress condition can be reproduced in the laboratory only with great difficulty. Figure 14.1 represents a pavement element showing the stresses which occur in it when a wheel load is approaching. This is, of course, very much a simplified representation. In practice the stresses are applied three-dimensionally: horizontal and shear stresses occur on planes which are at right angles to those shown in figure 14.1. As the wheel passes over the element, these stresses change with time, as shown in figure 14.2. To reproduce this complex stress regime accurately in the laboratory has yet to be done; the repeated load triaxial test comes close, but does not reproduce the shear reversal effect. Hence, simplified tests have been introduced which may reproduce certain aspects of the *in-situ* behaviour. These tests can be divided into three groups:

- fundamental:
  repeated load triaxial test;
  unconfined static uniaxial creep compression test;
  repeated load indirect tensile test;
  dynamic stiffness and fatigue tests.
- simulative:
  wheel-tracking test.
- empirical:
  Marshall test.

14.1 Fundamental tests

14.1.1 Repeated load triaxial test

As stated above, this test most closely reproduces the stress conditions in the pavement. It is extensively used in research to study permanent deformation as well as elastic properties of bituminous materials[1]. It has the advantage that both vertical and horizontal stresses can be applied at the levels predicted in the pavement, as shown in figure 14.3. Deformation monitoring devices to measure vertical and horizontal strains induced by the imposed stresses are mounted on the specimens. In view of the complex nature of the apparatus and testing procedures this type of test is only suitable for research investigations.

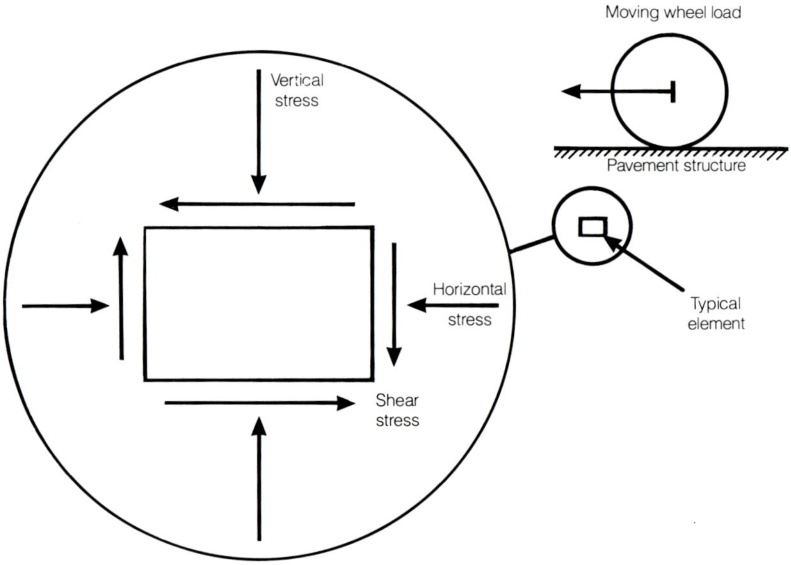

Figure 14.1 – Stresses on an element in a pavement[5]

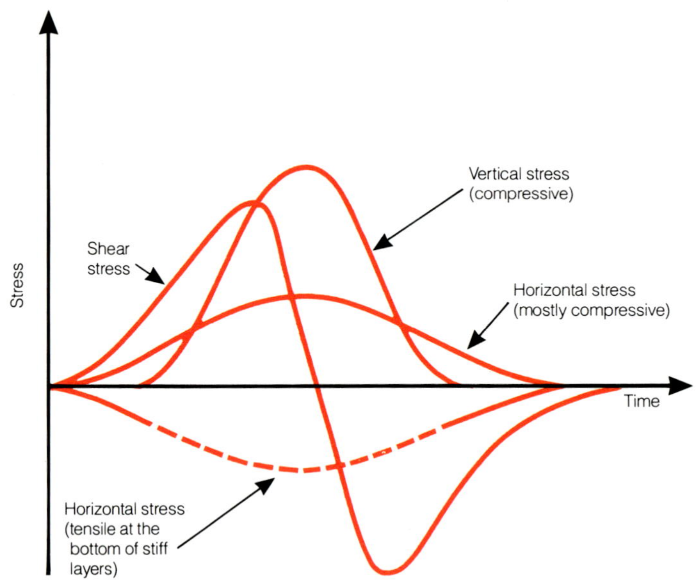

Figure 14.2 – Stresses induced by a moving wheel load[5]

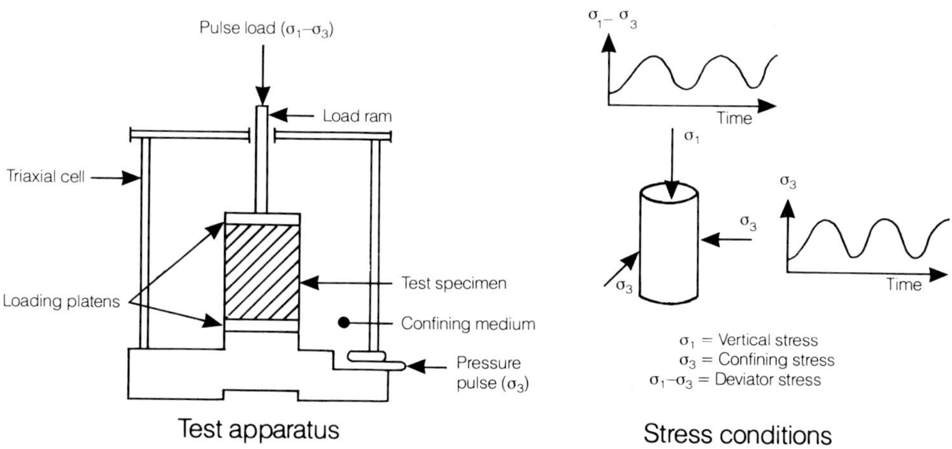

Figure 14.3 – The repeated load triaxial test[5]

14.1.2 Unconfined static uniaxial creep compression test

The simple creep test is used in Europe as a method of assessing the resistance of bituminous materials to permanent deformation[2]. In the test a static axial stress, σ, of 1 x 10^5 Pa is applied to a specimen for a period of 1 hour at a temperature of 40°C. These test conditions were standardised following a colloquium in Zurich in 1977[3]. During the test, axial deformation is measured as a function of time. Thus knowing the initial length of the specimen, the axial strain, ε, and therefore the stiffness modulus, S_{mix}, can be determined:

$$S_{mix} = \frac{\text{Applied stress}}{\text{Axial strain}} = \frac{\sigma}{\varepsilon}$$

Photographs 14.1 and 14.2 show a creep test apparatus where the axial load is applied by a ram actuated by compressed air. The resultant axial deformation is measured continuously by displacement transducer which is connected to a chart recorder.

14.1.3 Repeated load indirect tensile test

The repeated load indirect tensile test was developed in the U.S.A.[4] from the indirect tension test originally developed for concrete. In this test a pulse load is applied to a cylindrical specimen along the vertical diametrical plane, thereby generating indirect horizontal tensile stresses as shown in figure 14.4. By measuring the resulting horizontal deformation the elastic stiffness can be determined.

This non-destructive test has the advantage that elastic stiffness can be determined on either laboratory prepared specimens, eg Marshall specimens, cores from a slab or relatively thin cores (down to 35 mm thick) taken from the road.

This test is being developed with the aim of producing a relatively simple automated apparatus providing fundamental information on mix properties producing input data for pavement design and evaluation[5].

Photograph 14.2 – The creep test

14.1.4 Dynamic stiffness and fatigue tests

For the measurement of the dynamic stiffness and fatigue characteristics of bituminous mixes researchers have developed a number of tests which are illustrated diagrammatically in figure 14.5. Bending tests using beams or cantilevers involving repeated applications of load have been used to determine both dynamic stiffness and fatigue resistance[6,7,8,9]. With bending tests the maximum stress occurs at a point on the surface of the specimen and its calculation, using standard beam bending formula, depends on the assumption of linear elasticity.

In the U.S.A. a repeated load uniaxial test involving only compressive stresses has been adopted by ASTM as a standard method[10] for evaluating the dynamic stiffness or 'resilient modulus' of bituminous materials.

Photograph 14.3 shows a three-point vibration test where a beam of asphalt 200 mm long by 40 mm wide by 30 mm deep is held at its ends and the mid-point is vibrated at relatively low strains or deflections. By measuring the force required to achieve the level of strain the dynamic stiffness of the material can be calculated using the following standard beam bending formula:

$$\delta = \frac{WL^3}{48EI} \Rightarrow E = \frac{WL^3}{48\delta I}$$

where E = dynamic stiffness

δ = deflection

W = applied load

L = length of beam

I = second moment of area

$= \frac{bd^3}{12}$

(b = breadth, d = depth)

Most laboratory fatigue tests are carried out under uniaxial conditions, either bending or direct loading[11]. Fatigue tests are carried out by applying a load to a specimen in the form of an alternating stress or strain and determining the number of load applications required to induce 'failure' of the specimen. 'Failure' is not where the sample literally fails but is an arbitrary end point. In a constant strain test the sample is usually deemed to have 'failed' when the load required to maintain that level of strain has fallen to 50% of its initial value. Because of the scatter of test results associated with fatigue testing, it is normal to test several specimens at each stress or strain level and the results are plotted as stress or strain v cycles to failure on a log-log graph as shown in figure 14.6[12]. Controlled stress relates to thick construction and controlled strain to thin surfacings.

Photograph 14.3 – The three-point vibration test

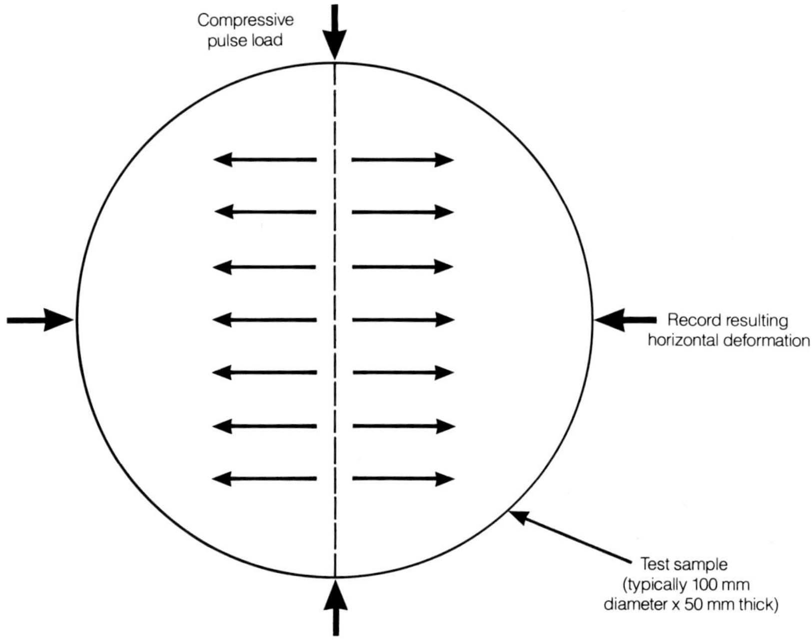

Figure 14.4 – The repeated load indirect tensile test

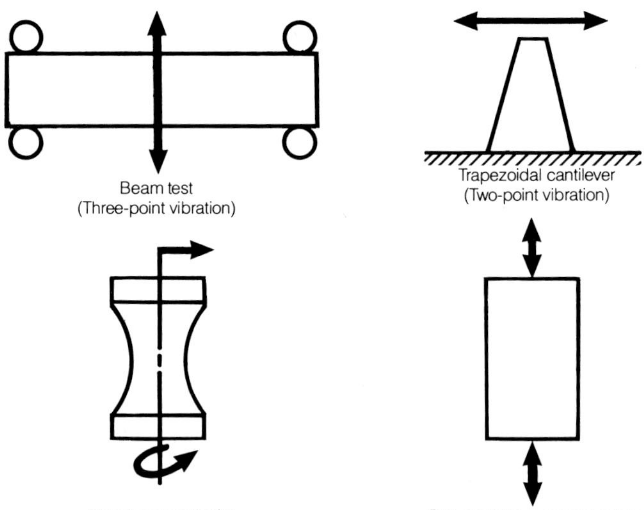

Figure 14.5 – Tests for measuring dynamic stiffness and fatigue

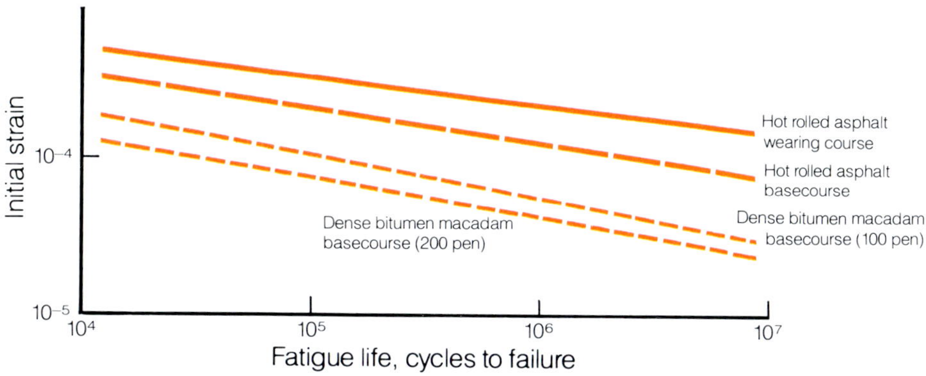

Figure 14.6 – Fatigue properties of different mixes (controlled stress)[12]

Using the apparatus shown in photograph 14.3 a constant strain fatigue test can be carried out immediately after determination of the dynamic stiffness simply by increasing the level of applied strain. The number of cycles to induce failure is automatically recorded on the HP controller when the test is complete.

14.2 Simulative tests

As the stress conditions in a pavement loaded by a rolling wheel are extremely complex and cannot be precisely replicated in a laboratory test on a sample of a bituminous mix, simulative tests have been used to compare the performance of different materials.

14.2.1 Wheel-tracking tests

The equipment ranges from extremely large full-scale pavement test facilities such as that at the Transport and Road Research Laboratory (TRRL)[13] through the mid-scale pavement test facility at the University of Nottingham,[14] both of which can be used to test complete pavement structures, to the smaller laboratory-scale wheel-tracking test (see photograph 14.4) for evaluating the resistance of a bituminous mix to plastic deformation[15]. The laboratory-scale test is usually carried out at 45°C, the wheel applying a load of 525 N to the surface of the sample. The performance of the material is assessed by measuring the resultant rut depth after a given number of passes or rate of tracking in mm/h. Work carried out by the TRRL has shown that a good correlation exists between plastic deformation on open stretches of road and the performance in the laboratory-scale wheel-tracking test[16].

14.3 Empirical tests

The current U.K. empirical pavement design procedure specifies three tests for the evaluation of the mechanical properties of pavement materials: the Marshall test, which forms part of the bituminous mix design procedure detailed in chapter 16; the California Bearing Ratio (CBR) test (see chapter 17, section 17.3.1), which is used for soils and unbound granular material; and the uniaxial compression test, used for the determination of the crushing strength of cement-bound materials.

Photograph 14.4 – The wheel-tracking test

The first two tests (Marshall and CBR) are tests to failure under continuous loading. In both instances the complex stress system set up in the material is quite unrelated to the actual pavement condition *in-situ* under traffic loading. Neither of the two tests can give basic information on the stress-strain characteristics of the material because each has its origin in an empirical method – mix design as far as the Marshall test is concerned, and pavement thickness in the case of the CBR test.

14.3.1 The Marshall test

The concepts of the Marshall test were developed by Bruce Marshall, formerly Bituminous Engineer with the Mississippi State Highway Department. In 1948 the U.S. Corps of Engineers improved and added certain features to Marshall's test procedure and ultimately developed mix design criteria(17). Since 1948 the test has been adopted by organisations and government departments in many countries, sometimes with modifications either to the procedure or to the interpretation of the results. Asphaltic concrete mixes for airfield runways and taxiways have been designed using the Marshall test for the last 20 years and the option to design the mortar fraction of hot rolled asphalt was introduced into BS 594(18) in 1973 and this was extended to an option for designing the whole mix in the most recent revision of this standard in BS 598 : Part 3 : 1985(19).

**Photograph 14.5 –
The Marshall compaction hammer**

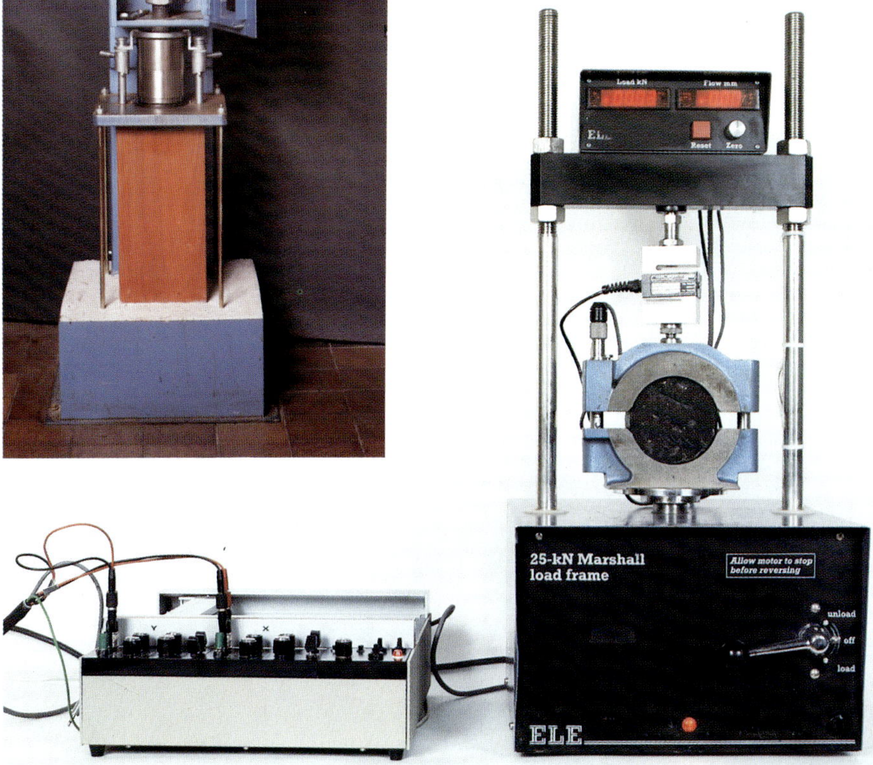

Photograph 14.6 – The Marshall test

The Marshall test consists of the manufacture of cylindrical specimens 102 mm in diameter x 64 mm high by the use of a standard compaction hammer and a cylindrical mould (see photograph 14.5). The specimens are tested for their resistance to deformation at 60°C at a constant rate of 50 mm/min in a test rig as shown in photograph 14.6. The jaws of the loading rig confine the majority but not all of the circumference of the specimen, the top and bottom of the cylinder being unconfined. Thus the stress distribution in the specimen during testing is extremely complex. Two properties are determined: the maximum load the specimen will carry before failure – this is known as the Marshall stability – and the amount of deformation of the specimen before failure occurred – this is known as Marshall flow. The ratio of stability to flow is known as the Marshall quotient, a sort of pseudo stiffness, which is a measure of the material's resistance to permanent deformation.

Although the Marshall test is very widely used, it is important to recognise its limitations. Research at the University of Nottingham[20] compared the mechanical properties of various mixes, using repeated load triaxial tests, triaxial creep tests, uniaxial unconfined creep tests and Marshall tests. The result of this work, shown in figure 14.7, suggests that the Marshall test is a poor measure of resistance to permanent deformation and may not be able to rank mixes in order of their deformation resistance, compared with more realistic repeated load triaxial tests. Figure 14.7a shows that the triaxial creep test is able to distinguish the dense bitumen macadams from the rest of the mixes. However, this test exhibits a very large range of stiffness due to two particularly high results. The unconfined creep test results shown in figure 14.7b can distinguish only between the conventional rolled asphalt basecourses and the other materials of better performance. However, the gap in the rankings is very clear and similar to that from the repeated load test. It would seem on the basis of this data that the unconfined creep test is able to identify good mixes from poor ones in general terms. The Marshall test data in figure 14.7c reveal rankings which are the opposite of those in the repeated load triaxial test. 'Good' mixes are not distinguished from 'poor' ones. In fact, the rolled asphalts all performed better in the Marshall test than any of the dense bitumen macadams.

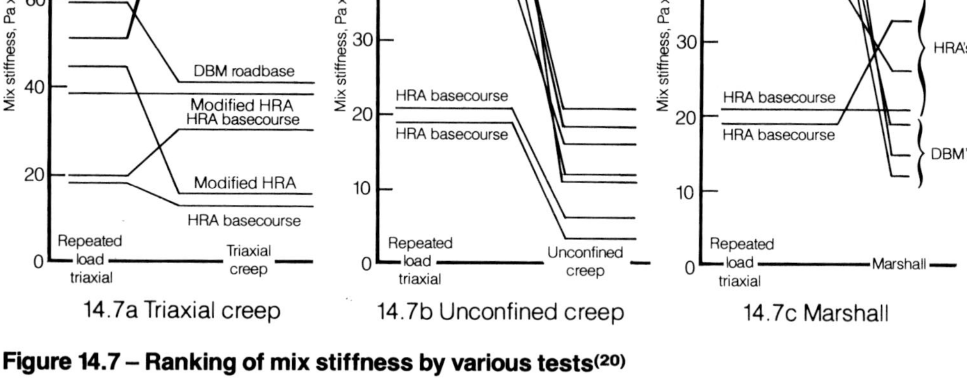

Figure 14.7 – Ranking of mix stiffness by various tests[20]

14.4 References

(1) Brown, S F and Cooper, K E: *The mechanical properties of bituminous materials for roadbases and basecourses*. Proceedings of the Association of Asphalt Paving Technologists, vol 53, pp 413-439, 1984.

(2) Bolk, H J N A: *The creep test*: SCW Record 5, Study Centre for Road Construction, Arnhem, 1981.

(3) *Recommendation for the performance of unconfined static creep test in asphalt specimens.* Proceedings of the international symposium on plastic deformability of bituminous mixes, pp 335-359, Zurich, 1977.

(4) Schmidt, R J: *A practical method of measuring the resilient modulus of asphalt-treated mixes.* Highway Research Record No 404, Highway Research Board, Washington D.C., pp 22-32, 1972.

(5) Pell, P S: *Bituminous pavements: materials, design and evaluation.* Residential course at The University of Nottingham, section E, *Laboratory test methods*, pp E1-E17, April 1988.

(6) Brown, S F: *Practical mechanical tests for the design and control of asphaltic mixes.* RILEM third international symposium on the testing of hydrocarbon binders and materials, Belgrade, 1983.

(7) Monismith, C L, Epps, J A and Finn, F N: *Improved asphalt mix design.* Proceedings of the Association of Asphalt Paving Technologists, vol 54, pp 347-391, 1985.

(8) Bonnaure, F, Gest, G, Gravois, A and Uge, P: *A new method of predicting the stiffness of asphalt paving mixtures.* Proceedings of the Association of Asphalt Paving Technologists, vol 46, pp 64-104, 1977.

(9) Cooper, K E and Pell, P S: *The effect of mix variables on the fatigue strength of bituminous materials.* Transport and Road Research Laboratory, Report LR 633, 1974.

(10) Annual book of the American Society for Testing and Materials Standard: Part II, bituminous materials, soil and rock, skid resistance, 1976.

(11) Pell, P S: *Characterisation of fatigue behaviour.* Highway Research Board, Special Report 140, pp 49-64, 1973.

(12) Pell, P S: *Bituminous pavements: materials, design and evaluation.* Residential course at The University of Nottingham, section J, *Bituminous materials – fatigue cracking*, pp J1-J10, April 1988.

(13) Transport and Road Research Laboratory: *Full-scale pavement testing.* Report by the OECD Scientific Expert Group, Paris, 1985.

(14) Brown, S F and Brodrick, B V: *The Nottingham pavement test facility.* TRB Record 810, pp 67-72, 1981.

(15) Road Research Laboratory: *The wheel-tracking test.* Leaflet LF 50, issue 2, 1971.

(16) Szatkowski, W S and Jacobs, F A: *Dense wearing courses in Britain with high resistance to deformation.* Proceedings of the international symposium on plastic deformability of bituminous mixes, pp 63-76, Zurich, 1977.

(17) The Asphalt Institute: *Mix design methods for asphalt concrete and other hot-mix types.* Manual Series No 2 (MS-2), 1988.

(18) British Standard 594: 1973: *Specification for rolled asphalt (hot process) for roads and other paved areas.*

(19) British Standard 598: Part 3: 1985: *Sampling and examination of bituminous mixtures for roads and other paved areas. Part 3. Methods for design and physical testing.*

(20) Brown, S F, Cooper, K E and Pooley, G R: *Mechanical properties of bituminous materials for pavement design.* Proceedings of the second Eurobitume symposium, pp 143-147, Cannes, October 1981. Reprinted in a special issue of Asphalt Technology, January 1982.

Chapter 15 – The mechanical properties of bituminous mixes

The analytical design of bituminous pavements requires that two aspects of material properties are examined. Firstly, the load-deformation or stress-strain characteristics which are used to analyse critical stresses and strains in the structure. Secondly, the performance characteristics of the materials themselves which will determine the likely mode of failure. The two prinicipal structural distress modes are fatigue cracking and excessive permanent deformation, the former applying to bituminous and cement bound materials, the latter to all the materials in the structure, both bound and unbound.

15.1 Dynamic stiffness of bituminous materials

As discussed previously, asphalt mixes are visco-elastic, ie their mechanical properties depend on the time of loading and on the temperature at which the load was applied. The complexity of the problem is increased by the heterogeneity of the mix components, the bitumen being responsible for the visco-elastic properties whereas the mineral skeleton influences elastic properties. The great diversity of the nature of the components and of the mix compositions make predicting mix properties extremely difficult.

There are two categories of stiffness, namely, elastic stiffness under conditions of low temperatures or short times of loading, and viscous stiffness at high temperatures or long times of loading. The former is used to calculate critical strains in the structure in analytical design. The latter is used to assess the resistance of the material to deformation.

Stiffness at a particular temperature and time of loading can be measured by a variety of methods, such as bending or vibration tests on a beam specimen (see chapter 14, section 14.1.4 and photograph 14.3) or direct uniaxial or triaxial tests on cylindrical specimens. Different types of loading can be used; for elastic stiffness relevant to moving traffic, sinusoidal or other repeated loading is most appropriate.

15.1.1 The prediction of mix stiffness

If the measurement of stiffness is not feasible, then the stiffness of a particular mix at any temperature and time of loading can be estimated to an accuracy acceptable for most design purposes.

In 1977 Shell produced the nomograph shown in figure 15.1 for predicting the stiffness of a bituminous material[1]. The inputs required for this nomograph are:

the stiffness modulus of the bitumen, Pa;
the volume of bitumen, %;
the volume of the mineral aggregate, %.

The University of Nottingham have also developed a method for calculating mix stiffness[2]. The inputs required for the Nottingham method are:

the stiffness modulus of the bitumen, Pa x 10^6;
voids in mixed aggregate, (VMA), %.

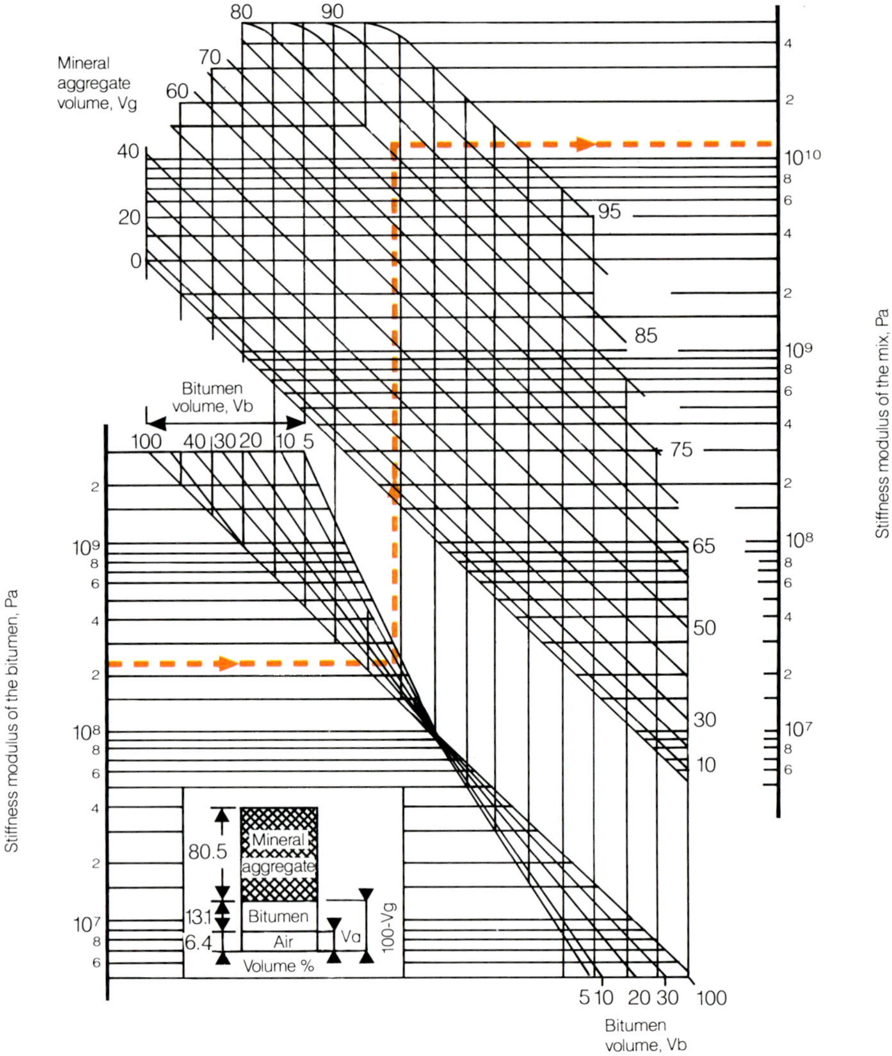

Example

Stiffness modulus of the recovered bitumen, 2.2×10^{8} Pa Vb: Volume of bitumen, 13.1% Vg: Volume of mineral aggregate, 80.5%	Stiffness modulus of the mix 1.1×10^{10} Pa

Figure 15.1 – Nomograph for predicting the stiffness modulus of bituminous mixes[(1)]

These two procedures can be applied only when the stiffness of the bitumen exceeds 5 x 10^6 Pa, ie under high stiffness conditions appropriate to moving traffic and, for the Nottingham method, values of VMA between 12% and 30%. These two methods assume that the grading, type and characteristics of the aggregate affect only the elastic stiffness of the mix insofar as they influence the packing characteristics of the aggregate and thus the state of compaction of the material.

15.2 Permanent deformation of bituminous materials

To determine the permanent deformation resistance of a bituminous mix, the low stiffness response of the material must be analysed, ie its response at high temperatures or long loading times. When the stiffness of the bitumen is <5 x 10^6 Pa, mix behaviour is much more complex than it is in the elastic zone. Under these conditions the stiffness of the mix not only depends on that of the bitumen and the volume of aggregate and bitumen, but also on a variety of other factors. These include the aggregate grading, its shape, texture and degree of interlock, and the method and degree of compaction. This is illustrated in figure 15.2[3].

The simplest test used to study the deformation behaviour of bituminous mixes is the creep test which is described in chapter 14, section 14.1.2 (see photographs 14.1 and 14.2). Figure 15.3a[4] shows the deformation behaviour of a mix tested at different temperatures. If these are plotted as mix stiffness (S_{mix}) and bitumen stiffness (S_{bit}) the test results form one continuous 'mastercurve' for the mix as shown in figure 15.3b[4], the influence of the temperature change being incorporated into S_{bit}. Similarly the effect of using different grades of bitumen or different stress levels can be combined.

Thus, the S_{mix} v S_{bit} curve represents a basic material characteristic for assessing resistance to permanent deformation which is independent of arbitrarily chosen test variables. For example, figure 15.4 shows the deformation characteristics of two asphalt mortar mixes having identical

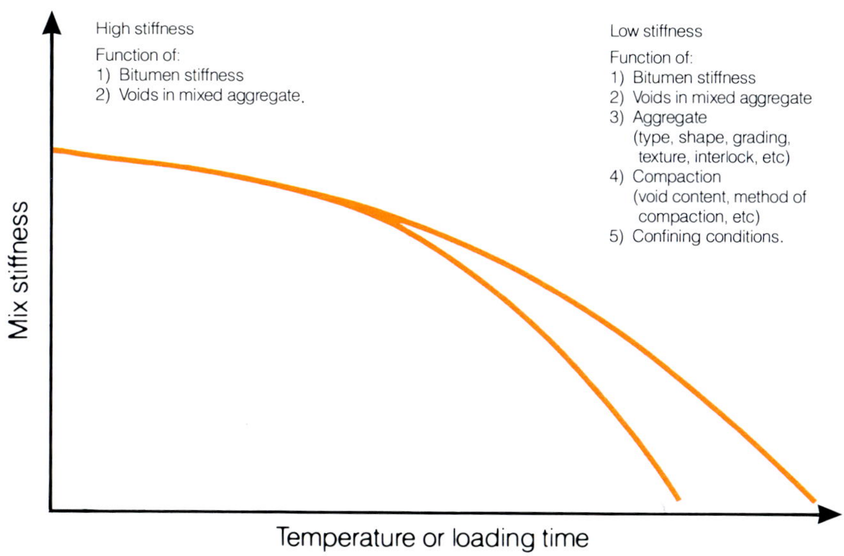

Figure 15.2 – Mix stiffness as a function of temperature or loading time[3]

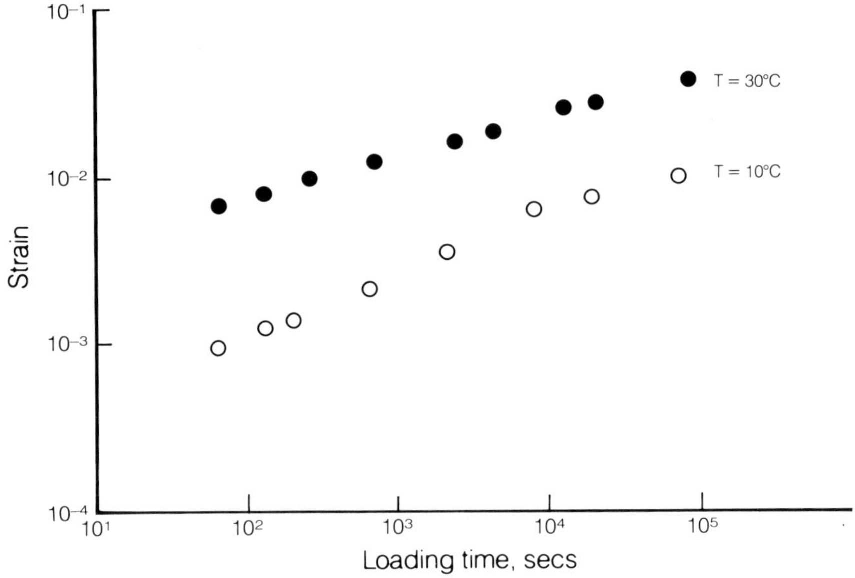

Figure 15.3a – Effect of temperature on deformation(4)

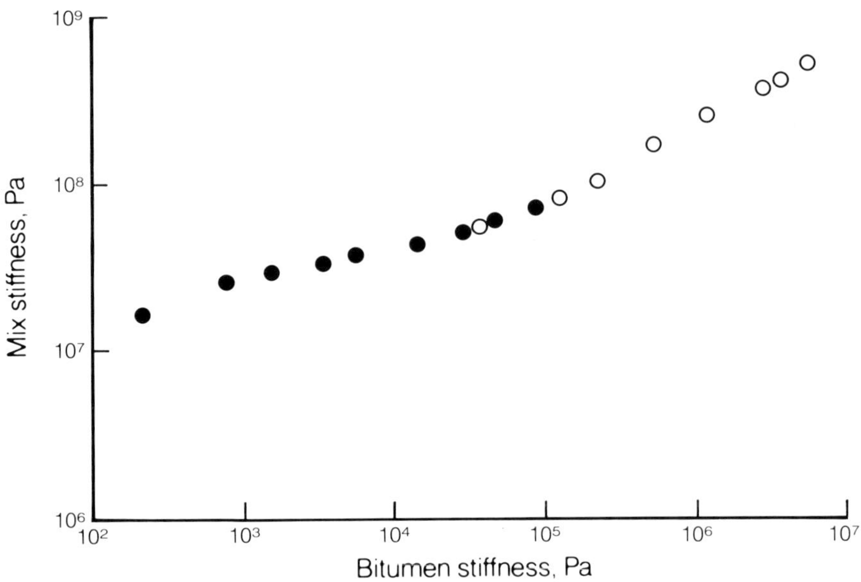

Figure 15.3b – The 'mastercurve' for the mix(4)

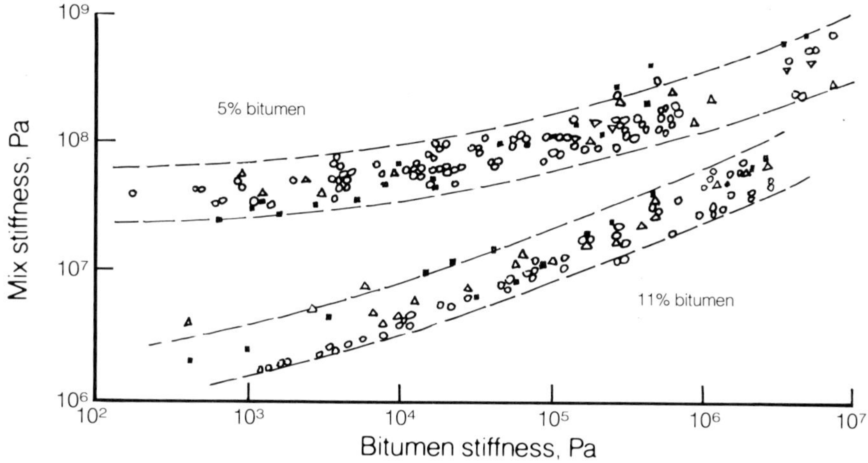

Figure 15.4 – Effect of bitumen content on creep properties[5]

aggregate gradings but containing 5% and 11% bitumen[5]. The leaner mix shows that S_{mix} tends to level out with decreasing S_{bit} whereas the richer mix continues to decrease. Clearly, maintaining a high value of S_{mix} when S_{bit} is decreasing, is a desirable characteristic for long term resistance to permanent deformation. Similarly figure 15.5 shows the effect of aggregate shape on an asphalt mortar having the same aggregate grading and bitumen content[6]. As one would expect, crushed aggregate increases aggregate interlock and this results in increased S_{mix} and higher resistance to permanent deformation.

The resistance of mixes to permanent deformation can be determined by tests such as the unconfined creep test and the repeated load uniaxial or triaxial test; their results are plotted as S_{mix} v S_{bit}. While these tests more accurately reproduce stress conditions on the road, they require more expensive and sophisticated equipment which is available at only a few research establishments.

15.2.1 The prediction of permanent deformation

When attempting to calculate the deformation on a road, account has to be taken of varying wheel load spectra, varying contact areas and pressures, lateral distribution of the wheel loads and varying ambient temperatures and temperature gradients within the asphalt layers. Clearly the situation is very complex. The final step in the Shell Pavement Design Manual[6,7], when the layer thicknesses have been determined, is to predict the rut depth from:

$$\text{Rut depth} = C_m \times h \times \frac{\sigma_{ave}}{S_{mix}}$$

where C_m = empirical correction factor

h = thickness of layer

σ_{ave} = average stress

S_{mix} = stiffness of the mix.

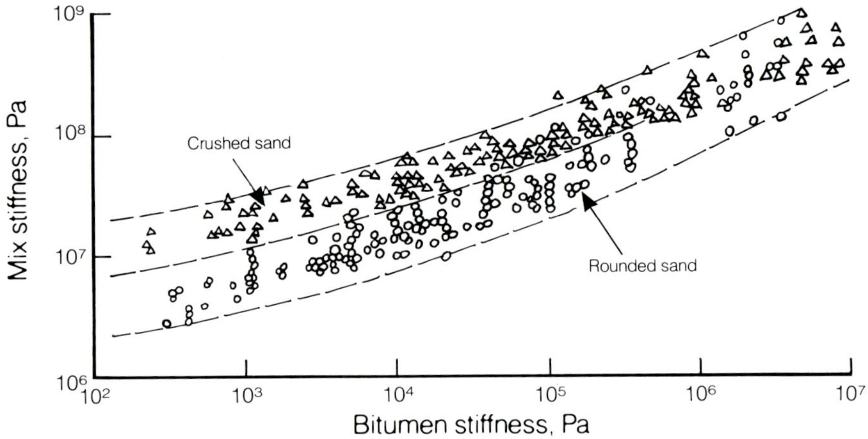

Figure 15.5 – Effect of aggregate shape on creep properties[5]

However, as temperature gradients occur within the asphalt layers and different mix types are often used in the wearing course, basecourse and roadbase, the total asphalt thickness must be subdivided and each sub-layer considered individually. The rut depth is the sum of the deformations in each of these sub-layers.

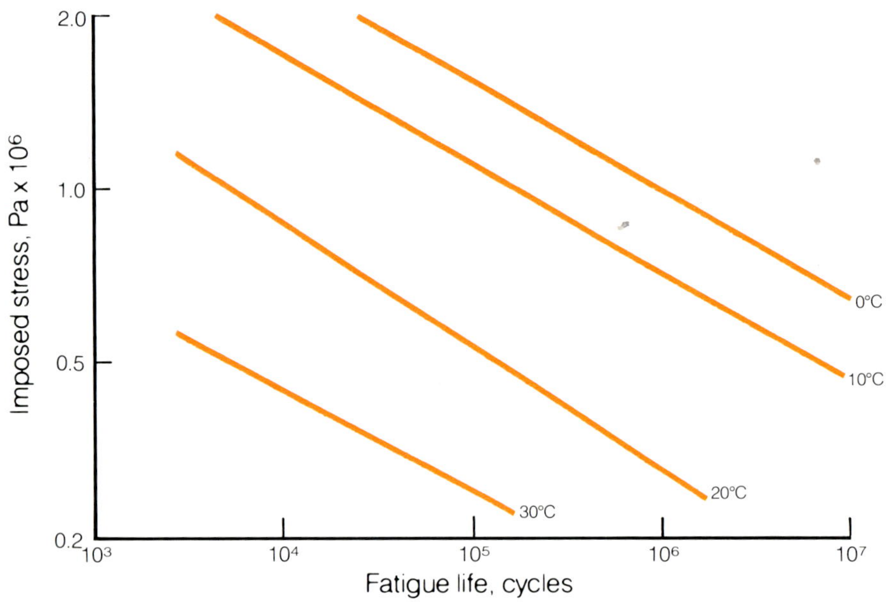

Figure 15.6 – Typical fatigue lines – stress criterion[8]

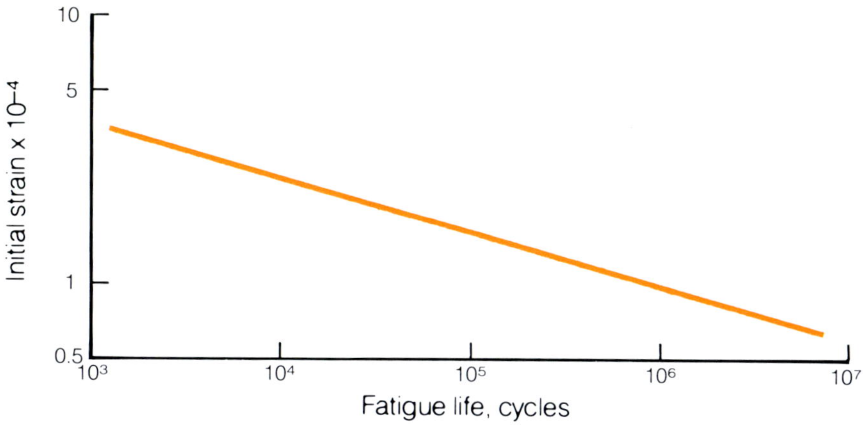

Figure 15.7 – Fatigue lines from figure 15.6 – strain criterion(8)

15.3 Fatigue characteristics of bituminous mixes

Fatigue can be defined as: "The phenomenon of fracture under repeated or fluctuating stress having a maximum value generally less than the tensile strength of the material"(8).

Under traffic loading the layers of a flexible pavement structure are subjected to continuous flexing. The magnitude of the strains is dependent on the overall stiffness and nature of the pavement construction, but analysis confirmed by *in-situ* measurements has indicated that tensile strains of the order of 30 to 200 x 10^{-6} for a standard wheel load occur. Under these conditions the possibility of fatigue cracking exists(8).

As described in chapter 14, section 14.1.4, dynamic bending or dynamic push-pull tests are normally used to measure the fatigue characteristics of bituminous mixes. The constant-stress fatigue life characteristics are shown in figure 15.6 for the same mix at different temperatures. The lines are essentially parallel and show longer fatigue lives at lower temperatures. If the tests are carried out at a different frequency the result is similar – fatigue lives are longer at high frequency. If one replots the fatigue results in terms of constant strain, ε_t as shown in figure 15.7, it is found that results from different stiffnesses coincide; this shows that the criterion of failure is strain and that temperature and time of loading affect stiffness. This effect is known as 'the strain criterion'(9). The general relationship defining the fatigue life is as follows:

$$N_f = c \times \left(\frac{1}{\varepsilon_t}\right)^m$$

where N_f = number of load applications to initiate a fatigue crack

ε_t = maximum value of applied tensile strain

c and m = factors depending on the composition and properties of the mix; m is the slope of the strain-fatigue life line.

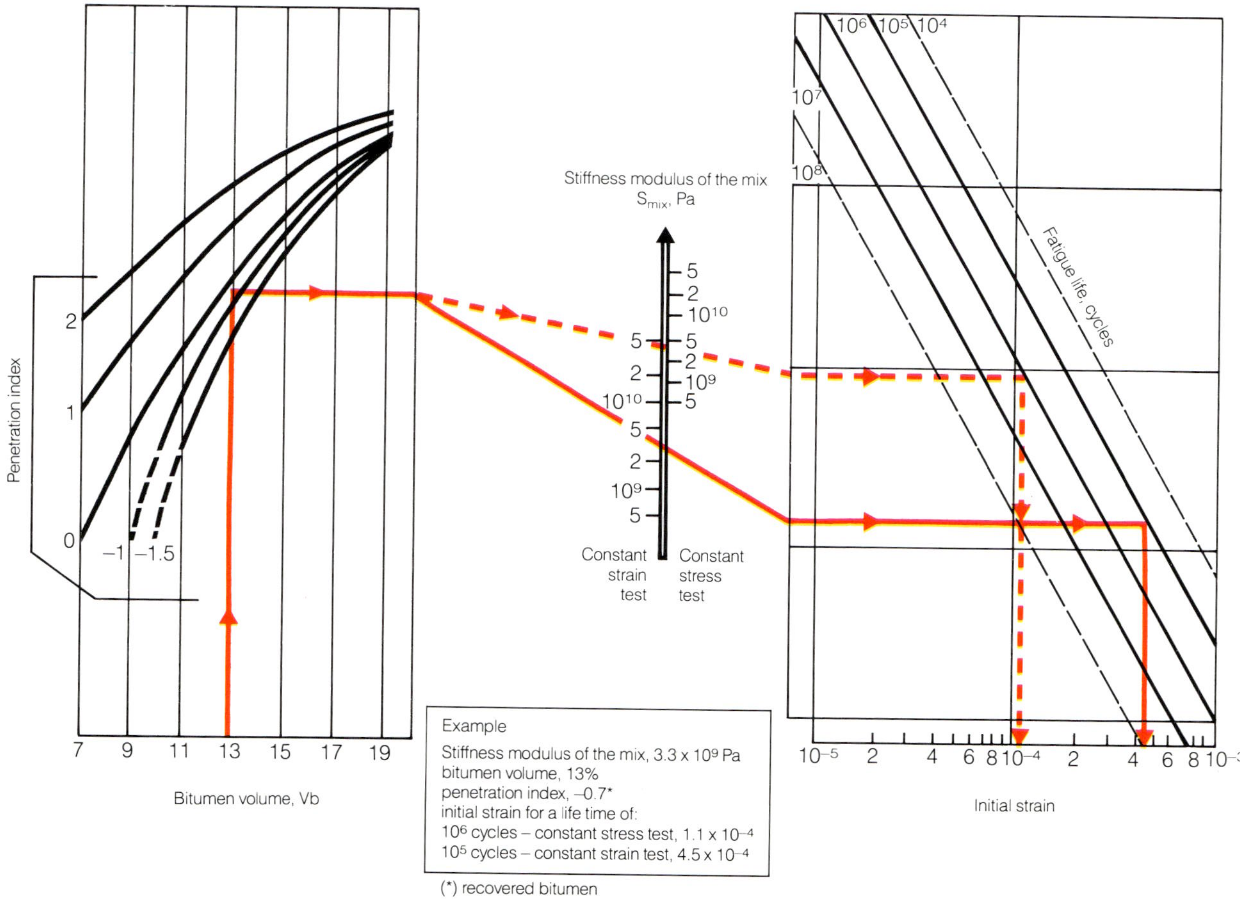

Figure 15.8 – Nomograph to predict the laboratory fatigue performance of asphalt mixes(11)

The fatigue characteristics of a mix can be influenced significantly by its composition. The latter also influences mix stiffness and it is important to distinguish between the two. For example adding filler or reducing the void content will increase mix stiffness, resulting in increased fatigue life at a given level of stress since the resulting strain is smaller; in other words, a point has been reached lower down on the strain-fatigue life line. At a certain strain level, however, it has been found[10] that the entire line moves to the right, ie the fatigue life is improved, if, for example, the bitumen content is increased. Basic fatigue performance is influenced by other mix variables only insofar as these affect the volume of bitumen in the mix. If rounded aggregate is used, for instance, a denser mix with a lower void content will result and this will have the effect of increasing the relative bitumen content and thus improve fatigue performance.

15.3.1 The prediction of fatigue performance

To establish the fatigue line for a particular mix requires somewhat involved and specialised testing equipment and clearly a simpler procedure is desirable to predict the fatigue performance with sufficient accuracy for pavement design purposes. A method for predicting the fatigue life of bituminous mixes has been developed by Shell[11] using the nomograph shown in figure 15.8. The inputs required are:

the volume of bitumen, %;

the penetration index of the bitumen;

the stiffness modulus of the mix, Pa;

the initial strain level.

The application of laboratory-determined fatigue lives to predict actual pavement performance in practice is a complex problem which is likely to yield conservative results, as simple continuous cycles of loading neglect the beneficial effects of rest periods which occur in practice between axle loads. Longer lives in practice are also likely because of the lateral distribution of wheel loads in the wheel track, and a degree of crack propagation will have to occur before pavement performance is adversely affected. Because of these problems, laboratory fatigue lives have to be calibrated to correlate with actual pavement performance and the calibration factor is likely to depend on environmental and loading conditions.

15.4 References

(1) Bonnaure, F, Gest, G, Gravois, A and Uge, P: *A new method of predicting the stiffness of asphalt paving mixtures*. Proceedings of the Association of Asphalt Paving Technologists, vol 46, pp 64-104, 1977.

(2) Brown, S F and Brunton, J M: *An introduction to the analytical design of bituminous pavements*. Third edition. The University of Nottingham.

(3) Brown, S F: *Bituminous pavements: materials, design and evaluation*. Residential course at The University of Nottingham, section I, *Bituminous materials: elastic stiffness and permanent deformation*, pp I1-I13, April 1988.

(4) Hills, J F, Brien, D and Van de Loo, P J: *The correlation of rutting and creep tests on asphalt mixes*. Journal of the Institute of Petroleum, Paper IP74-001, 1974.

(5) Hills, J F: *The creep of asphalt mixes*. Journal of the Institute of Petroleum, vol 59, No 570, November 1973.

(6) *Shell Pavement Design Manual*. Shell International Petroleum Company Limited, 1978.

(7) Claessen, A I M, Edwards, J M, Sommer, P and Uge, P: *Asphalt pavement design – the Shell method*. Proceedings of the fourth international conference on the structural design of asphalt pavements, vol 1, pp 39-74, 1977.

(8) Pell, P S: *Bituminous pavements: materials, design and evaluation*. Residential course at The University of Nottingham, section J, *Bituminous materials – fatigue cracking*, pp J1-J10, April 1988.

(9) Pell, P S and Taylor, I F: *Asphaltic road materials in fatigue*. Proceedings of the Association of Asphalt Paving Technologists, vol 38, pp 371-422, 1969.

(10) Pell, P S and Cooper, K E: *The effect of testing and mix variables on the fatigue performance of bituminous materials*. Proceedings of the Association of Asphalt Paving Technologists, vol 44, pp 1-37, 1975.

(11) Bonnaure, F, Gravois, A and Udron, J: *A new method for predicting the fatigue life of bituminous mixes*. Proceedings of the Association of Asphalt Paving Technologists, vol 49, pp 499-529, 1980.

Chapter 16 – The design and specification of bituminous mixes

Prior to 1945 the main requirements of a road surface were often held to be that it should be low-cost and durable. At this time the majority of county roads and many trunk roads were surfaced with British Standard (BS) 1621 macadams, which proved to be less and less durable in the inter war years as traffic increased. Highway engineers recognised that improved durability would be achieved using dense, impermeable mixes. The gradual increase in the use of BS 594[1,2,3] rolled asphalt for trunk roads and the development of dense bitumen macadam, which first appeared in the 1961 edition of BS 1621[4], reflected this awareness.

Since the mid sixties both the volume and axle weight of vehicles on the roads in the U.K. have increased dramatically, and in the early seventies it was realised that recipe mixes, which had hitherto given long and satisfactory performance, were deforming under the increasing numbers of heavier vehicles.

Thus, bituminous mixes have to fulfil a wide range of requirements for today's traffic, in particular the ability to:

- resist permanent deformation;
- resist fatigue cracking;
- be workable during laying, enabling the material to be satisfactorily compacted with the available equipment;
- be impermeable, to protect the lower layers of the road from water;
- be durable, resisting abrasion by traffic and the effects of air and water;
- contribute to the strength of the pavement structure;
- be easily maintained and, most importantly, this must be cost-effective.

In addition to the above, wearing course materials must also fulfil the following tyre/pavement interaction requirements:

- provide a skid-resistant surface under all weather conditions;
- have an acceptable level of rolling resistance;
- provide a surface which, under trafficking, produces an acceptable level of tyre/road noise;
- provide a surface of acceptable riding quality.

This is a formidable list of requirements demonstrating the need for optimal mix design to ensure that the required performance is achieved without wasting valuable and limited resources. Thus, the use of traditional recipe mixes is simply inappropriate for heavily trafficked roads. The need to produce materials which would fulfil the above criteria led to a re-appraisal of mix deisgn philosophy and ultimately to the inclusion of the Marshall design procedure in BS : 594 : 1973[5] which has since been superseded by BS : 598 : 1985[6].

The objective of mix design is to produce an economical material, making full use of local resources, that meet the engineering requirements, ie to design for optimal service and economic performance, not maximum achievable performance.

16.1 Traditional recipe specifications

A recipe specification defines a mix in terms of the aggregate grading, mix composition and the method by which the mix should be manufactured, placed and compacted. Recipe specifications are simple to use and are applicable to all types of mix. They are based on experience of known compositions which have performed successfully in practice and, provided there is widespread experience of different mix compositions in practice, it is not difficult for the customer to specify the mix he considers suitable. The supplier usually has little difficulty making the mix to the required composition and it is relatively easy to check compliance against the specification. As a result the recipe approach to mix design is still the basis of most specifications, particularly in the U.K.

However, recipe formulations have a number of severe limitations. The conditions of traffic, climate, etc, to which the mix is to be subjected, will almost certainly not be the same as those existing when the recipe formulation was being evaluated. Variations in the components of this mix may occur that are outside previous experience. If this necessitates modifications to the recipe, there is no means of assessing what these modifications should be or their effect.

Checking for compliance with recipe mixes is relatively simple; however, there is no means of assessing the seriousness or the practical implications of a minor failure to comply with the specification. While the importance of adhering to the relevant specification is clear in the interests of quality control, and since it is assumed that the particular specification represents the best current knowledge available from experience, there is perhaps too ready an assumption that any mix failing to meet the specification is defective in some respect. The fallacy of making this assumption irrespective of engineering properties is well illustrated by the recent development of heavy duty macadam (HDM), which is manufactured using 50 pen bitumen and filler contents of 7% to 11%, and dense bitumen macadam which is manufactured using 50 pen bitumen (DBM 50)[7,8] (see chapter 17, section 17.4). These mixes, which do not comply with the current, or recently superseded, editions of BS 4987 and would previously have been rejected out of hand as failing to comply, have been shown in extensive laboratory and field trials to exhibit superior load spreading ability and to be more cost-effective than normal dense bitumen macadam.

This illustration emphasises the weakness of reliance on recipe specifications alone, where a combination of consideration of the degree and nature of the deviation from the recipe together with recourse to some laboratory and/or field assessment of the engineering properties of the mix in question, could lead to the avoidance of some unnecessary and expensive removals of newly laid materials marginally failing to comply with the specification. In addition, recipe specification tend to be restrictive towards new developments and the use of cheap locally available materials, a restriction which may result in unnecessary expenditure. The basis of recipe specifications is the aggregate grading curve and as these curves cannot, except in a very general way, take account of the aggregate properties and packing characteristics, they are not very reliable, particularly when mixtures of aggregates are used.

16.2 The principles of mix design

The design of a bituminous mix involves the choice of aggregate type, aggregate grading, bitumen grade and the determination of a bitumen content which will optimise the engineering properties in relation to the desired behaviour in service. While the importance of the first three is not underestimated, the majority of discussion and research has been centred on the methods of arriving at a design bitumen content, whereas in recipe specifications the bitumen content is selected solely on the basis of experience. In 'design' methods the choice of bitumen content is based in whole or in part upon the results obtained in one or other laboratory test procedure.

While the Marshall test has been the most widespread in use, other test apparatus and procedures have also been the basis of bitumen content design, eg the Hveem method[9] and various methods involving use of triaxial or creep tests. In many of these the general approach to the selection of the design bitumen content has been to average the optimum bitumen contents giving peak values for certain mechanical properties, together with bitumen contents giving other desired properties relating to density and voids content. Although Bruce Marshall in his original method selected the design bitumen content as that giving the peak on the unit weight versus bitumen content curve, checking other properties at this bitumen content against desired standards, the customary practice in subsequent methods based on the Marshall test has been to follow the averaging approach mentioned above . Details of this approach are given in the following two sections. Some of the difficulties associated with this method of averaging, eg identification of the 'peak', the problem of double peaks and the risk of basing calculations on severely asymmetrical curves are discussed by Lees[10] who describes an alternative method whereby the ranges of bitumen content over which certain desired properties are satisfied are considered and the choice of design bitumen content based upon these considerations.

In order to design a bituminous mix by laboratory procedures, with the aim of producing the required density and engineering properties, it is desirable to prepare specimens which reproduce, as closely as possible, the material that will be used in practice. The packing characteristics of the aggregates significantly influence the mix properties which cannot be assessed from grading curves, as these assume uniformity of particle shape, surface roughness and specific gravity. Evaluation of compacted specimens takes into account these aggregate factors. Ideally specimens should be compacted to the same thickness as used in practice, by rolling, to a density similar to that attained under traffic during the service life. However, this is not done in any current mix design procedure, and specimens prepared for routine mix designs are compacted by hammer. Rolling tends to orientate elongated and flaky aggregate particles so that their longest dimensions are horizontal; thus the specimens tend to be more anisotropic than those prepared by hammer.

Knowing the masses and the specific gravities of the mix components and the density of the compacted mix, the volumetric composition can be determined. The samples are then tested to assess their relative resistance to deformation using either the Marshall test or the creep test.

16.3 The design of continuously graded mixes

The objective when designing continuously graded mixes is to achieve a dense mix of high stability, but with sufficient voids between the aggregate particles, ensuring that sufficient bitumen is added to achieve flexibility, durability and workability without sacrificing resistance to permanent deformation.

16.3.1 The Asphalt Institute design method

The most widely used procedure for designing continuously graded mixes is the Asphalt Institute method[9]. The basis of this method is described below.

The aggregate grading is designed to compact to a void content of between 16% and 19%. This ensures that there are sufficient voids to accommodate the bitumen but still leave some air voids. If the void content is too low, the aggregate will be unable to retain sufficient bitumen and the mix will lack durability. If the void content is too high, the material will be permeable and again lack durability. This could be overcome by adding more bitumen but this would be at the expense of deformation resistance and would be uneconomic. The maximum density grading curve can be determined using the Fuller curve[9] which is of the form:

$$P = 100 \left(\frac{d}{D}\right)^n$$

where D = maximum size of aggregate
d = size of aggregate in question
P = percentage of aggregate in question
n = 0.5.

Nijboer[11] made mixes with aggregate blended to grading curves of this form, but with a range of n values, and showed that the densest mixes were obtained with n between 0.4 and 0.5. Using n = 0.45 results in 60% stone content as the optimum for a 18 mm maximum size and 52% stone content for a 12 mm maximum size.

The first stage in the design procedure is to dry and grade the aggregate and blend it to achieve the desired grading. Then the bulk specific gravity, dry compacted density and voids in the mixed aggregate are determined.

The quantity of bitumen added to the aggregate should ideally result in a void content in the mix of between 3% and 5%. If too much bitumen is added, gradual secondary compaction by traffic will result and excess bitumen will be squeezed to the surface, making it slippery. It may also lead to hydrostatic pressure in the body of the mix, reducing the stability of the mix. If insufficient bitumen is added, then water and air will enter the mix, resulting in hardening of the bitumen and possibly stripping. The amount of bitumen to be added to the mix can be estimated from:

$$B = \frac{V_{ma} - V_a}{G_{sc}} \times G_b$$

where B = parts mass of bitumen per 100 parts mass of aggregate (abbreviated to pha)
G_{sc} = bulk specific gravity of the compacted aggregate
G_b = specific gravity of bitumen
V_{ma} = voids in the mineral aggregate
V_a = desired void content.

If the bitumen content is required in terms of the parts mass of the total mix, P_b, %m, is:

$$P_b = \frac{B}{B + 100} \times 100$$

Five mixes are usually made with bitumen contents at increments of 0.5 pha or 0.5% m each side of the estimated optimum bitumen content. Four specimens are made at each bitumen content. Compaction is carried out with the standard hammer, 35, 50 or 75 blows on each end of the specimen, according to the expected traffic. The specimens are weighed in air and water to determine the bulk specific gravity of the compacted mix. At each bitumen content the following mix properties are determined (average of the four specimens):

the bulk specific gravity of the compacted mix;

the bulk specific gravity of the compacted aggregate;

the theoretical maximum specific gravity of the mix;

the voids in the mixed aggregate;

the theoretical voids in the mix (assumes no bitumen is absorbed into the aggregate).

The specimens are then tested in the Marshall apparatus and the average Marshall stability and flow values determined for each bitumen content. Five graphs are plotted – stability, flow, specific gravity of the mix, void content of the mix and void content of the mixed aggregate, all against bitumen content as illustrated in figure 16.1[9]. From these graphs the three bitumen contents corresponding to maximum stability, maximum specific gravity and the desired value of void content are determined. The optimum bitumen content is taken as the average of these three values. The flow value and voids in mixed aggregate that correspond to the optimum bitumen content is determined and compared with permissible values. If all the requirements are not met, the procedure should be repeated with appropriate modification to the aggregate grading.

If all the properties are satisfactory, a correction to void content for bitumen absorption should be made. The 'theoretical' voids in the mix are determined using the relative density of the aggregate calculated on an oven dried basis. This assumes that there is no bitumen absorption into the aggregate. In some aggregates a significant amount of absorption occurs and one of the most reliable and direct methods of determining bitumen absorption is based on Rice's method[9]. Briefly this involves initially determining the maximum specific gravity of the mix. A sample of the mix is then warmed and broken into small pieces so that as much entrained air can escape as possible; the sample is then weighed in air and water. The difference between this value and the maximum specific gravity of the mix, calculated using the known amount of bitumen and aggregate and their respective specific gravities, is a measure of the amount of bitumen absorbed into the aggregate.

The Marshall criteria recommended by the Asphalt Institute for roads[9] and airfields[12] are summarised in tables 16.1 and 16.2 respectively.

16.3.2 The DoE (Airfields Branch) method[13]

In the U.K. continuously graded asphaltic concrete mixes are normally used only for airfield pavements and are designed using a variation of the Asphalt Institute method. The Department of the Environment (DoE) method differs in several minor but significant respects. In the DoE method the actual maximum theoretical density is not determined and instead of voids in mixed aggregate, voids filled with bitumen are determined. The other significant difference is that the bitumen absorption is estimated using the apparent relative density, the relative density on an oven dried basis and the water absorption value of the aggregate. If the water absorption value of the aggregate is less than 1%, all the calculations are made using the apparent relative density, ie assuming that the bitumen absorption is equal to the water absorption. If the water absorption is greater than 1% the average of the apparent relative density and the relative density on an oven dried basis is used, ie assuming that the bitumen absorption is only half of the water absorption.

Five graphs are plotted – stability, flow, maximum apparent relative density, voids in the mix and voids filled with bitumen, each against bitumen content. From these graphs the four bitumen contents corresponding to maximum stability, maximum apparent relative density and the desired values of voids in the mix and voids filled with bitumen are determined. The optimum bitumen content is taken as the average of these four values. The flow value that corresponds to this optimum value is determined from the graph and compared with the permissible values (see table 16.3).

Mix criteria	Traffic category		
	Light	Medium	Heavy
Number of compaction blows	2 x 35	2 x 50	2 x 75
Minimum Marshall stability, kN	3.3	5.3	8.0
Marshall flow, mm	2.0-4.5	2.0-4.0	2.0-3.5
Void content, %	3.0-5.0	3.0-5.0	3.0-5.0

	Maximum size of aggregate, mm	Minimum voids in mixed aggregate, %
Voids in mixed aggregate, %	25	13.0
	19	14.0
	12.5	15.0
	9.5	16.0
	4.75	18.0
	2.36	21.0
	1.18	23.5

Table 16.1 – The Asphalt Institute design criteria for roads[9]

Mix criteria	Category of airfield		
	Small	General	Air carrier
Number of compaction blows	2 x 50	2 x 75	2 x 75
Minimum Marshall stability, kN	2.2	4.5	8.0
Marshall flow, mm	2.0-5.0	2.0-4.0	2.0-3.5
Void content, %	3.0-5.0	3.0-5.0	3.0-5.0

Voids in mixed aggregate – see table 16.1

Table 16.2 – The Asphalt Institute design criteria for airfields[12]

Mix criteria	Wearing course	Basecourse
Minimum Marshall stability, kN	8.0	8.0
Marshall flow, mm	0-4.0	0-4.0
Void content, %	3.0-4.0	3.0-4.0
Voids filled with bitumen, %	76-82	67-77

Table 16.3 – The DoE design criteria for airfield pavements[13]

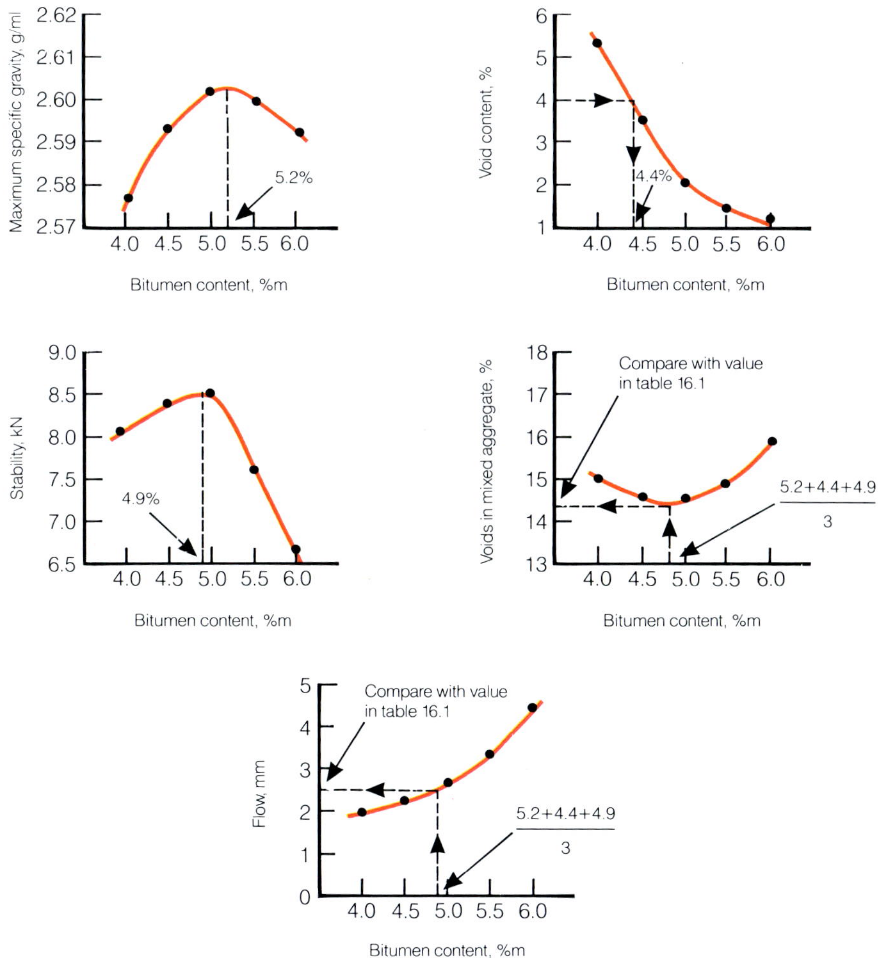

$$\text{Optimum bitumen content} = \frac{5.2+4.4+4.9}{3} = 4.8\%$$

At the optimum bitumen content:

1) The Marshall flow and void content should comply with the ranges given in table 16.1.
2) The Marshall stability and voids in mixed aggregate should exceed the minimum values given in table 16.1.

Figure 16.1 – Marshall design curves, the Asphalt Institute method[9]

16.4 The design of gap-graded mixes

Gap-graded mixes have been used for many years for surfacing major roads in the U.K. and traditionally their compositions have been specified in BS 594 by recipe. The 1973 edition of BS 594[5] continued the tradition of recipe specification, the coarse and fine aggregates being specified separately with the filler and bitumen specified in tables detailing alternative compositions.

16.4.1 The BS 594 : 1973 mortar design method

In addition to the recipe specification, BS 594 : 1973 introduced the option of using the Marshall test to estimate the bitumen content required to optimise the properties of the fine aggregate (sand) : filler mixture. These two components are blended in the ratio 6 : 1 by weight and Marshall specimens are prepared and tested over a range of bitumen contents, usually 8.0% to 12.0% at 0.5% increments. From the test results three graphs are plotted – stability, compacted density of the mix and compacted density of the aggregate, each against bitumen content. The three bitumen contents corresponding to the maximum values are determined and the mean of these three values is taken as the optimum bitumen content for that sand : filler blend. The sand and filler are then used with added stone to comply with one of the standard formulations and the bitumen content for the mix with the added stone is calculated using a mathematical formula. Limits are placed on the minimum bitumen content allowable.

The weaknesses of this method are as follows:

- a 6 : 1 sand : filler blend may not be optimal;
- the performance of a mortar may not be a sufficient indication of the behaviour of the corresponding stone filled rolled asphalt[14];
- calculating the bitumen content for the stone filled rolled asphalt empirically may not be optimal;
- the stability, mix density and compacted aggregate density measure only some of the requirements and do not give a true measure of the material's strength or voids which are vitally important.

In 1979 the Department of Transport issued Departmental Standard HD/2/79[15,16] which required that for all trunk roads and motorways carrying more than 250 commercial vehicles per lane per day, the mortar design method should be used. Again restrictions on minimum acceptable bitumen content for the mortar at 9.2% were imposed, reflecting fears about material durability at low bitumen contents. HD/2/79 also introduced the first guidance on the required levels of stability and flow for various traffic categories.

16.4.2 The BS 598 : 1985 design method

The 1985 edition of BS 594[17] broke with the traditional approach of specifying recipe in terms of fine and coarse aggregates separately and specifies an 'all-in' grading. The tabulation of alternative compositions and bitumen contents is retained. BS 598 : 1985[6] provides an alternative for designing the mix composition; this time the total mix including the coarse aggregate is evaluated. Briefly the method consists of testing three specimens at each bitumen content; a minimum of nine bitumen contents are examined at 0.5% increments. From the test results four graphs are plotted – stability, flow, mix density and compacted aggregate density, each against bitumen content. Idealised curves are shown in figure 16.2[6]. The three bitumen contents corresponding to the

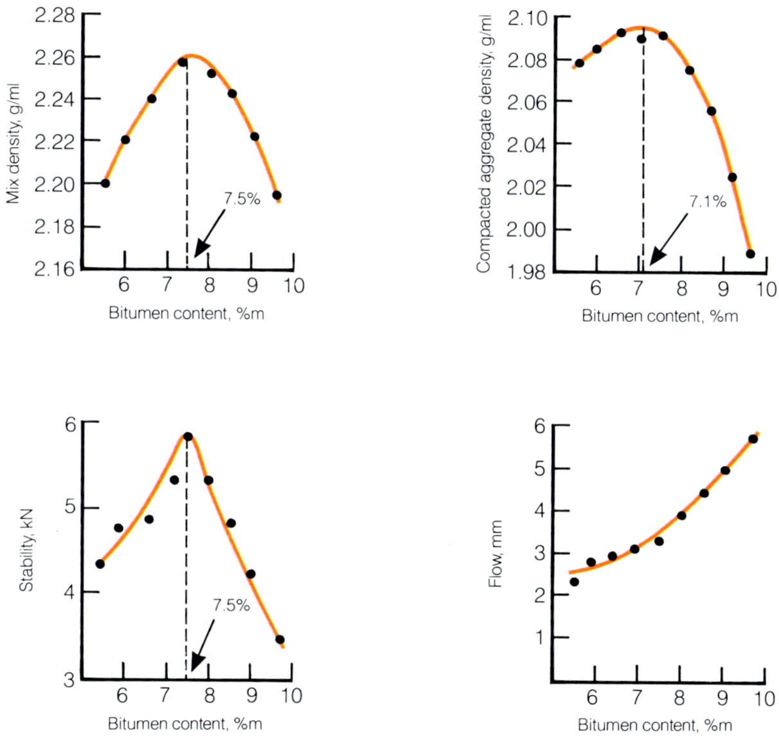

$$\text{Optimum bitumen content} = \frac{7.5 + 7.5 + 7.1}{3} = 7.4\%$$

For a 30% stone content mix the design bitumen content = 7.4 + 0.7 = 8.1%

The Marshall flow at the design bitumen content should not exceed the values given in table 16.4.

Figure 16.2 – Marshall design curves, the BS 598 : 1985 method(6)

maximum values of stability, mix density and compacted aggregate density are determined and the mean of these three values is termed the optimum bitumen content. An empirical factor is added to the optimum bitumen content to determine the 'design' bitumen content. This empirical addition is 0.7% for a mix containing 30% coarse aggregate. The bitumen content specified for the final mix composition is termed the target bitumen content. The target bitumen content is the design bitumen content plus (if deemed necessary) an unspecified quantity of bitumen to "allow for such properties as workability and durability". This extra bitumen typically ranges from 0% to 1.0%. Thus the actual target bitumen content may be as much as 1.7% above the optimum bitumen content determined by the Marshall design procedure.

The 1973 design method does not include provision for the measurement of void content; however, the 1985 method includes details of the calculation of void content relationships, but no guidance is given regarding their use.

In part 1 of the 1985 edition of BS 594[17] there is only limited guidance on design criteria, as detailed in table 16.4. It is left for the engineer to use his knowledge of all the available design criteria in asessment of mix designs.

Traffic flow, comm.veh./lane/day	Marshall properties	
	Stability, kN	Maximum flow, mm
less than 1500	2.0-8.0	5.0
1500-6000	4.0-8.0	5.0
Over 6000	6.0-10.0	5.0 or 7.0*

*For stabilities up to 8.0 kN the maximum flow is 5.0 mm,
for stabilities greater than 8.0 kN the maximum flow is 7.0 mm.

Table 16.4 – The BS 594 : 1985 design criteria for roads[17]

16.5 An alternative mortar design method

A system of more general applicability for the design of gap-graded mixes, not constrained by U.K. recipes, and based on sound principles, has been developed by Brien[18,19,20,21]. The dominant material in gap-graded mixes is the sand/filler/bitumen mortar, which controls the properties of the total mix. The Brien procedure consists of evaluating mortar mixes by Marshall quotient, ie Marshall stability divided by Marshall flow. The Marshall quotient can be increased by one or more of the following:

- the addition of coarse aggregate;
- the use of a more rugous sand;
- the use of a harder grade of bitumen;
- varying the composition of the mortar.

Initially a series of mixes are made using the available sand at a fixed sand : filler ratio, say 6 : 1, over a range of bitumen contents. The bitumen content required to give a maximum Marshall quotient is determined and the voids in the mortar of this mix are calculated. The above four factors are examined in turn to increase the measured Marshall quotient to that required.

The effect of the coarse aggregate on Marshall quotient[18,19,20] and voids in mortar is well documented[19,20] and shown in figures 16.3 and 16.4. These two figures can be used to assess the effect of adding different amounts of coarse aggregate.

Of all the components that form a gap-graded mix, the sand is probably the most important in controlling the deformation resistance of the final mix. The effect of bitumen content on the wheel-tracking rate of four 30% stone content hot rolled asphalt mixes, manufactured using 50 pen bitumen and four different asphalt sands, all complying with the grading envelope in BS 594 : 1973, is illustrated in figure 16.5[22,23]. The results clearly show that the optimum bitumen content for

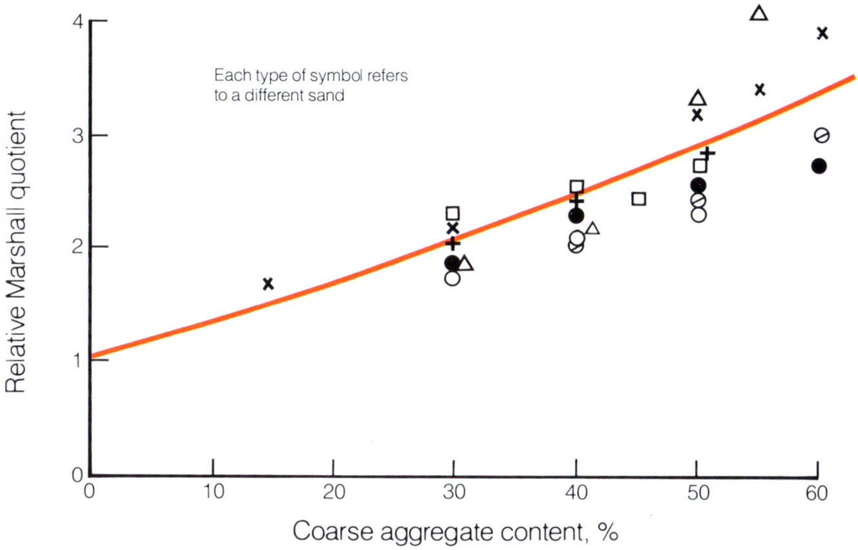

Figure 16.3 – Relationship between Marshall quotient and coarse aggregate content(18,19,20)

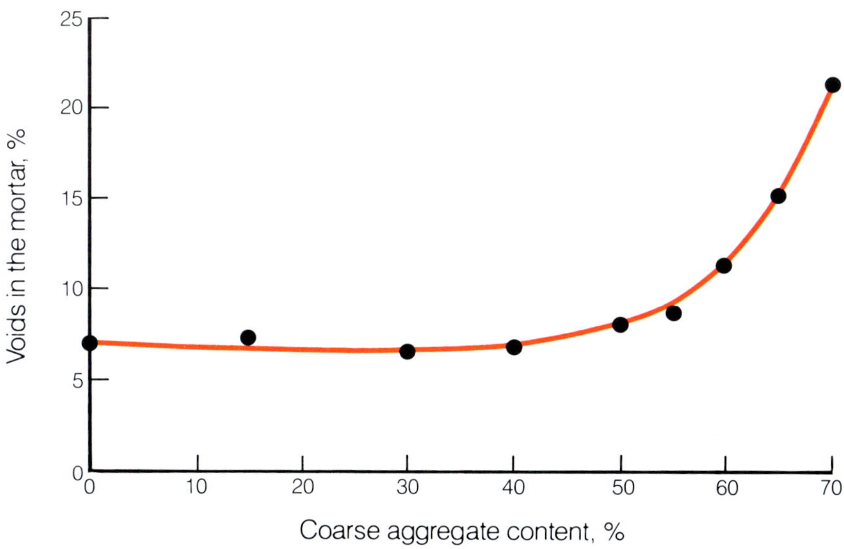

Figure 16.4 – Relationship between voids in the mortar and coarse aggregate content(19,20)

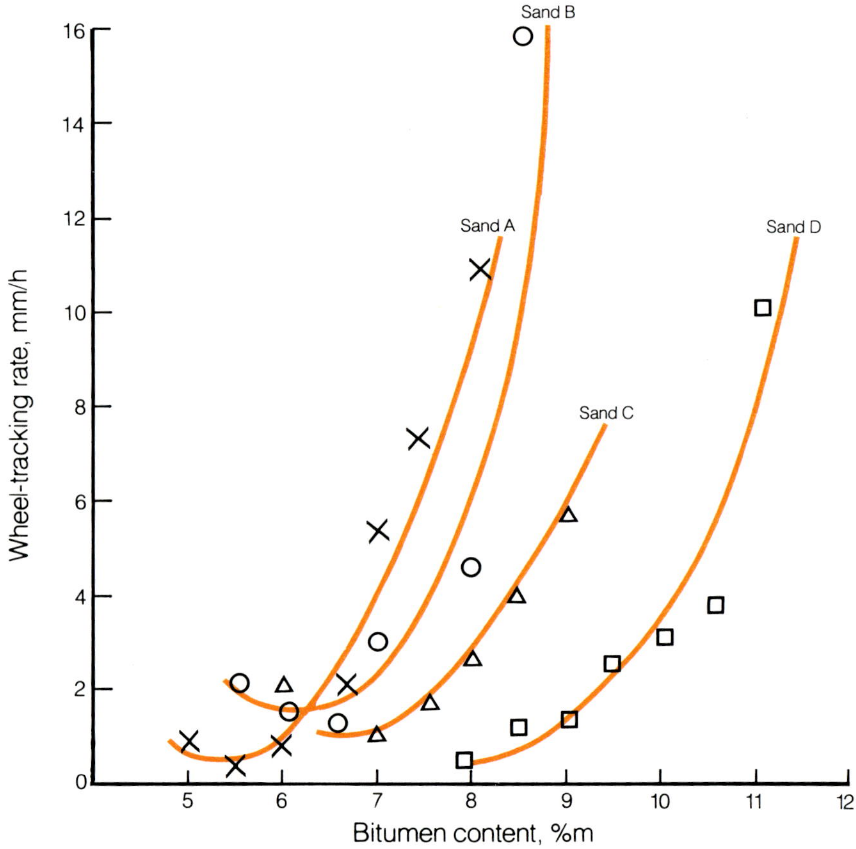

Figure 16.5 – The effect of bitumen content on resistance to deformation of four different sands[22,23]

different sands can vary widely. It also shows that increasing the bitumen content by 1% above the optimum reduces the resistance to deformation by a factor of 2 or more. The surface texture of the sand is also very important. Wheel-tracking tests on sands with a range of surface textures[19,24] have shown that the deformation resistance can be improved by as much as a factor of 4 by using a sand with a coarse surface texture. The addition of crushed rock fines to the sand generally improves deformation resistance but often the addition of this material reduces the workability of the asphalt significantly. One method of characterising the surface texture of sands is the dry viscosity test[18,19,20]. In this test, shown in figure 16.6, the time taken for a known weight of sand to flow through a 6 mm orifice is determined. The time for 100 grams to pass through the orifice is termed the dry viscosity of the sand. Values range from 13 seconds for smooth rounded particles to 18 seconds for angular particles. Figure 16.7 shows the relationship between the Marshall quotient and the dry viscosity of the sand.

The effect of bitumen grade on the mix properties is discussed at length in chapter 7, section 7.2.

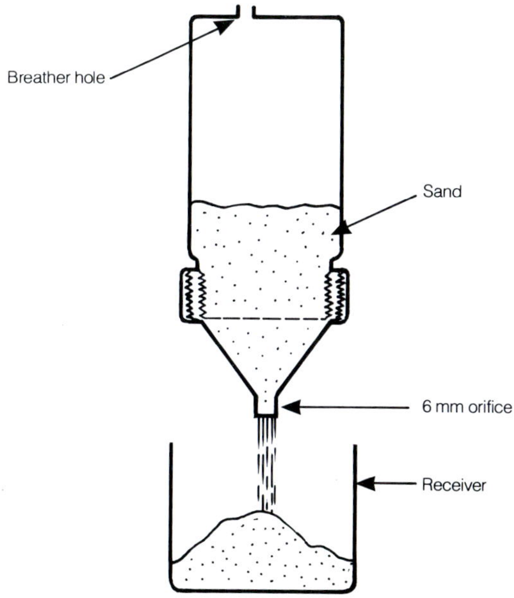

Figure 16.6 – The dry viscosity test

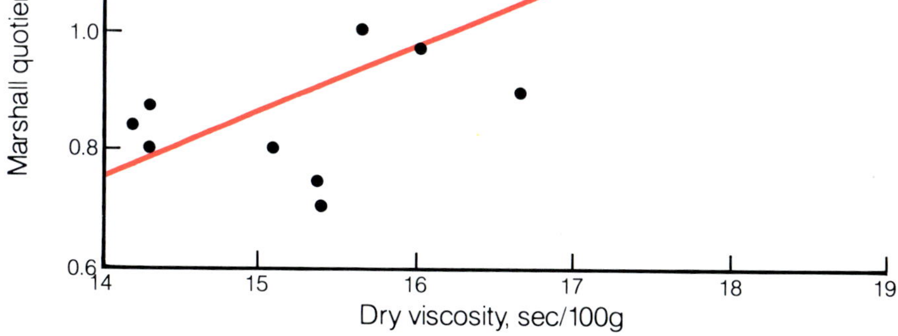

Figure 16.7 – Relationship between Marshall quotient and the dry viscosity of the sand[19,20]

Brien examined a series of mortars in which the filler and bitumen contents were varied systematically and the Marshall quotients of each mix were determined. The relationships between filler : bitumen ratio and Marshall quotient and void content are shown in figures 16.8a and 16.8b respectively. The co-ordinates of any point define a fixed composition and the lines are contours of equi-Marshall quotient. If mixes with a fixed filler content are considered, low values of quotient are obtained at the right hand side of the diagram, where the bitumen content is high, ie the voids are overfilled. At the left hand side of the diagram the volume of bitumen is too low and therefore inhibits adequate compaction. Between these two extremes, there is an optimum. Similarly, if mixes having constant bitumen contents are considered, low values are obtained at the bottom of the diagram, where the bitumen is relatively fluid because there is little filler to stiffen it, and there are also low values at the top, due to the highly stiffened bitumen preventing adequate compaction. Between these two extremes, an optimum value is again found. There is thus a 'ridge' along which the maximum quotient occurs. A similar pattern of results is obtained when considering other mix properties such as stability.

If the void content is treated in a similar way, as shown in figure 16.8b, the ridge is associated with a constant void content. This suggests that the durability of the mix that is associated with void content, ie porosity to air and water, will be unaffected if the mix composition is changed along the ridge. However, the fatigue life of the mix will be reduced at the lower bitumen contents and this must be taken into account. Thus if a new mortar is to be designed with a different filler content a new point is located on the ridge without the need to carry out a full investigation. Other mixes can then be made having compostions close to the ridge.

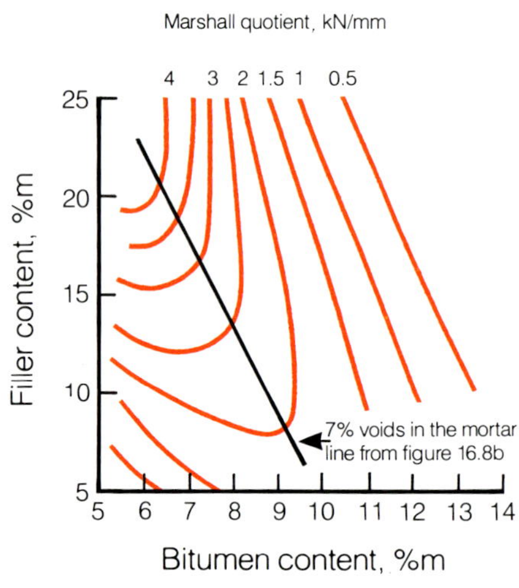

Figure 16.8a

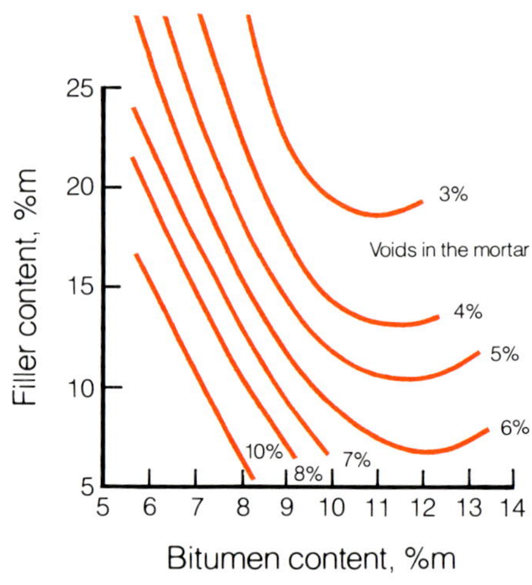

Figure 16.8b

Marshall quotient and voids in the mortar—filler : bitumen contours(18,19,20,21)

The importance of both bitumen and filler content on the deformation resistance of a 30% stone content hot rolled asphalt mix manufactured using 50 pen bitumen is illustrated in figure 16.9[23]. At the optimum bitumen content varying the filler content by ± 2.0% has only a small effect on deformation resistance. However, if the bitumen and filler contents are increased by 0.6% and 1.5% respectively, ie the upper tolerance limits in BS 594 : 1973, the resistance to deformation is reduced by a factor of 2.

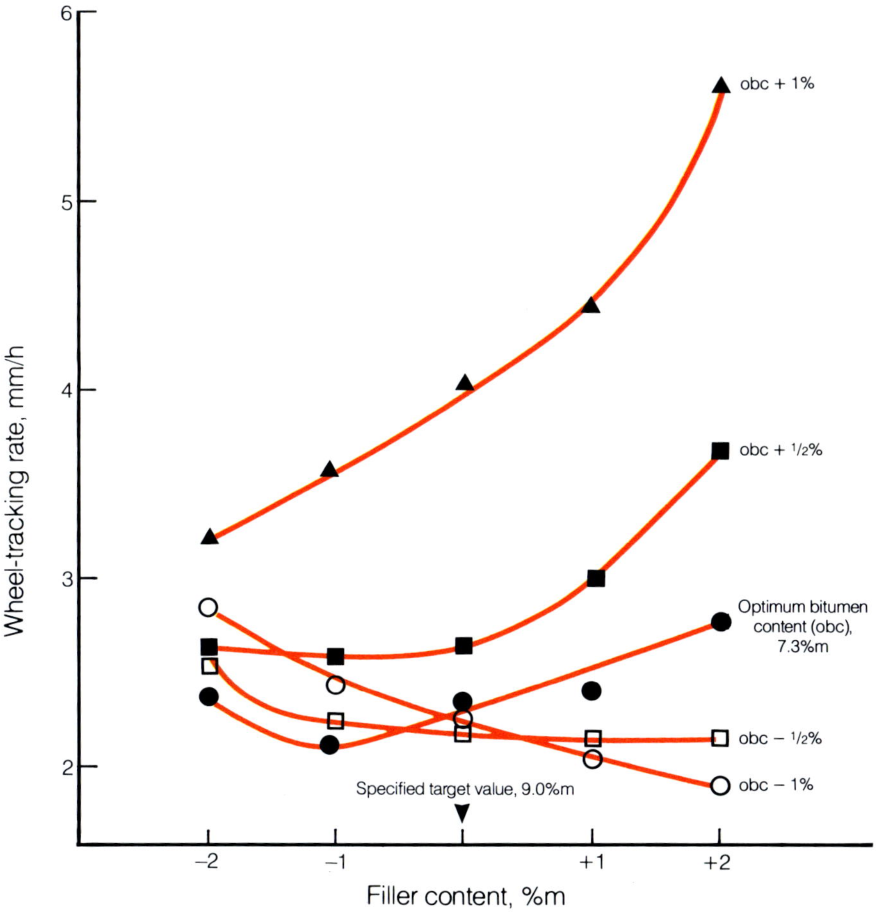

Figure 16.9 – The effect of bitumen and filler content on resistance to deformation[23]

16.6 An alternative approach to specifying Marshall properties

Marshall stability is the maximum load a specimen can accept before failure and the Marshall flow is the maximum diametral compressive strain measured at the instance of failure. The Marshall quotient, calculated as the ratio of stability to flow and thereby representing an approximation of the ratio of load to deformation under the particular conditions of the test, can be used as a measure of the material's resistance to permanent deformation in service[20,25,26,27]. However, accepting this, it is important to recognise that one drawback of using Marshall quotient to evaluate mixes is the precision, ie repeatability and reproducibility, of this parameter. As Marshall quotient is the resultant of dividing stability by flow, the precision of the quotient must be poorer than the precision of the stability and flow individually.

BS 594 : 1985 specifies maximum values of Marshall stability for each named traffic category, in addition to the customary specification of minimum values. It is sometimes said that mixes with a high stability are difficult to compact and lack flexibility. However where this is true it is because high stability has been accompanied by high mix stiffness. If Marshall stability is recognised for what it is, namely the ultimate load or strength of the mix, is there a logical reason for proposing a maximum value for any traffic category? However, there may be very sound reasons for placing limits on the mix stiffness, the most appropriate parameter for which in the Marshall test procedure is the quotient. In other words, if the objective of imposing a maximum Marshall stability is to ensure adequate workability during compaction, then in the absence of a reliable direct means of measuring workability, the Marshall quotient may give a more meaningful indication of mix stiffness during compaction. Similarly if the objective of specifying a maximum stability is to ensure an appropriate level of flexibility, where the structural and traffic conditions indicate this is required, then the Marshall quotient again is probably the most relevant parameter. Finally if a further control against excessively brittle behaviour is required then the specification of a minimum Marshall flow could be a more appropriate measure.

If it is accepted that the four relevant structural properties required of a bituminous mix are:

- resistance to permanent deformation;
- adequate stiffness as a contribution to pavement structural properties;
- adequate flexibility and freedom from brittleness in order to accommodate pavement deflections and avoid excessive cracking;
- resistance to fatigue;

then for a structurally thick pavement, as defined below, Lees[10] has proposed that these can be achieved by specification of minima for Marshall stability, flow and quotient, as shown graphically in figures 16.10 and 16.11. Figure 16.10 compares this proposal with the existing published methods of the Asphalt Institute and the U.K. Property Services Agency (PSA) in which no limits are placed upon quotient but which include specification of an upper limit on flow. Since flow is just the accumulated increments of deformation up to the point of failure, ie the stability load, it is reasoned that high stability mixes (which necessarily must accumulate more of these increments of deformation) are unfairly penalised by imposition of a maximum flow value. Insurance against excessively low stiffness and consequent risk of deformation, which removal of the flow maximum might otherwise permit, is obtained by inclusion of the specification of a minimum quotient. In figure 16.10 it is shown that by reliance upon a maximum flow criterion without consideration of the desirable Marshall quotient the present Asphalt Institute/PSA specifications reject some mixes which should be accepted and accept others which should be rejected, when judged in terms of their desirable engineering properties.

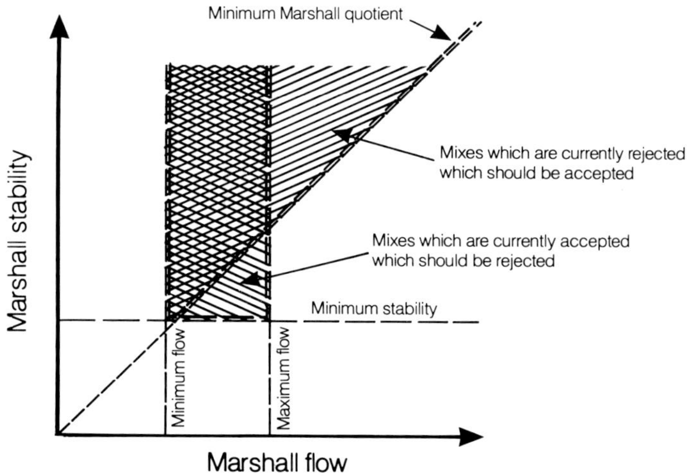

Figure 16.10 – Acceptance/rejection criteria from Marshall test parameters(10)

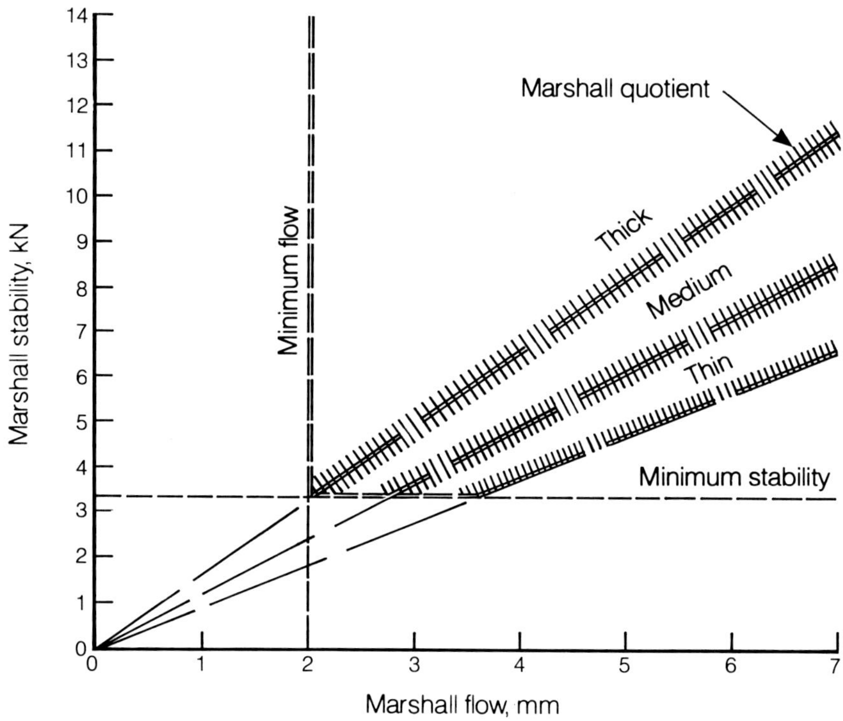

Figure 16.11 – Proposed Marshall test parameters for different pavement constructions, medium traffic category(10)

The current BS 594 : 1985 specification is given in table 16.4. The absence of a minimum Marshall flow specification permits the acceptance of mixes with relatively low strain at failure, ie this specification does not exclude brittle mixes. The inclusion of a higher maximum flow value for the highest traffic category acknowledges that the Marshall stability and flow values are related, but for whatever reason, BS : 594 : 1985 does not take the approach of adopting a minimum Marshall quotient.

Lees[10] recommends a combination of the U.K. and U.S.A. approach with further recommendations in relation to fatigue resistance. In controlled stress fatigue tests the longest fatigue life is obtained with the stiffest mixes, these having the lowest strain for the given repeated load applications, whereas, in controlled strain fatigue tests the stiffest mixes have the shortest fatigue lives. As a useful aid to practical design, Lees, following the work of Pell[28], and Monismith and Deacon[29], suggested that an asphalt layer greater than 100 mm thick could be compared in fatigue behaviour to the controlled stress test, in that the strain resulting from the applied stress will be a function of the stiffness of the mix. On the other hand with thin asphalt layers, say less than 50 mm thick, the deflection induced by a wheel load will be governed by the structural response of the underlying materials and will be largely independent of the stiffness of the thin bituminous layer. Under this circumstance the behaviour compares with the laboratory controlled deflection test where the most flexible materials, ie those of lower stiffness, show the longest fatigue life.

It follows from the above that for a thick bituminous pavement there is no conflict between deformation resistance and fatigue because both are optimised by high stiffness. Thus in terms of Marshall parameters only a minimum Marshall quotient need be specified. However, for a thin bituminous construction the *in-situ* condition more closely resembles the controlled strain test, Thus, in Marshall parameters a minimum Marshall quotient is required for resistance to deformation and a maximum Marshall quotient for flexibility. For pavements of intermediate thickness a compromise solution involving assessment of the relative magnitudes of the contributory factors discussed above is necessary.

This approach would result in three graphs of the form shown in figure 16.11, for light, medium and heavy traffic conditions. This approach forms a realistic basis for design criteria, both for road and airfield construction.

16.7 References

(1) British Standard 594 : 1950 : *Rolled asphalt, asphaltic bitumen and fluxed lake asphalt (hot process).*

(2) British Standard 594 : 1958 : *Rolled asphalt, asphaltic bitumen and fluxed lake asphalt (hot process).*

(3) British Standard 594 : 1961 : *Rolled asphalt hot process.*

(4) British Standard 1621 : 1961 : *Bitumen macadam with crushed rock or slag aggregate.* (Replaced by British Standard 4987 : 1973).

(5) British Standard 594 : 1973 : *Specification for rolled asphalt (hot process) for roads and other paved areas.*

(6) British Standard 598 : Part 3 : 1985 : *Sampling and examination of bituminous mixtures for roads and other paved areas. Part 3. Methods for design and physical testing.*

(7) Powell, W D : *Collaborative research on improved bitumen macadams.* A paper given at the ACMA seminar, Sutton Coldfield, pp 25-36, 1987.

(8) Nutt, P E: *The implementation of heavy duty macadam.* A paper given at the ACMA seminar, Sutton Coldfield, pp 37-47, 1987.

(9) The Asphalt Institute: *Mix design methods for asphalt concrete and other hot-mix types.* Manual series No 2, 1988.

(10) Lees, G: *Asphalt mix design for optimum structural and tyre interaction purposes.* Proceedings of the sixth international conference on the structural design of asphalt pavements, Ann Arbor, Michigan, vol 1, pp 414-417, 1987.

(11) Nijboer, L W: *Plasticity as a factor on the design of dense bituminous road carpets.* Elsevier Publishing Company, London, 1948.

(12) The Asphalt Institute: *Mix design methods for asphalt concrete and other hot-mix types.* Manual series No 2, 1984.

(13) Property Services Agency (Airfields Branch): *Standard specification clauses for airfield pavement works.* PSA specification, part 4. Bituminous surfacing, 1979.

(14) Akeroyd, F M L, Chipperfield, E H and Hoban, T W S: *Mix design of rolled asphalt.* Asphalt Technology, No 27, pp 42-50, July 1979.

(15) Department of Transport: *Specification for road and bridgeworks. Rolled asphalt wearing course.* Departmental Standard HD/2/79, London, 1979.

(16) Department of Transport: *Specification for road and bridgeworks. Rolled asphalt wearing course.* Advice Note HA/3/79, London, 1979.

(17) British Standard 594: Part 1: 1985: *Hot rolled asphalt for roads and other paved areas. Part 1. Specification for constituent materials and asphalt mixtures.*

(18) Brien, D: *Asphalt mix design.* Chapter 3 from Developments in Highway Pavement Engineering, Part 1, Applied Science Publishers, 1978.

(19) Brien, D: *The design of improved asphalt road mixtures.* A paper presented to the Society of Chemical Industry, London, October 1976.

(20) Brien, D: *Research in the design of asphalt.* The Highway Engineer, vol 24, pp 14-21, October 1977.

(21) Brien, D: *A design method for gap-graded asphalt mixes.* Shell Bitumen Review 56, pp 9-13, September 1977.

(22) Daines, M E: *The design of rolled asphalt: A full-scale experiment on A33 Winchester bypass.* Transport and Road Research Laboratory, Supplementary Report 22UC, 1972.

(23) Szatkowski, W S: *Rolled asphalt wearing course with high resistance to deformation.* Proceedings of an Institution of Civil Engineers conference on the performance of rolled asphalt road surfacings, pp 107-122, October 1980.

(24) Duthie, J L: *The Marshall test in the design of hot rolled asphalt wearing courses.* A paper presented to the Society of Chemical Industry, London, January 1974.

(25) Fabb, T R J and Heyes, J V: *Deformation of asphalt wearing courses.* Proceedings of the Canadian Asphalt Technical Association, pp 334-361, 1979.

(26) Choyce, P W, Lammiman, K A and Taylor, I F: *Resistance to deformation of hot rolled asphalt.* Highways and Transportation, vol 31, pp 28-32, January 1984.

(27) Jacobs, F A: *Hot rolled asphalt: effect of binder properties on resistance to deformation.* Transport and Road Research Laboratory, Report LR 1003, 1981.

(28) Pell, P S: *Fatigue of bituminous materials in flexible pavements.* The Highway Engineer, vol 8, pp 17-23, 1971.

(29) Monismith, C L and Deacon, J A: *Fatigue of asphalt paving mixtures*. Proceedings of the American Society of Civil Engineers 95(TE2), pp 317-346, 1969.

Chapter 17 – The design of bituminous pavements

The objective of pavement design is to produce an engineering structure that will distribute traffic loads efficiently, whilst minimising the whole-life cost of the pavement, ie both initial construction and maintenance costs. Pavement design is essentially a structural evaluation process, ensuring that the traffic loads are so distributed that the stresses and strains developed at all levels in the pavement and the subgrade are within the capabilities of the materials at those levels. It involves the selection of materials for the different layers and the calculation of the required thickness. The load-carrying capacity of a pavement is a function of both the thickness of the material and its stiffness. Consequently, the mechanical properties of the materials comprising the layers in a pavement are important for designing the structure. Climatic conditions influence the performance of the whole pavement. Moisture affects the subgrade, sub-base or unbound base and temperature affects the bitumen-bound materials; therefore it is essential that design methods take account of climatic conditions.

17.1 The structural elements of a bituminous pavement

Bituminous pavements have essentially three components as shown in figure 17.1:

- the pavement foundation;
- the roadbase;
- the surfacing.

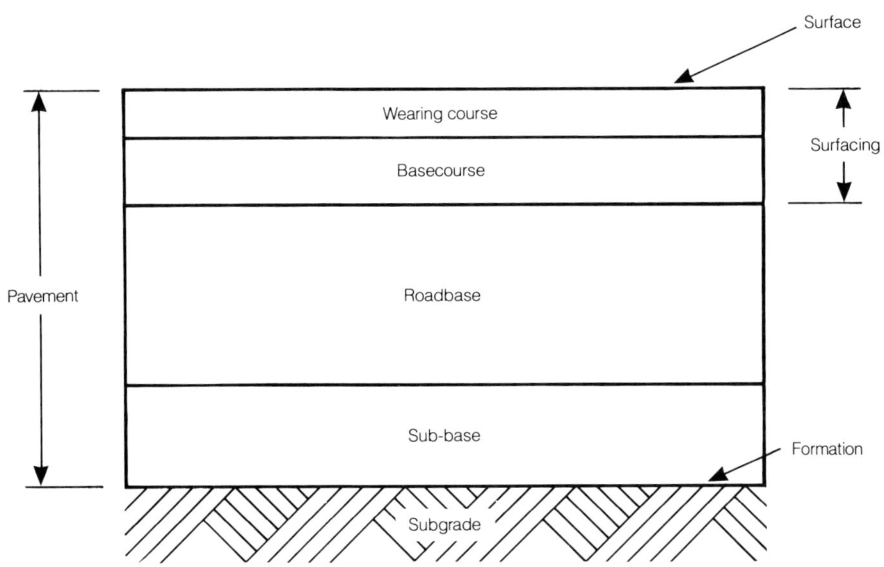

Figure 17.1 – Structural layers in a bituminous pavement

17.1.1 The pavement foundation

The foundation of a pavement is essentially comprised of two layers. The upper layer is termed the sub-base and is usually formed of good quality granular material. The lower section, the subgrade, is the natural soil or fill material which provides the surface upon which the pavement is constructed. The interface between these two layers is called the formation.

Additional layers are sometimes necessary. In cases where a thick sub-base is required it can be more economic if it is partially replaced by a thicker layer of intermediate quality. This is known as a capping layer. The capping layer may be provided by treating the natural soil with lime or cement, or by placing a layer of coarse aggregate.

The sub-base performs four main functions:

- it provides a structural layer which distributes loads to the subgrade;
- it provides a working platform for construction traffic and a compaction platform onto which bituminous materials can be laid and compacted;
- it acts as an insulating layer protecting the subgrade from frost;
- it provides a drainage layer to remove water from the pavement.

Provision of a free draining layer may not, however, be consistent with a layer that is dense and can withstand and transmit construction and traffic stresses. It is seldom necessary for material at this location to be free draining and therefore normal practice is to opt for strength through the provision of a well-graded, properly compacted sub-base.

17.1.2 The roadbase

The roadbase is the main structural element in the pavement. Its function is to distribute the imposed loading so that the underlying materials are not overstressed. It must resist permanent deformation and cracking caused by fatigue through repeated loading; it must also be capable of withstanding stresses induced by temperature gradients through the structure. The roadbase is usually a dense bituminous material, but wet lean concrete or cement bound material may also be used. For less heavily trafficked situations an unbound granular base may be adequate.

17.1.3 The surfacing

The surfacing comprises two layers. The lower layer is termed the basecourse. Its purpose is to provide a well shaped surface on which to construct the wearing course, and also to distribute the traffic loads over the roadbase. The basecourse is usually 60 mm thick.

The upper layer of the surfacing is termed the wearing course. This layer has to fulfil a formidable list of requirements:

resist deformation by traffic;

be impervious, therefore protecting the lower layers of the pavement. The exception to this is pervious macadam which is designed to be permeable (see chapter 11, section 11.4);

be durable, resisting the effects of weather, abrasion by traffic and fatigue;

provide a skid-resistant surface;

contribute to the strength of the pavement structure;

provide a surface of acceptable riding quality.

In most bituminous pavements the stiffness of the material in each layer is greater than that in the layer below and less than that in the layer above. An exception to this is the use of wet lean concrete or cement bound material as a roadbase. The stiffness of such a layer in its initial condition will be greater than that of the bituminous surfacing layers.

17.2 Factors involved in pavement design

17.2.1 Pavement loading

Loads are applied to the pavement through contact between vehicle tyres and the surface of the wearing course. Traffic loading is one of the major factors affecting the design and performance of pavements and it is both the magnitude and the number of loadings which contribute to overall damage to the pavement. The tyre pressure primarily affects the stresses (and strains) developed at the surface and within the surface layers of the pavement. (In road pavement design a typical commercial vehicle tyre pressure would be 0.5 MPa, whereas aircraft tyres use pressures up to 3.0 MPa, which has a significant influence on the design of the materials used as the surfacing of road and airfield pavements). The structural effects of multiple wheel configurations are also primarily confined to the upper layers of the pavement.

Although much publicity is given to the physical size and gross weight of commercial vehicles, it is the double wheel or axle load that is critical in pavement design and performance. Heavier vehicles are usually carried on a larger number of axles, thereby maintaining or reducing the axle load. The importance of traffic loading is recognised by most countries by the imposition of statutory limits for axle loads, for example(1):

Austria, Denmark, West Germany, Japan, Netherlands and Sweden	10.00 tonnes (98 kN)
U.K.	10.17 tonnes (100 kN)
European Community proposed limit	11.50 tonnes (113 kN)
Italy	12.00 tonnes (118 kN)
Belgium, France, Greece, Luxembourg	13.00 tonnes (128 kN)

However, these limits are difficult to enforce and are regularly and flagrantly ignored.

Normal traffic on conventional roads is mixed in composition and axle loads, and it is therefore necessary for design purposes to simplify the real situation by converting actual axle loads to an 'equivalent' loading system. This idea originated with the American Association of State Highway Officials (AASHO) road test in the U.S.A. thirty years ago(2) and was adopted in the U.K. with the publication of the third edition of Road Note 29 in 1970(3). The mixed axle load spectrum is converted to an equivalent number of standard axles, the standard adopted in the U.K. being 80 kN. This is based on equivalent damage to the pavement. Although the relative damage done by wheel loads of different magnitudes depends on a variety of factors, for design calculations normally the 'fourth power law' is used, viz:

$$\text{Equivalence factor} = \left(\frac{W}{80}\right)^4$$

where W = axle load in kN.

It has been shown(4) that on weak pavements the power could be as high as 6 if trafficked by heavy axle loads at or above the current legal limit. However, the value of 4 is a reliable figure for normal conditions.

Thus under normal conditions it takes 17 individual 40 kN axle loads to cause the same damage as a single application of an 80 kN axle load and one individual application of a 160 kN axle load induces 16 times the damage of the standard 80 kN axle load.

17.2.2 Environmental effects

The two environmental effects that influence pavement design are moisture, in particular the water table which influences subgrade performance, and temperature, which strongly affects the properties of bituminous materials.

The properties of subgrade soils should be determined for critical conditions existing after the construction of the pavement. The relationship between elastic modulus and moisture content of the subgrade is important. In addition frost can significantly affect soils and unbound pavement materials. By choosing materials having low susceptibility to frost and ensuring that any which might be so affected are located below the depth to which frost is likely to penetrate, frost damage can be minimised.

The effect of temperature on the stiffness of bituminous materials has been discussed in depth in previous chapters.

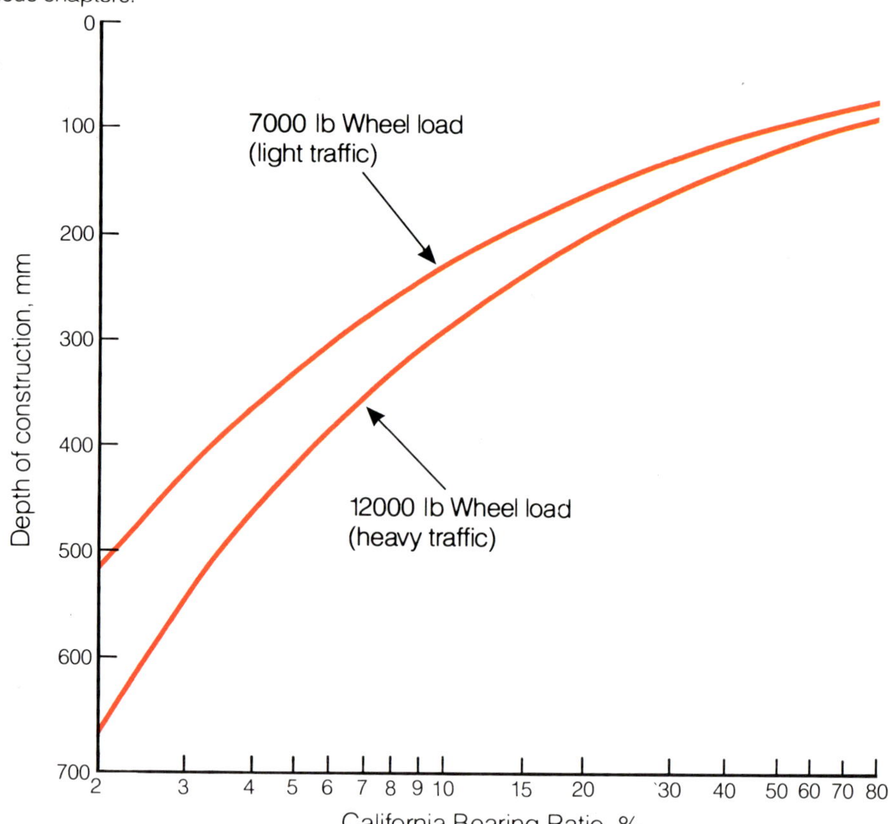

Figure 17.2 – California Bearing Ratio design curves

17.3 Pavement design

There are basically two approaches to pavement design: empirical and analytical. Most design methods in current practice around the world are empirical, being based on experience accumulated in practice and from specially constructed test sections. This approach to pavement design is somewhat removed from engineering principles and cannot cope with circumstances beyond those included in the trial section on which the method is based. In contrast the analytical approach uses theoretical analysis, mechanical properties of bituminous materials and is capable in principle, of dealing with any design situation.

17.3.1 Empirical and semi-empirical design methods

Design methods falling within this category relate the pavement thickness to the strength of the subgrade, proposed materials and anticipated traffic during the life of the pavement. One of the most widely used methods in this category is the California Bearing Ratio (CBR) method in which the strength of the subgrade and unbound granular materials is measured by the CBR test. In this test a circular piston 1932 mm^2 in area is driven into the material at a specified rate. The load acting on the piston is recorded and that corresponding to a penetration of 2.5 mm is determined. The ratio of that load to 1360 kg (the value obtained with a standard crushed stone sample) expressed as a percentage is the CBR value of the material. The total thickness of the pavement is related to the CBR of the subgrade for particular wheel loads by curves of the type shown in figure 17.2.

The Road Note 29(3) design procedure produced by the Transport and Road Research Laboratory (TRRL) is of the CBR type, the thicknesses of the layers of the pavement being related to the CBR of the subgrade and the type of the bituminous materials used. The results of the TRRL's many full-scale road trials have been used in the derivation of the design charts and tables contained in Road Note 29(5,6). Until the introduction of analytical design methods (as described below), most pavements in the U.K. were designed using Road Note 29. The method allows a pavement to be designed for a life of a selected number of years by assessing the cumulative number of commercial vehicles to be carried. This number is obtained by determining the traffic intensity at the time of construction and making an allowance for the anticipated traffic growth. A series of graphs given in Road Note 29, each of which is valid for one particular growth rate, gives the cumulative number of commercial vehicles in the slow lane. This value is converted to its equivalent number of 80 kN standard axle loads using load-equivalence factors from the Department of Transport's (DTp) technical memorandum H6/78(7) which have superseded the values given in Road Note 29.

The thickness of the sub-base is related to the CBR of the subgrade and to the cumulative number of standard axle loads expected during the design life of the pavement in a series of graphs of the type illustrated in figure 17.3. The thickness of the roadbase depends upon both the cumulative number of standard axles to be carried and upon the type of material to be used. The required thickness is obtained from graphs as shown in figure 17.4. A range of different materials can be used with charts provided for each material. The thickness of the surfacing depends upon both the type of roadbase material and upon the number of standard axles to be carried.

Increases in the damaging power of traffic necessitated addenda to Road Note 29: technical memorandum H6/78(7) in 1978 and TRRL report LR 910(8) in 1979. Further increases in the damaging power of commercial vehicles using major roads have occurred resulting in the most heavily trafficked roads being designed to carry up to 200 million standard axles. Road Note 29 was developed from observations on pavements that had carried less than 10 million standard axles, although it is generally agreed that Road Note 29 designs are satisfactory for pavements carrying up to 40 million standard axles.

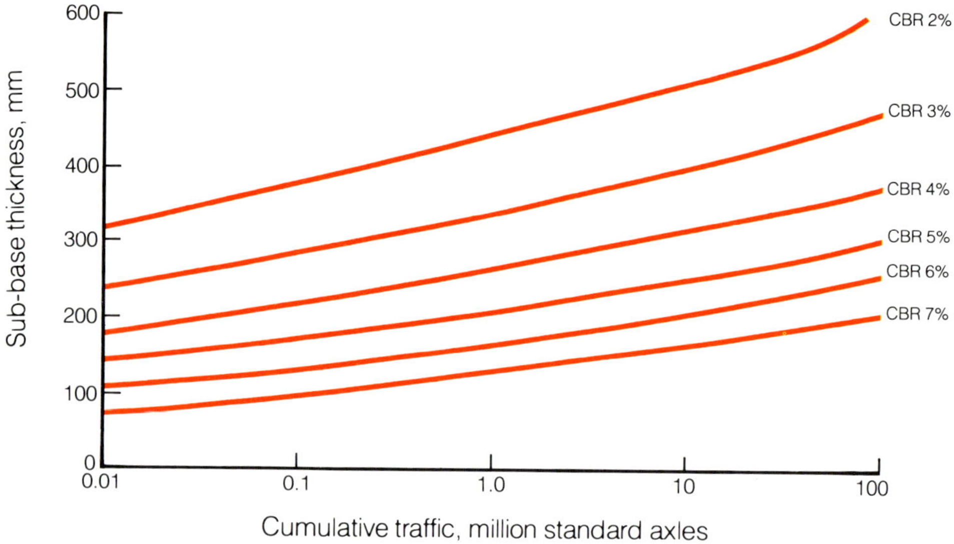

Figure 17.3 – Design curves for sub-base thickness(3)

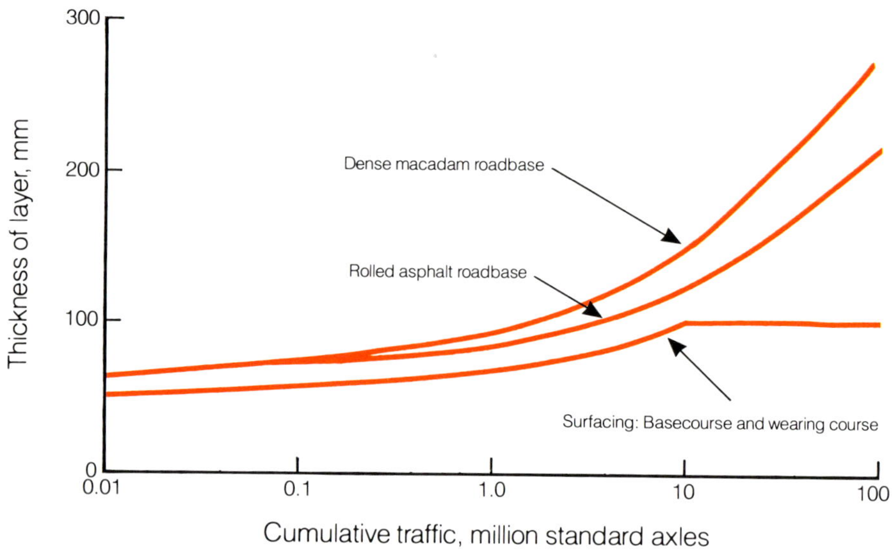

Figure 17.4 – Design curves for roadbase and surfacing(3)

Thus the empirical method of design used to develop Road Note 29 does not provide a satisfactory basis upon which to produce the pavement designs for traffic levels that are so far in excess of those observed on the experimental roads. In addition, this empirical method inhibits the use of new or improved materials or innovation in methods of construction. Road Note 29 does not include designs for reconstruction where the available depth may be limited and where the foundation is unlikely to be the same as in normal designs for new construction. However, it is in designing for heavy traffic and for reconstruction that new and improved materials are most likely to be needed.

To overcome these inadequacies the TRRL developed a design method that used the databank of information on the road experiments, but analysed the information using mathematical models of pavement behaviour and tested road materials and pavements in a systematic re-analysis of the behaviour of the road experiments. This work culminated in the publication of LR 1132(9) in 1984 which is the basis of the DTp's new design standard HD14/87(10,11).

LR 1132 contains design recommendations, derived from the structural performance of experimental roads, for pavements with different types of roadbase. The extrapolation from about 20 million standard axles carried by full scale road trials to 200 million standard axles has been based on analysis. The basic designs provide reference structures, having an 85% probability of survival, which can be modified, using the analytical approach, to cater for new materials or alternative designs. Thus the LR 1132 approach to pavement design recognises the variability in the road structure and the uncertainties of traffic forecasts. The concept of probability of survival certainly offers a challenge to the engineer who is aiming to achieve a minimum whole-life cost.

The design life of a road is the length of time between the beginning of the use of the road and the time when the road has deteriorated to a predetermined condition. In Road Note 29 the end of the design life was a rut depth of 20 mm or severe cracking. Strengthening of such a pavement did not always result in acceptable subsequent performance. In LR 1132 critical conditions are 10 mm rut depth or when cracks are first observed. Overlaying at the onset of critical conditions is recommended. A 20 year initial design period is normally recommended on the basis of minimum discounted costs.

In LR 1132 the estimation of the cumulative number of equivalent standard 80 kN axles to be carried during the design life is similar to the modified methods used as part of Road Note 29, but allowances are made for further increases in vehicle damage factors.

Subgrades that become wet during construction subsequently achieve lower equilibrium strengths than those which are kept dry. More precise allowances are made for drainage conditions. The subgrade is characterised for two functions. Firstly, so that the sub-base can be designed to carry construction traffic and secondly to establish the long-term, *in-situ* service conditions for design of the final pavement.

The thickness of the sub-base is based on the properties of the subgrade during construction. During the service life of the pavement, the sub-base makes some structural contribution and its thickness is taken into account in the final design stage, when the long-term *in-situ* properties of the subgrade are used, rather than the properties during construction. When working on subgrades with a CBR of less than 5% at the time of construction, a capping layer is normally required.

The re-assessment of the data collected from experimental roads has suggested that the overall performance of dense roadbase macadam and rolled asphalt roadbases, made to conventional recipe specifications, was very similar. The design curve adopted, shown in figure 17.5, provides an 85% probability that neither excessive surface deformation nor cracking will occur during the design period. The effect of design CBR on the thickness requirements is shown in figure 17.6 and this can be used to make adjustments between roadbase and sub-base thicknesses in those cases when the design CBR is not 5%.

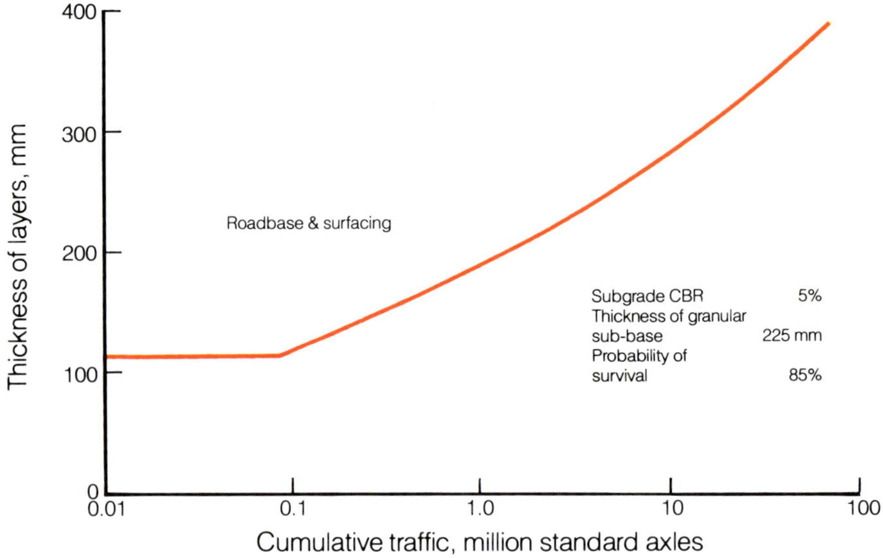

Figure 17.5 – Design curve for roads with bituminous roadbase[9]

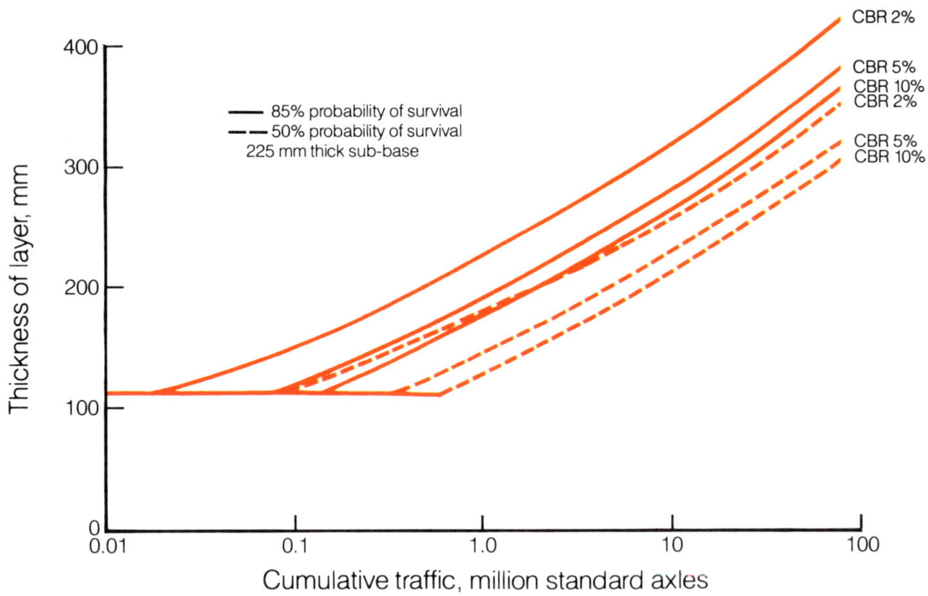

Figure 17.6 – Design curves for dense bitumen macadam and hot-rolled asphalt roadbase[9]

The standard designs illustrated in figures 17.5 and 17.6 cover most normal conditions, but for new situations the bituminous designs can be adapted using the analytical approach. One example of this is the use of rolled asphalt roadbase placed between the sub-base and upper roadbase as illustrated in figure 17.7, the objective being to place a mix with high resistance to cracking at the point of maximum tensile strain.

The approach used in LR 1132 is a combination of data from full scale road experiments and analysis, in which the field data was used to calibrate the theory; it is a semi-empirical design approach. LR 1132 suggests that the design criteria established may be used to modify the standard designs for alternative materials. It does not provide full details of the analytical model or full properties of some of the materials.

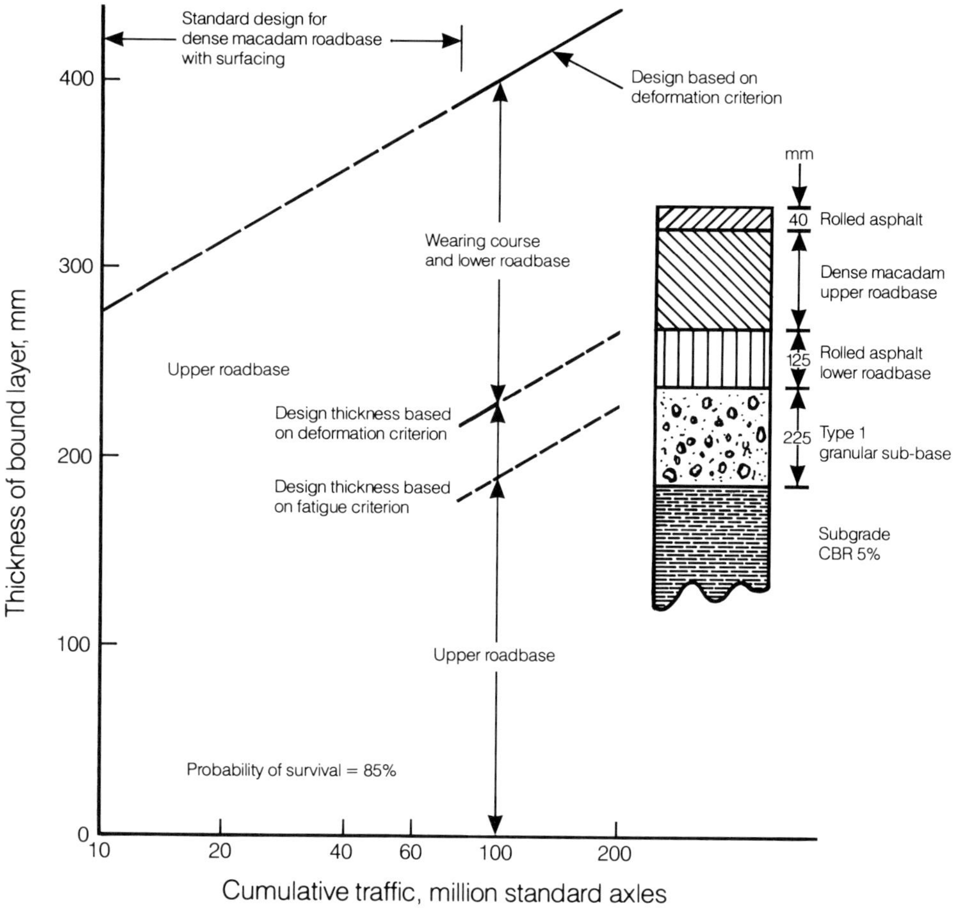

Figure 17.7 – Design for traffic in excess of 80 msa[9]

17.3.2 Analytical design methods

The philosophy of the analytical approach to pavement design is that the structure should be treated in the same way as other civil engineering structures. The basic procedure is as follows[12]:

(a) assume some form for the structure which will usually involve rational simplification of the actual structure to facilitate analysis;

(b) specify the loading;

(c) estimate the size of the components;

(d) carry out the structural analysis to determine the stresses, strains and deflections at critical points in the structure;

(e) compare these values with the maximum allowable values to assess whether the design is satisfactory;

(f) adjust the materials or geometry repeating steps (c), (d) and (e) until a satisfactory design is achieved;

(g) consider the economic feasibility of the result.

The development of analytical techniques has been the subject of considerable research effort in the last 25 years. Methods based on the use of theoretical analysis take several forms, those based on charts such as the Shell Pavement Design Manual[12,13,14] or computer based procedures such as those developed at the University of Nottingham[15].

In 1963 Shell International Petroleum Company Limited published a set of design charts[16,17] which were developed from structural analysis of a range of options. This was superseded in 1978 by the Shell Pavement Design Manual[13] (SPDM) and Addendum[18]. Although based on computer models of pavement structures these two documents also present the design method in the form of graphs, charts and tables so that it is accessible to engineers in the field who do not have access to sophisticated computing aids.

The SPDM is based on a three-layered structure consisting of an asphalt pavement on a base of unbound granular material overlying the existing subgrade (see figure 17.8). For convenience, the traffic loads imposed on the structure are converted to an equivalent number of standard 80 kN axle loads. The life of the structure is expressed as the number of standard axle loads applied to the structure before a given level of damage occurs. In this context the design method distinguishes between structural damage, ie fatigue failure of the asphalt and permanent deformation resulting from deformation of the subgrade (see figure 17.9), and plastic deformation of the bituminous layers.

The principal criterion for asphalt fatigue is the horizontal strain at the bottom of the bituminous layer, for subgrade deformation it is the vertical strain at the top of the subgrade. Cracking of the asphalt layer arises from repeated tensile strains, the maximum value of which occurs at the bottom of the bituminous layer, as shown in the figure. The crack, once initiated, propagates upwards causing gradual weakening of the structure. The development of the rut arises from the accumulation of permanent strain throughout the structure. The calculation of rut depth using analytical procedures is a complicated process; however, if the vertical strain in the subgrade, ε_z, is kept below a certain level, experience has shown that excessive rutting will not occur, unless the material is poorly designed or inadequately compacted.

The design problem, therefore, is to proportion the pavement structure so that, for the chosen material, the critical levels of strain will not be exceeded in the design life. To achieve this, the mechanical properties of the materials are required to analyse the structure. The simplified three-layered structure used as part of the SPDM is regarded as a linear elastic multi-layered system in

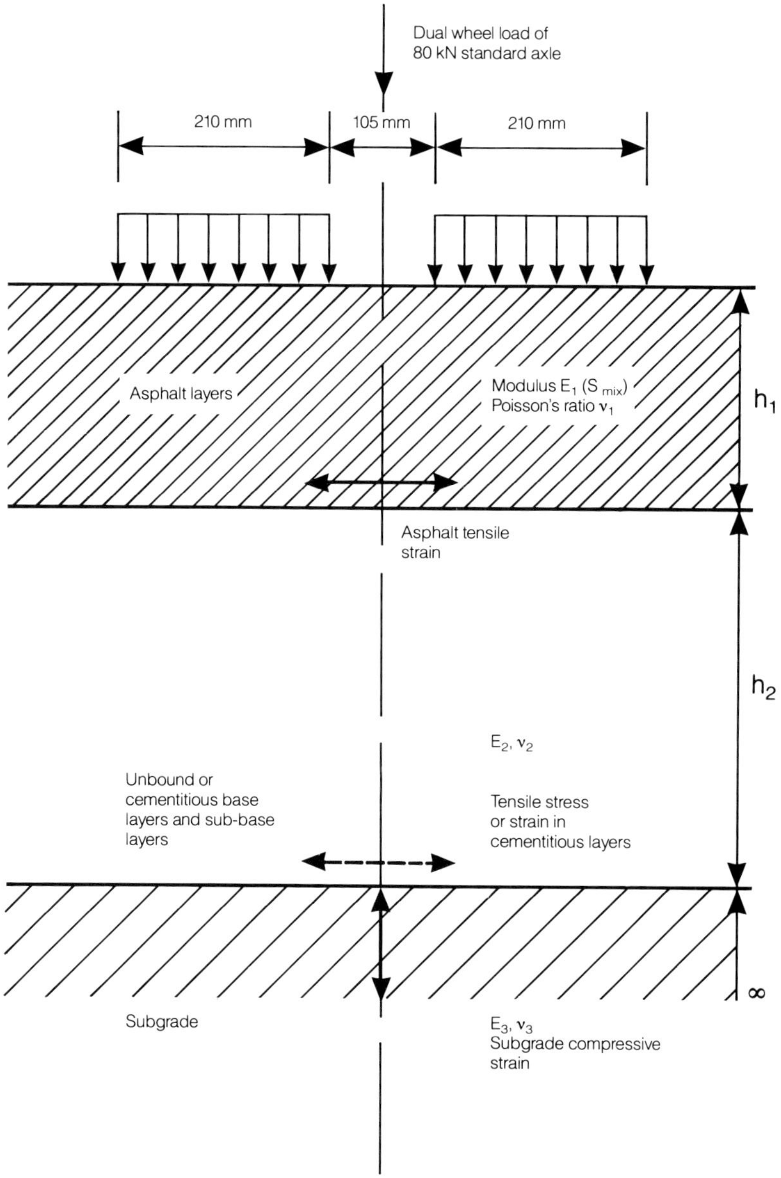

Figure 17.8 – Structural layers, with analytical design criteria[13]

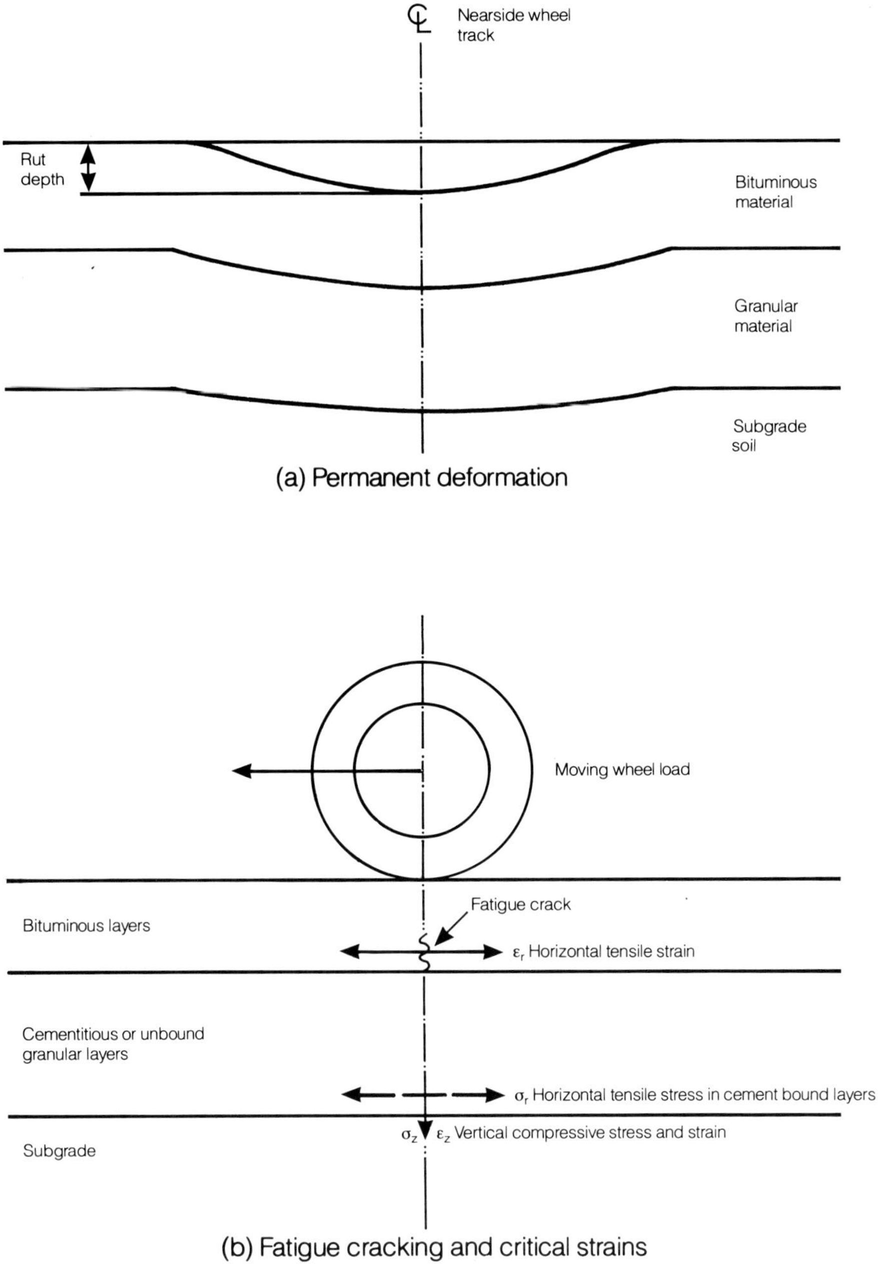

Figure 17.9 – Failure modes and critical strains in asphalt pavements[12]

which the materials are characterised by Young's modulus of elasticity and Poisson's ratio. The materials are assumed to be homogeneous and isotropic. It was on the basis of this model that in the late sixties Shell Research Limited developed the BISTRO (BItumen STructures in ROads) program which calculates stresses, strains and displacements at any position in a multi-layer system. This program has been enhanced and renamed BISAR[19,20] (BItumen Stress Analysis in Roads). Recent advances in computer technology have produced personal computers that are powerful enough to solve problems which until recently could be handled only on a mainframe computer. As a result Shell Research Limited have developed BISAR-PC[21] which is a truncated version of the mainframe program. BISAR-PC will calculate maximum tensile and compressive stresses and strains in the selected positions. The mainframe version of BISAR has the facility for handling a greater number of pavement layers and points where the stresses and strains can be calculated. However, BISAR-PC, should provide sufficient information for everyday pavement design and analysis. Using BISAR-PC the horizontal tensile strain at the bottom of the asphalt layer and the vertical compressive strain at the surface of the subgrade can be determined for various pavement configurations and these two values compared with allowable values.

Considerable effort has been made to reduce the number of variables involved in the analysis of pavements to keep the process as simple as possible. It has been shown that the modular ratio between a granular layer and its supporting subgrade is approximately 2.5, thus eliminating the granular modulus as an independent variable. The thick sub-bases specified in Road Note 29 for use on weak subgrades are not considered necessary for pavements having bituminous bound bases. Hence a standard sub-base thickness may be specified for all soils and is usually taken as 200 mm. Materials testing has shown that Poisson's ratio does not vary greatly and a value of 0.4 can be assumed for the asphalt layer and subgrade and 0.3 for the sub-base. These simplifications leave only three variables for any particular system, ie:

- elastic modulus of the bituminous layer;
- thickness of the bituminous layer;
- elastic modulus of the subgrade.

17.4 Improved roadbase and basecourse materials

In the U.K. improved roadbase and basecourse materials are required, particularly for the reconstruction of heavily trafficked roads where the available construction depths are often restricted by bridge clearance and drainage levels. To improve the performance of the pavement without increasing the depth of construction, improved materials are required. Appropriate design curves for these materials can be derived by the measurement of their properties in relation to those of the standard dense bitumen macadam roadbase. The most important property is that of the stiffness modulus, which strongly affects the critical strains generated in both the roadbase and subgrade.

One improved dense bitumen macadam roadbase is heavy duty macadam[22,23,24] (HDM). In this material the 100 pen bitumen is replaced by 50 pen bitumen and the target filler content is increased from 5% to 8%. A three-fold increase in stiffness is obtained by using the 50 pen bitumen and a further, but smaller, benefit resulted from increasing the filler content. Laboratory fatigue tests showed that the harder bitumen and higher filler content had an insignificant effect on fatigue life in comparison with the effect of increased stiffness, which reduces the strains induced at the bottom of the roadbase where fatigue cracking is assumed to originate. Tests using a pneumatic tyred wheel-tracking machine at 30°C showed that adding 3% more filler reduced deformation by about 40% in properly compacted materials.

During 1986, a number of proving trials were carried out on heavily trafficked roads using this material to identify any problems that might arise during manufacture or laying. The only significant difficulty experienced was that in some types of mixing plant longer mixing times were required than with conventional dense macadam due to the higher filler content. This reduced the output and increased the cost. Experience with this material has shown that increasing the bitumen content slightly can do much to overcome this problem without significantly affecting the performance of the mix.

Appropriate design thicknesses have been determined by the TRRL(23). For a given design life, the introduction of HDM in place of the conventional material leads to a reduction in design thickness, typically between 15% and 20%. This curve is included in the DTp's Departmental Standard HD 14/87(10). A specification for HDM is included in clause 930 of the DTp specification (25).

17.5 Bituminous substitution

In situations where construction depth is limited, for example during reconstruction, there may be an economical and structural advantage if some or all the sub-base is replaced by a structurally equivalent but thinner bituminous roadbase. Alternatively, an equivalent thickness could be used, thereby increasing the structural strength of the pavement. This form of construction is often referred to as 'full-depth asphalt', but may be more correctly described as asphalt substitution or bituminous substitution, as it may be neither necessary nor desirable to replace all of the granular sub-base.

Analytical techniques have been used to establish thicknesses of sub-base and bituminous roadbase that give equivalent performance. From this analysis it has been estimated that, for a given design life, 10 mm of dense roadbase macadam is equivalent to about 30 mm of type 1 sub-base material. This equivalence varies little with thickness and it generally agrees well with the empirical data collected from full-scale experiments. More authoritative designs will result from additional full-scale road trials which are currently underway.

One of the major concerns about total bituminous substitution has been the problem of achieving satisfactory compaction on relatively weak subgrades. However, work carried out by the TRRL(26,27) has shown that thick lifts of bituminous roadbases, particularly rolled asphalt, can be satisfactorily compacted even on a subgrade as weak as 3% CBR, which is too weak to support a paver or supply vehicles. In practice on subgrades weaker than 5% CBR a capping layer is normally applied or some other subgrade improvement carried out. Thus, there should be no difficulty compacting roadbase materials on CBRs greater than 5% or on capping material. One of the objectives of the current road trials is to assess the practical problems of carrying construction traffic on weak subgrades and of compacting the lower pavement layers under difficult and variable site conditions. The use of polymer geotextiles or geogrids appears to improve resistance to deformation both during construction and in service; however, the practicality and economic justification for using these materials has yet to be proven.

17.6 References

(1) Rhodes, A H: *Flexible pavements and bituminous materials.* Residential course at the University of Newcastle upon Tyne, section A, *Traffic loading*, pp A1-A21, September 1989.

(2) Liddle, W J: *Application of AASHO road test results to the design of flexible pavement structures.* Proceedings of the first international conference on the structural design of asphalt pavements, Ann Arbor, Michigan, pp 42-51, 1962.

(3) Road Note 29: *A guide to the structural design of pavements for new roads*. Third edition, HMSO, London, 1970.

(4) Addis, R R and Whitmarsh, R A: *Relative damaging power of wheel loads in mixed traffic*. Transport and Road Research Laboratory, Report LR 979, 1981.

(5) Leigh, J V and Croney, D: *The current design procedure for flexible pavements in Britain*. Proceedings of the third international conference on the structural design of asphalt pavements, London, pp 1039-1048, 1972.

(6) Thompson, R, Croney, D and Currer, E W H: *The Alconbury Hill experiment and its relation to flexible pavement design*. Proceedings of the third international conference on the structural design of asphalt pavements, London, pp 920-937, 1972.

(7) Department of Transport: *Road pavement design*. Technical Memorandum H6/78, 1978.

(8) Currer, E W H and O'Connor, M G D: *Commercial traffic: its estimated damaging effect 1945-2005*. Transport and Road Research Laboratory, Report LR 910, 1979.

(9) Powell, W D, Potter, J F, Mayhew, H C and Nunn, M E: *The structural design of bituminous roads*. Transport and Road Research Laboratory, Report LR 1132, 1984.

(10) Department of Transport: *Structural design of new road pavements*. Departmental Standard HD 14/87, London, 1987.

(11) Department of Transport: *Structural design of new road pavements*. Advice Note HA 35/87, London, 1987.

(12) Brown, S F and Brunton, J M: *An introduction to the analytical design of bituminous pavements*. Third edition, 1986.

(13) Shell International Petroleum Company Limited: *Shell Pavement Design Manual*, 1978.

(14) Mobil Oil Company Limited: *Asphalt pavement design manual for the U.K.*, 1985.

(15) Brown, S F and Brunton, J M: *Computer programs for the analytical design of asphalt pavements*. Highways and Transportation, vol 31, pp 18-27, 1984.

(16) Shell International Petroleum Company Limited: *Shell design charts for flexible pavements*, 1963.

(17) Dormon, G M and Edwards, J M: *Shell 1963 design charts for flexible pavements. An outline of their development*. Shell International Petroleum Company Limited, April 1964.

(18) Shell International Petroleum Company Limited: *Addendum to the Shell pavement design manual*, 1985.

(19) Peutz, M G F, Van Kempen, H P M and Jones, A: *Layered systems under normal surface loads*. Highway Research Record No 228, pp 34-45, 1968.

(20) De Jong, D L, Peutz, M G F and Korswagen, A R: *Computer program BISAR. Layered systems under normal and tangential surface loads*. Koninklijke/Shell Laboratorium Amsterdam, external report AMSR.0006.73, 1973.

(21) Koole, R C: *BISAR-PC – A personal computer version of Shell's multilayer pavement model*. Shell Bitumen Review 63, pp 9-11, April 1988.

(22) Hingley, C E, Peattie, K R and Powell, W D: *French experience with grave-bitume, a dense bituminous roadbase*. Transport and Road Research Laboratory, Supplementary Report 242, 1976.

(23) Leech, D: *A dense coated roadbase macadam of improved performance*. Transport and Road Research Laboratory, Report LR 1060, 1982.

(24) Nunn, M E, Rant, C J and Schoepe, B: *Improved roadbase macadams: Road trials and design considerations*. Transport and Road Research Laboratory, Report RR 132, 1987.

(25) Department of Transport: *Specification for Highway Works*. Part 3, series 900. Sixth edition, August 1986.

(26) Leech, D and Nunn, M E: *Substitution of bituminous roadbase for granular sub-base*. Third Eurobitume symposium, The Hague, pp 249-255, 1985. Also reprinted in Asphalt Technology, February 1986.

(27) Leech, D and Nunn, M E: *Substitution of bituminous roadbase for granular sub-base*. Transport and Road Research Laboratory, Report RR 58, 1986.

Chapter 18 – Surface dressing

Surface dressing is a well established, economical and highly effective method of maintaining the surface of roads. It is without doubt the most extensively used form of surface treatment, with on average approximately 170 million m^2 of road surface dressed per annum[1] in the U.K. County councils in England spend on average almost 25% of their structural maintenance budget on surface dressing. Surface dressing can be used successfully on roads of all types, from the country lane that carries only an occasional vehicle to trunk roads carrying tens of thousands of vehicles per day. It is a simple and inexpensive process; its very simplicity may, however, lead to inadequate planning, control and supervision, with the result that failures can and do occur. Although adverse weather conditions remain the major cause of failure, many of the problems encountered could be alleviated if the recommendations of Road Note 39[2] were more widely implemented. If the industry is to make surface dressing more 'socially acceptable' to motorists, it is essential that dressings are properly designed, and laid by skilled operatives (paying particular attention to traffic control and aftercare) to reduce as far as practicable the problem of flying chippings.

Normal surface dressing consists of spraying a thin film of binder onto the road surface, followed by the application of a layer of stone chippings (see photograph 18.1). The chippings are then rolled to promote contact between the chippings and the binder and to initiate the formation of an interlocking mosaic. The Marshall Report[3] stated that the object of surface dressing "is to seal porous surfaces against the entry of water, to bind surfaces at the onset of deterioration due to stone loss etc, to delay surface deterioration or to restore loss of skidding resistance (which may be due to loss of texture

Photograph 18.1 – Surface dressing *(Photograph reproduced by kind permission of Colas Roads Limited)*

depth, polishing of the aggregate or fatting-up)." This statement neatly encapsulates the three main purposes of a surface dressing, viz:

- to provide a durable, skid-resistant road surface;
- to seal the surface of the road against ingress of water;
- to arrest disintegration of the road surface.

However, it should be understood that surface dressing will not restore evenness to a deformed road, nor will it significantly contribute to the structural strength of the road structure.

18.1 Factors which influence the performance of surface dressings

The principal factors which influence the performance and the service life of a surface dressing are:

- *traffic*; the volume of commercial vehicles per lane per day governs the rate of embedment of chippings into the old road surface; a commercial vehicle is defined as a vehicle of unladen weight greater than 1.5 tonnes; vehicles of lower axle weight have no significant effect on embedment;
- *existing road surface*; the degree to which chippings will penetrate into the old road surface is related to the hardness of the old road surface and to the number of commercial vehicles in the traffic lane under consideration; where an old surface is porous, allowance must be made for the effective reduction in binder film thickness caused by penetration of binder into the underlying surface; severe cracks in the old road surface will ultimately reflect through the surface dressing;
- *size and type of chipping*; chippings should not be so small that they are rapidly embedded into the underlying surface nor so large that they may be dislodged by traffic; chippings should have strength characteristics and resistance to polishing appropriate to the road being surface dressed;
- *binder*; the function of the binder is to seal cracks and bind the chippings to the underlying surface; the viscosity of the binder must be such that it can wet the chippings adequately at the time of application, prevent dislodgement of chippings when the road is opened to traffic, and not become brittle during periods of prolonged low temperature;
- *rate of spread of binder*; the rate of spread of binder has to be such that there is sufficient binder to retain chippings during the early life of a dressing, without filling a proportion of voids in the layer of chippings in excess of that required to ensure adequate surface texture depth during the life of the dressing;
- *environmental conditions*; these may be related to location, climate or exceptional traffic conditions; the rate of cure of binders can be influenced by whether the surface dressing is laid in an exposed or sheltered area, eg under trees or bridges; the time of year in which surface dressing is carried out and the weather conditions during the early life of a surface dressing may have an important influence on performance; traffic may impose exceptional stresses due to braking, acceleration or turning movements at junctions or roundabouts, which increase the risk of chippings being dislodged, or by additional loading on long inclines, which increases embedment of chippings.

18.2 Design of surface dressings

18.2.1 The Road Note 39 design approach[2]

The guiding design principle in Road Note 39 is to select the size of chipping to offset the embedment caused by traffic into road surfaces of different degrees of hardness. The latest edition of Road Note 39 has refined the empirical approach used previously, which was based on experience gained from extensive road trials. This empirical approach allows the engineer to design the surface dressing for a particular site. Figure 18.1 shows the Road Note 39 design parameters in the form of a flow diagram; these comprise

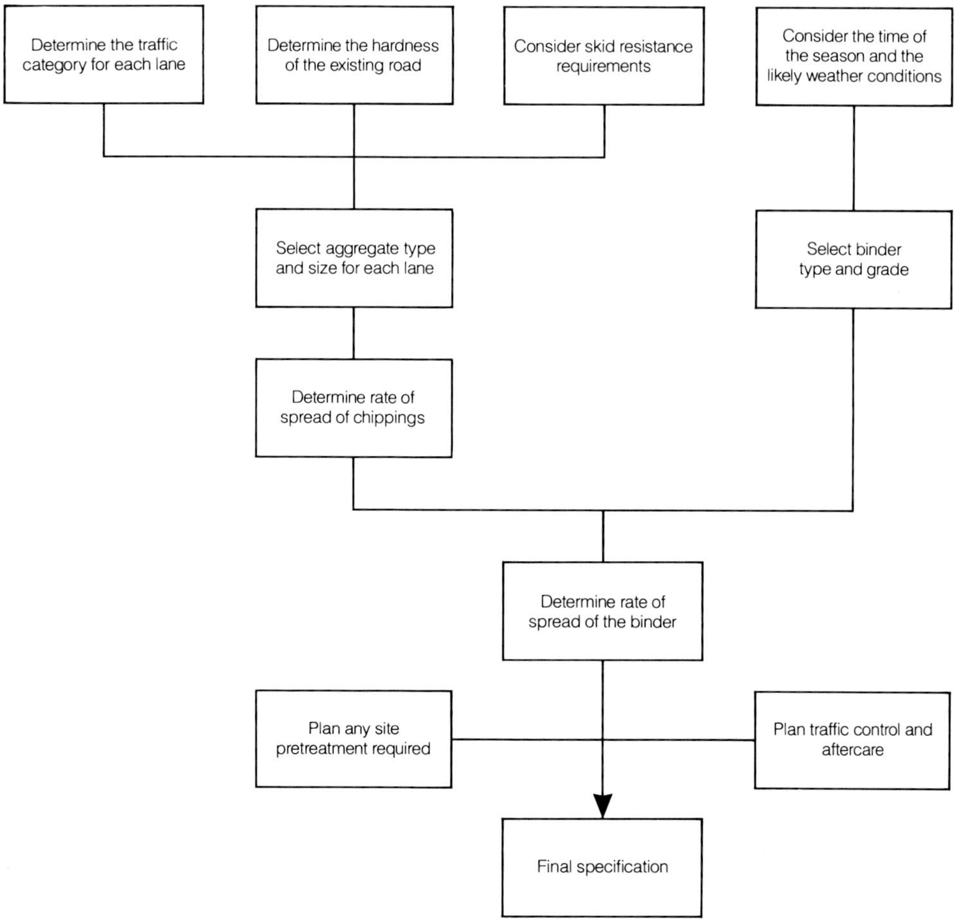

Figure 18.1 – Flow diagram for the planning and specification of surface dressings[2]

- relating the number of commercial vehicles (see table 18.1) and the hardness of the road surface (see table 18.2) to the size of chipping required; this is illustrated in figure 18.2;
- ensuring that both microtexture and macrotexture are maintained during the life of the surface dressing by selecting a chipping of appropriate size, polished stone value (PSV) and aggregate abrasion value (AAV); additional guidance on the selection of suitable aggregate to provide appropriate levels of skidding resistance for various classifications of road is given in the Transport and Road Research Laboratory (TRRL) Report LR 510[4];
- ensuring that the rate of spread of chippings is sufficient to cover the binder film after rolling, without excess chippings lying free on the surface of the dressing;

Category	Approximate number of commercial vehicles carried per day in the lane under consideration
1	2000 – 4000
2	1000 – 2000
3	200 – 1000
4	20 – 200
5	Less than 20

Table 18.1 – Categories of traffic[2]

Hardness category	Penetration at 30°C, mm	Classification of surface
Very hard	0-2	Surfaces such as concrete or exceptionally lean bituminous mixtures with dry stony surfaces into which there will be negligible penetration of chippings under very heavy traffic
Hard	2-5	Surfaces containing some hard bituminous mortar into which chippings will penetrate only slightly under heavy traffic
Normal	5-8	Surfaces into which chippings will penetrate moderately under heavy and medium traffic
Soft	8-12	Surfaces into which chippings will penetrate considerably under heavy and medium traffic
Very soft	>12	Surfaces into which even the largest chippings will be submerged under heavy traffic; such surfaces are usually rich in binder

Table 18.2 – Categories of road surface hardness[2]

- choosing a binder type and viscosity appropriate to the time of year, traffic category and degree of stress likely to be encountered; the application rate appropriate to the traffic category, road hardness and nominal size of chipping is determined from tables: tables 18.3a and 18.3b show the target rates of spread for cutback bitumens and emulsions respectively; these tables are only guides and must not be looked upon as specifications.

Type of surface	Lane traffic category									
	1		2		3		4		5	
	Nominal size of chipping, mm	Binder rate, l/m²	Nominal size of chipping, mm	Binder rate, l/m²	Nominal size of chipping, mm	Binder rate, l/m²	Nominal size of chipping, mm	Binder rate, l/m²	Nominal size of chipping, mm	Binder rate, l/m²
Very hard	10	1.1	10	1.1	6	1.1	6	1.2	6	1.4
Hard	14	1.0	14	1.1	10	1.1	6	1.1	6	1.2
Normal	20	1.0	14	1.0	10	1.1	10	1.1	6	1.1
Soft			Not recommended		14	1.0	14	1.1	10	1.1
Very soft	Conditions not suitable for surface dressing				20	1.0	14	1.0	10	1.0

Table 18.3a – Target rates of spread for cutback bitumen(2)

Type of surface	Lane traffic category							
	2		3		4		5	
	Nominal size of chipping, mm	Binder rate, l/m²	Nominal size of chipping, mm	Binder rate, l/m²	Nominal size of chipping, mm	Binder rate, l/m²	Nominal size of chipping, mm	Binder rate, l/m²
Very hard	Not recommended		6	1.3	6	1.5	6	1.6
Hard	14	1.5	10	1.3	6	1.3	6	1.4
Normal	14	1.4	10	1.3	10	1.3	6	1.3
Soft	20	1.3	14	1.2	14	1.3	10	1.3
Very soft	Not recommended		20	1.2	14	1.2	10	1.2

Table 18.3b – Target rates of spread for bitumen emulsions(2)

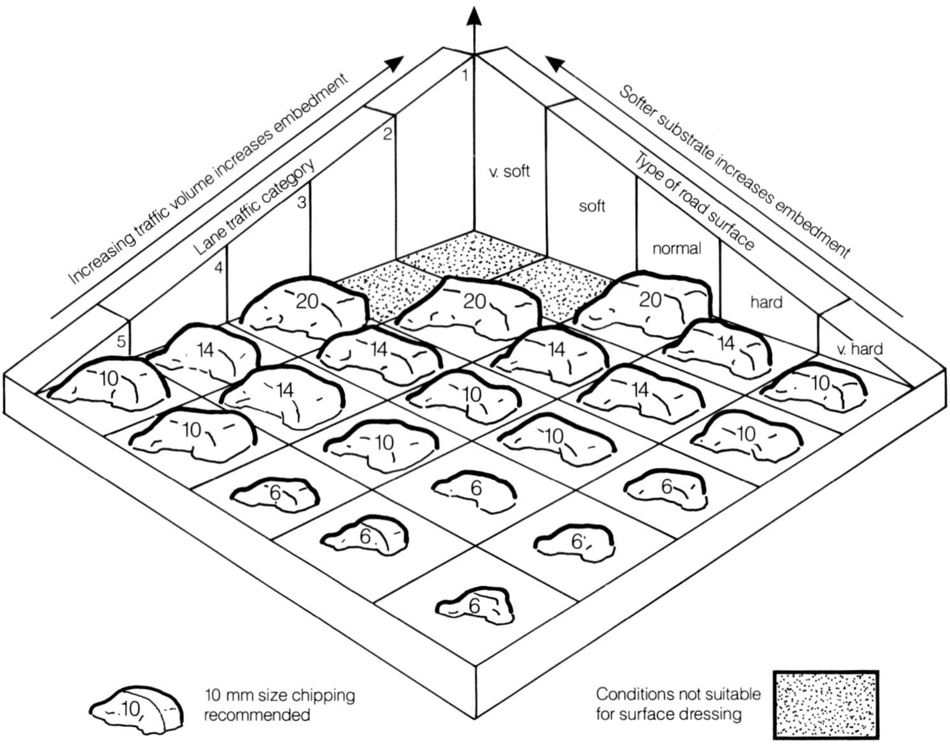

Figure 18.2 – The use of different size chippings to offset embedment as a result of traffic on substrates of differing hardness(2)

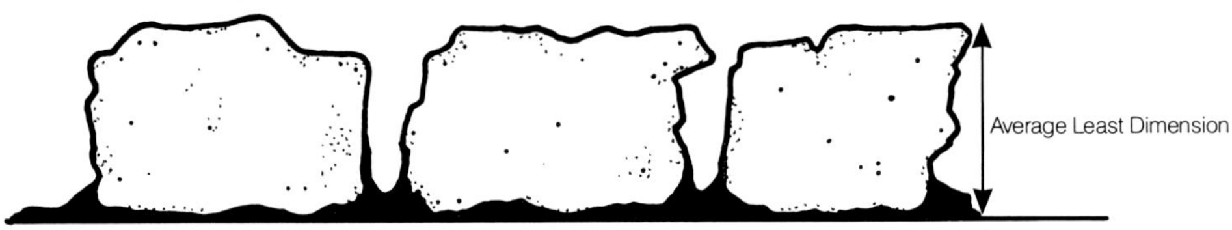

Figure 18.3 – Orientation of chippings after trafficking

Empirical design by definition is based on previous experience with particular conditions and therefore the method in Road Note 39 only allows the engineer to design conventional treatments for application on average sites. This approach may not produce the optimum and most cost-effective solution. To obtain the best value-for-money solution it is necessary for the engineer to appreciate and consider all the factors that affect the performance of the treatment for particular conditions and requirements. This requires the application of a more comprehensive rational design approach which must be based upon reliable data regarding site conditions, material properties, traffic characteristics and design objectives. It must also include financial analysis of alternatives, based upon comparison of the cost per unit area per annum of serviceable life.

18.2.2 An engineering approach to the design of surface dressings

In 1935 a New Zealand engineer, F M Hanson, developed an engineering approach to the selection of optimum rates of spread of binder and chippings for surface dressings[5]. Hanson's design procedure considers the volume of voids between the chippings when they are spread and rolled, and the orientation which the chippings adopt after trafficking. Hanson observed that, when trafficked, chippings tend to adopt a position in which their least dimension is vertical, thus giving rise to the concept of *average least dimension* (ALD) (see figure 18.3).

The ALD of a chipping can be determined by using the nomograph shown in figure 18.4. Two parameters are required for this nomograph – firstly the flakiness of the chipping and secondly the median size of the chipping. This latter value is the theoretical sieve size through which 50% of the

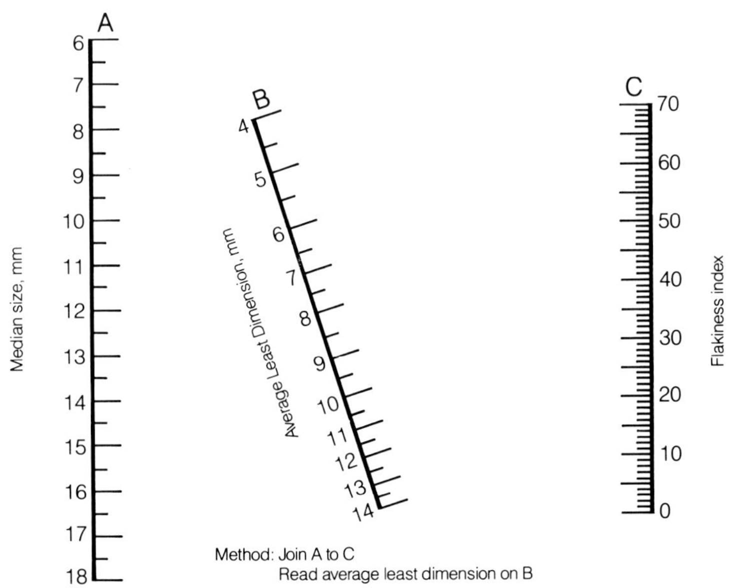

Figure 18.4 – Determination of Average Least Dimension (ALD)[5]

aggregate would pass. Alternatively the ALD can be measured directly by averaging the least dimension of 200 representative chippings.

The value of ALD is related to the nominal chipping size, viz:

Nominal chipping size, mm	*Typical ALD*, mm
20	11 – 15
14	8 – 10
10	5 – 6.5
6	2 – 4

Generally, higher ALDs mean greater packing together of the chippings, higher binder demand and possibly greater texture depth depending upon embedment.

Some fracturing of the chippings under rolling and traffic does not invalidate the ALD theory, as such fractures generally take place in a vertical direction, so that, even if a chipping breaks in two or three places, its depth on the road is not decreased. Clearly this is not the case with very soft stone which tends to pulverise, but this type of stone should not be used in surface dressing.

The space between the chippings after rolling, expressed as a percentage of the ALD volume of chippings spread on a smooth surface shoulder to shoulder, is dependent on the ALD of the chippings. ALD may therefore be used as a measure of the ability of the chippings to pack together. This space provides room for the binder volume and the design texture depth after allowance for embedment and chipping wear. Typically the space available varies from 15% to 40% of the ALD volume.

The concept of ALD has been further developed by Jackson(6) to include:

intensity of traffic;

condition of the existing road surface;

climatic conditions;

type of chippings.

Jackson allocated factors for each of the above four categories to characterise each site. These factors are given in table 18.4. The design binder and chipping application rates are determined by summing the four factors from table 18.4 and entering this value in figure 18.5. The intersection point of the ALD and the factor line gives the design rate of spread of binder (bottom scale). The intersection point of the ALD and the line AB gives the application rate for the chippings (top scale).

Traffic level		Type of chipping		Existing surface		Climate	
Vehicles/day	Factor	Description	Factor	Description	Factor	Description	Factor
0 – 100	+ 3	Round	+ 2	Untreated	+ 6	Wet and cold	+ 2
100 – 500	+ 1	Dusty	+ 2	Primed base	+ 6	Wet and hot	+ 1
500 – 1000	0	Cubical	0	Very lean	+ 4	Temperate	0
1000 – 3000	– 1	Flaky	– 2	Lean	0	Dry and hot	– 1
3000 – 6000	– 3	Pre-coated	– 2	Average	– 1	Very dry and very hot	– 2
6000 +	– 5			Rich	– 3		

Table 18.4 – Factors to be applied using the Jackson method for designing surface dressings(6)

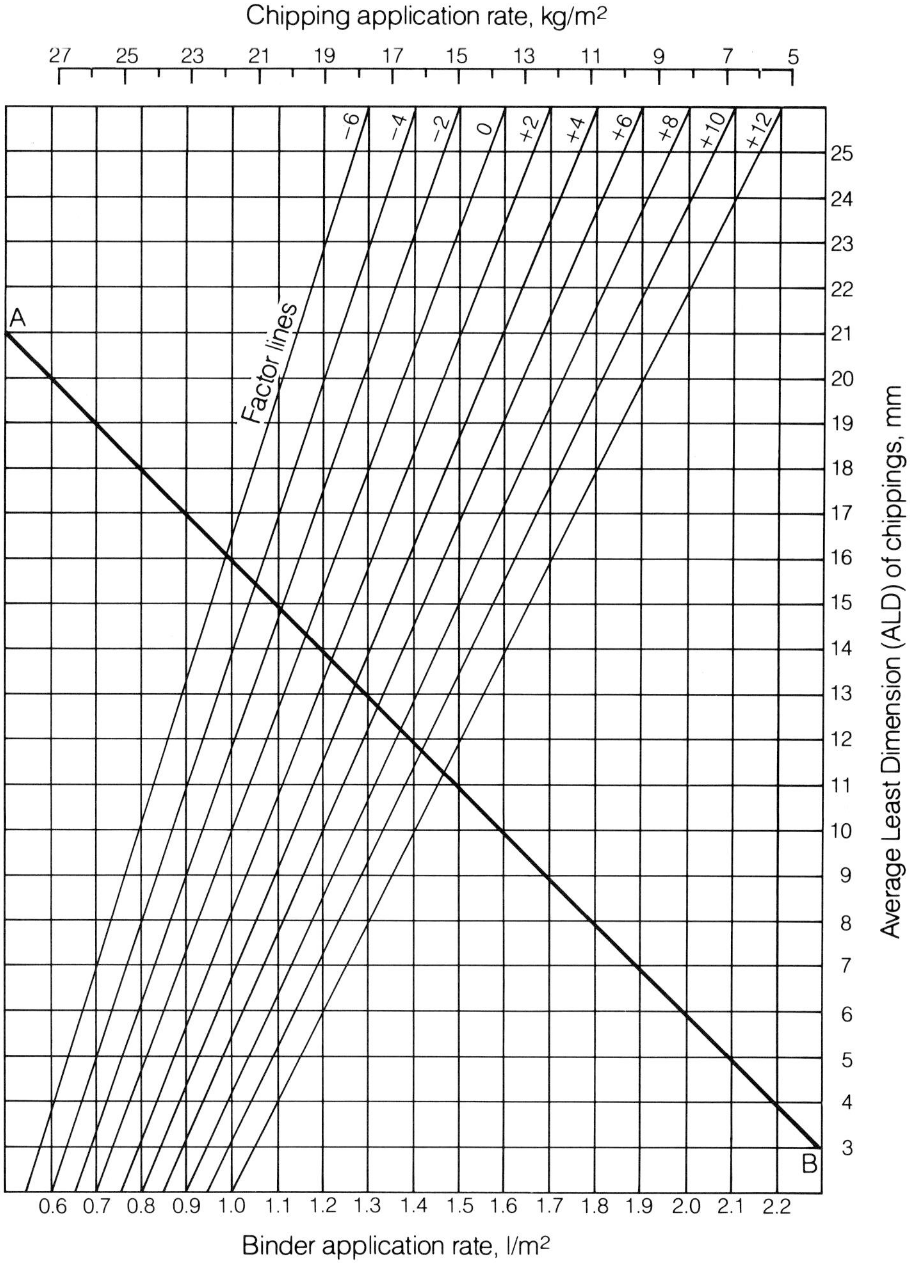

Figure 18.5 – Determination of design binder and chipping application rate – Jackson method[(6)]

The above principles form the basis of most rational design methods. However, there is a number of other factors which should be considered:

18.2.2.1 Embedment of chippings[7]

Embedment of chippings is a complex property that depends on a large number of variables, viz:

- *surface hardness*; the degree to which chippings will penetrate into the road surface is related to the hardness of the surface; the hardness of the road surface influences the choice of nominal size of chippings which must be used; the use of chippings which are too small will result in early embedment of the chippings into the underlying road surface leading to a loss of texture depth and in extreme cases fatting-up; the hardness of the road surface also influences the rate of spread of binder required for a given size of chipping; the rate of spread must be decreased where the road surface is soft, in order to compensate for the greater penetration of chippings into the surface under the action of traffic.

  The categories of hardness of existing road surfaces for the purposes of surface dressing design are set out in table 18.2[2]; in addition to describing each category of hardness, the table gives the ranges of penetration measured by the CTRA road hardness probe[8] which correspond to each category of hardness;

- *traffic loading*; it is possible to consider the effect of cars, but heavy commercial vehicles are by far the most significant factor; most chipping embedment occurs in the wheel-tracks and this is where surface hardness readings should be taken; as less embedment occurs outside the wheel-tracks, more voids are available for binder – typically 20% more;

- *road surface temperature*; if a section of road is colder throughout the year, less chipping embedment will occur; this explains the need for more binder in areas of shadow such as under trees or under bridges; in these locations typically 10% to 20% more binder is required;

- *vehicle speed*; the slower the vehicle the longer the chippings are loaded and as a result more embedment occurs; therefore less binder is required on steep inclines; typically the application rate on these locations is reduced by 10%;

- *tyre contact area*; this depends partly on tyre pressures, but also as embedment occurs a point is reached where the tyre touches the binder; this increases contact area and as a result the rate of embedment decreases.

Clearly chipping embedment is a complex property that depends on a large number of factors. A study by the TRRL of loss of texture depth on the M40 High Wycombe bypass[9] has shown that 60% of chipping embedment occurs in the first two years. The loss of texture on the site over twelve years is shown in figure 18.6.

18.2.2.2 Existing road surface texture[7]

This should be measured to allow assessment of binder required to saturate the existing surface. Surfaces that are dry and beginning to fret are an example of where this is important. The sand patch test can be used to provide a measure of the existing surface texture. Typically, binder application rates should be increased by approximately 30% above the texture depth, eg for a texture depth of 0.5 mm the application rate should be increased by 30% of 0.5, ie 0.15 l/m^2.

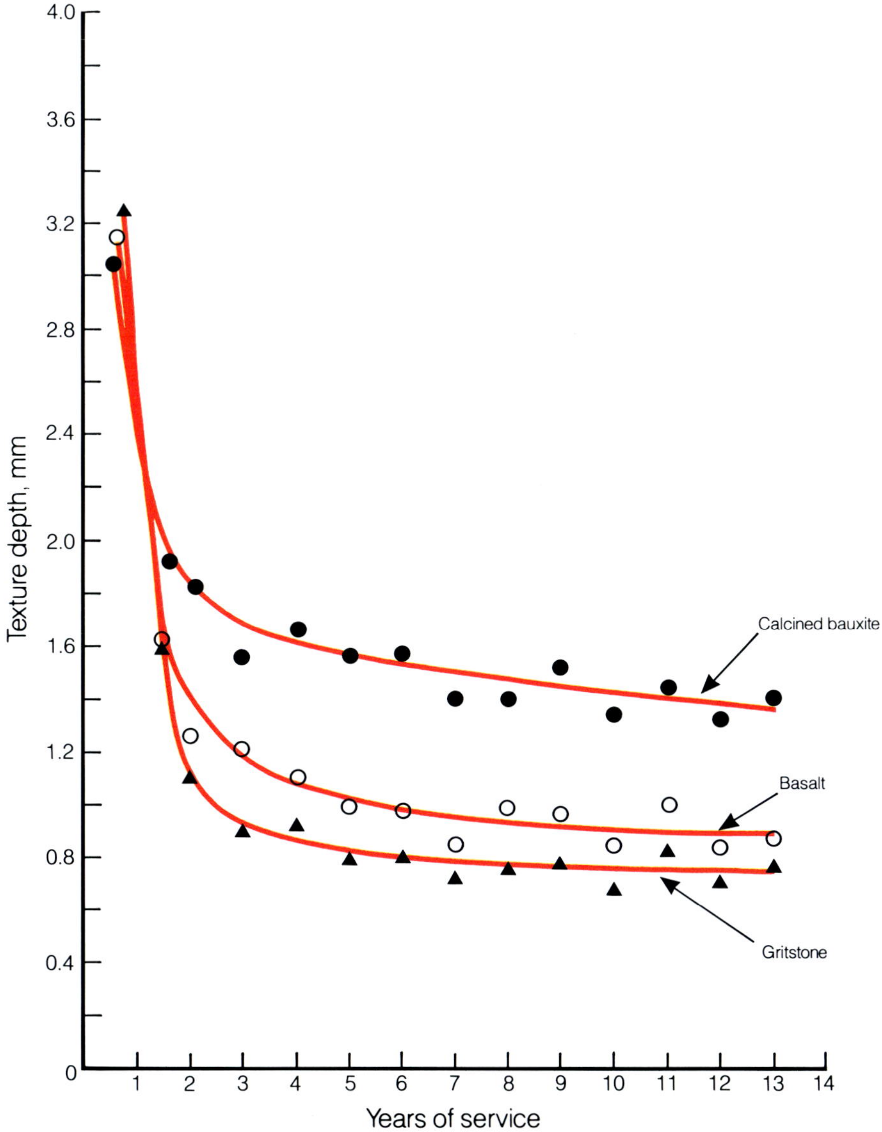

Figure 18.6 – Effect of aggregate type on the development of texture depth[9]

18.2.2.3 Surface texture and skid resistance

The skidding resistance of a road surface is determined by two basic characteristics; microtexture and macrotexture (see figure 18.9). Microtexture is the surface texture of the aggregate; high microtexture is necessary to enable the vehicle tyre to penetrate thin films of water and thus achieve dry contact. Macrotexture is the overall texture of the road surface; high macrotexture is necessary to provide drainage channels for the removal of bulk water between the tyre and the road surface.

At low vehicle speeds (up to 50 km/h) microtexture is the dominant factor; however, as speed increases, macrotexture controls skid resistance. To ensure that skid resistance at both high and low traffic speeds is maintained throughout the life of the dressing, the engineer specifies texture depth requirements to maintain macrotexture and PSV requirements to maintain microtexture.

Having taken account of all the factors and the relative economics, the engineer is then able to produce the most cost-effective solution:

- *chipping size*; based on expected embedment and required texture depth during the life of the treatment;
- *chipping application rate*; based on the shape of chippings (ALD) and how they are packed together; graphs are available to determine the chipping rate for chippings of particular ALDs;
- *binder rate*; based on the space available between the chippings and the existing road surface texture.

With the introduction of the micro-processor, it is possible to automate the design procedure. Such a method has been developed in Denmark and includes the automatic measurement of road surface hardness, surface temperature, and transverse vehicle distribution. These measurements can then be recorded on computer tape and fed into a computerised design process which takes account of the design factors previously discussed. The specification can then be fed into a programmable sprayer. Clearly the application of this type of technology adds to the cost of the surface dressing; however, if this reduces the risk of failure and produces dressings with longer lives, the expenditure is easily justified.

18.3 Types of binders used in surface dressing

The essential function of the bitumen is to seal the road surface and provide the initial bond between the chippings and the road surface. The subsequent role of the binder is dependent on the traffic stresses applied to the surface dressing from low stressed, low traffic country roads, to highly stressed sites carrying large volumes of traffic. Clearly a range of binders is necessary if optimal performance is to be achieved under these widely differing circumstances. Three broad ranges of bitumen-based surface dressing binders are available:

- traditional cutback and emulsified bitumens;
- proprietary polymer modified bitumens;
- epoxy-resin thermoset binders.

18.3.1 Traditional cutback and emulsified bitumens

18.3.1.1 Bitumen emulsions

The principal surface dressing binder is K1-70 bitumen emulsion. This is a cationic bitumen emulsion containing approximately 70% of 200 pen or 300 pen bitumen modified by the addition of approximately 1% by weight of kerosine. The relatively low viscosity of bitumen emulsions enables them to be sprayed at 75°C to 85°C through swirl jets. Emulsions are able to coat satisfactorily most aggregates, even through water films. However, dusty aggregates should be avoided because the dust can prevent the coalescing globules of bitumen from penetrating and wetting the aggregate. The relatively low viscosity of K1-70 allows the binder to flow on the road surface; thus it is important that the chippings are applied to the sprayed emulsion film as soon as possible. Once the chippings have been applied to the emulsion film, the time taken for the emulsion to break is a function of several variables: ambient temperature, humidity, aggregate water absorption, wind speed, the physio-chemical properties of both the emulsion and the aggregate and finally the mechanical forces applied to the surface dressing by rolling and traffic. Pre-coated chippings must not be used with emulsions. The chemistry and properties of emulsified bitumens are described in detail in chapter 4.

18.3.1.2 Cutback bitumen

This is usually 100 pen or 200 pen bitumen which is diluted with kerosine to comply with an STV viscosity specification[10], usually either 50 seconds, 100 seconds, or 200 seconds. The suffix X on the SHELPHALT range of cutback bitumens, eg SHELPHALT 100X indicates that they have been doped with a specially formulated thermally stable passive adhesion agent. This additive assists 'wetting' of the aggregate and the road surface during application and resists stripping of the binder from the aggregate in the presence of water. The effectiveness of this additive can be demonstrated by two laboratory tests; the immersion tray test[2] and the total water immersion test.

In the immersion tray test six clean 14 mm chippings are lightly pressed into a 1.5 mm film of cutback bitumen which is submerged in 25 mm of demineralised water at a temperature of 25°C. The chippings are left in contact with the binder for 10 minutes and are then carefully removed from the binder film and the percentage of binder retained on the aggregate is assessed visually.

Photograph 18.2 shows the percentage of binder retained on the aggregate when a conventional undoped cutback bitumen is used. This clearly shows that a significant proportion of the aggregate has not been 'wetted' by the binder and consequently the binder has not adhered to the aggregate. This photograph also shows that when the same test is carried out using SHELPHALT 100X, over 95% of the contact face is coated with binder, clearly demonstrating that the additive promotes 'wetting' of the aggregate even in the presence of water.

In the total water immersion test, developed by Shell Research, clean 14 mm aggregate is mixed with a quantity of cutback bitumen until 100% coating of the aggregate is achieved. After 30 minutes curing at room temperature the mix is submerged in demineralised water at 40°C. After 3 hours the pecentage of binder retained on the aggregate is assessed visually.

Photograph 18.3 shows three samples of coated aggregate:

top – coated aggregate before 3 hour soak;

middle – conventional cutback bitumen after 3 hour soak;

bottom – SHELPHALT 100X after 3 hour soak.

The photograph clearly shows that after soaking a substantial amount of the conventional cutback bitumen has been detached (or stripped) from the aggregate, whereas, with SHELPHALT 100X only a small amount of detachment has occurred, demonstrating that the additive not only promotes 'wetting' of the aggregate but also assists the binder to resist stripping in the presence of water.

However, it should be emphasised that if surface dressing is being carried out in marginal weather conditions or if the aggregate is wet it is recommended that an active wetting agent is added to SHELPHALT cutback bitumens immediately prior to spraying.

Photograph 18.2 – The immersion tray test

Photograph 18.3a – Coated aggregate before testing

Photograph 18.3b – Undoped cutback after testing

Photograph 18.3c – Doped cutback after testing

Photograph 18.4 – A modern surface dressing sprayer with rear wheel steering *(Photograph reproduced by kind permission of Colas Roads Limited)*

Cutback bitumens are sprayed through swirl or slot jets at temperatures between 140°C and 160°C. During spraying between 10% and 15% of the kerosine evaporates and a further 50% evolves from the surface dressing in the first few years after application.

18.3.1.3 Selection of binder

The choice between these two conventional binders is a balance between cost, initial adhesion, performance and ease of spraying. Excellent performance is obtained with both binders if they are applied correctly.

Generally, bitumen emulsions are more vulnerable during the early life of the dressing; however, they are more tolerant to damp chippings and as a result they are often preferred at the beginning and at the end of the surface dressing season, ie during April/May and August/September when the weather is variable and not suitable for the use of cutback bitumens. Thus, the day-to-day demand for emulsions during these two periods is also less predictable. As a result emulsion suppliers may increase emulsifier contents to improve storage stability. This, together with dusty chippings and cool weather conditions, may result in a semi-cured emulsion state often known as the 'cheesy' state. Here the bitumen droplets have agglomerated but not coalesced and the dressing is extremely vulnerable. Thus it is essential that the emulsion is completely broken before uncontrolled traffic is allowed on the dressing. This can be a serious disadvantage on heavily trafficked roads.

While previous editions of Road Note 39 had given little advice on application rates for bitumen emulsions, the 1981 edition contains detailed advice. However, after nine years experience with Road Note 39, it is widely believed that the rates of spread of binders given in the tables are about 10% low. When this increase has been implemented this should result in a reduction of failures due to insufficient binder. It is now accepted that K1-70 emulsion is suitable for category 2 roads, ie roads carrying between 1000 and 2000 commercial vehicles per lane per day.

Cutback bitumens have benefits in terms of immediate cohesive 'grip' of the chips, but generally the residual binder properties are more variable because of the differences in evolution of the diluent.

The choice between these two traditional binders is simply a case of 'horses for courses'.

18.3.2 Proprietary polymer modified binders

In Road Note 39 conventional surface dressings are considered suitable for sites carrying up to 4000 commercial vehicles per day. This level of traffic imposes high demands on the dressing and conventional systems have not always been sufficiently durable to withstand the stresses imposed. Although epoxy-resin systems are suitable for this intensity of traffic, the economic justification exists only for the most difficult sites, such as high-speed corners or approaches to traffic lights.

Thus there exists a 'gap' between these two alternatives which polymer modified bitumens can bridge[11]. The relationship between this gap and traffic category, type of chipping, etc, is shown in figure 18.7[11,12]. This gap in the surface dressing market has, at least in part, been the incentive for suppliers to develop proprietary polymer modified binders.

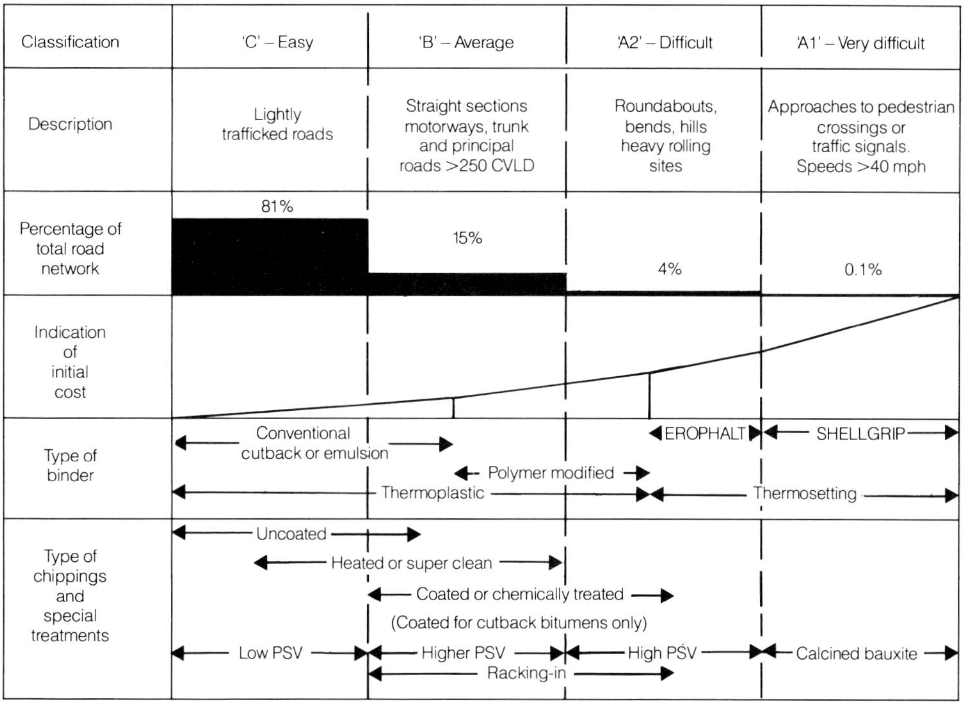

Figure 18.7 – Relationship between binder type and performance[11, 12]

In general the aim is to:

increase the number of sites suitable for surface dressing;

improve tolerance to climatic variations;

improve the temperature susceptibility characteristics of the binder;

improve the initial cohesive 'grab' of the binder to prevent flying chippings during early trafficking of the dressing;

reduce the risk of failure by construction or design shortcomings.

However, regardless of the improved properties of a particular binder, it is the performance of the aggregate/binder system and application mode which dictate the success or otherwise of the dressing.

The addition of polymers not only improves the initial cohesive strength of the binder but also lowers its temperature susceptibility, thus improving resistance to bleeding in warm weather and chipping loss in cold weather. The higher initial cohesive strength of the binder also allows a more rigorous suction sweeping procedure soon after rolling, which further reduces the chances of loose chippings and windscreen breakage.

The development of polymer modified surface dressing binders to complement conventional and epoxy-resin systems offers to the highway engineer a range of binders and binder/aggregate systems suitable for all categories of site (see figure 18.7). However, a rational approach is required for assessing the suitability of each for a particular site.

The University of Ulster[13] is currently investigating adapting the Sideway Force Coefficient Routine Investigation Machine (SCRIM)[14] to categorise sites, using variable wheel angles to assess changes in stress induced by turning, slewing, gradient changes and straight manoeuvres. From the data generated a 'stress contour map' of a site will provide highway engineers with the information necessary to make an engineering decision on which type of binder or binder/aggregate system is required.

The model site shown in figure 18.8[13] shows the site sub-divided into four sections, viz:

Section Ⓐ – evenly graded, straight road;

Section Ⓑ – steeply graded, including crawling and braking stresses;

Section Ⓒ – severe bend, causing slewing stresses;

Section Ⓓ – junction, turning braking and slewing stresses.

Obviously, section Ⓐ is subjected to much lower stresses than the other sections, which are subjected to shear stresses because of turning or braking stresses.

A solution to the widely differing stresses placed on the dressing at these different locations would be to select the binder or binder/aggregate system to match the imposed stresses, eg:

Section Ⓐ – conventional surface dressing;

Section Ⓑ – conventional surface dressing with increased binder application and where possible a larger chipping;

Section Ⓒ – polymer modified binder with racked-in chippings;

Section Ⓓ – epoxy-resin system.

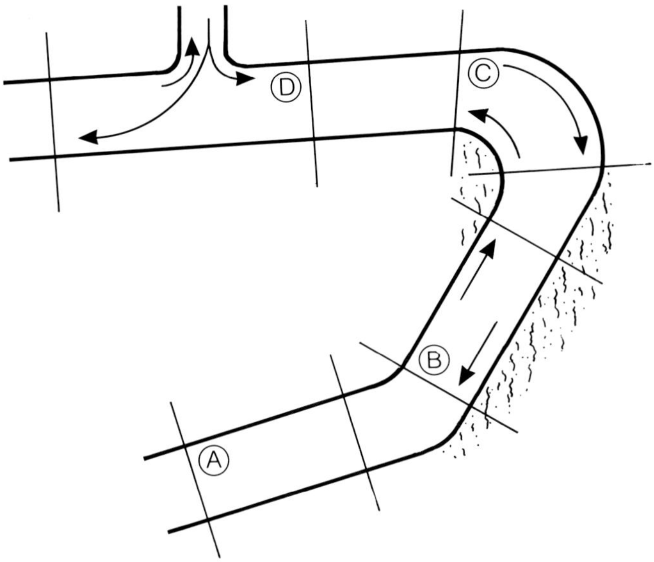

Section	*Description*
Ⓐ	Evenly graded, straight road
Ⓑ	Steeply graded, straight road
Ⓒ	Severe bend
Ⓓ	Junction with both braking and turning

Figure 18.8 – A model surface dressing site[13]

18.3.3 Epoxy-resin thermoset binders[15]

These are bitumen extended epoxy-resin binders used with high PSV aggregates such as calcined bauxite, for example SHELLGRIP and SPRAYGRIP. Epoxy-resin based binders are high performance systems that are fully capable of resisting the stresses imposed by traffic on the most difficult sites, for example roundabouts or approaches to traffic lights. These binders are classified as 'thermosetting', as the epoxy-resin components cause the binder to cure by chemical action and harden and it is not subsequently softened by high ambient temperatures nor by spillage of fuel. Thus the dressing acts as an effective seal against ingress of oil and fuel, which is particularly important on roundabouts where regular fuel spillages occur. With this type of binder very little embedment of the chippings into the road surface takes place and the integrity of the surface dressing is largely a function of the cohesive strength of the binder.

The extended life of this binder justifies the use of an exceptionally high-PSV and durable aggregate. The initial cost of this surface dressing system is high compared with conventional binders, but its exceptional wear resistant properties and the ability to maintain the highest levels of skid resistance throughout its life make it a cost-effective solution for very difficult sites by significantly reducing the number of skidding accidents.

Further work has been aimed at reducing the cost of the binder by using less resin and this has resulted in the development of EROPHALT. The rheological properties and curing rate of this binder differ from SHELLGRIP, and less expensive high-PSV aggregates are used instead of calcined bauxite. This binder shows great promise where high risk category sites justify the increased costs and channellised heavy traffic results in differences of embedment across traffic lanes.

18.4 Aggregates and chippings

18.4.1 Chipping selection

As stated earlier, the guiding design principle of surface dressing is to ensure that the size of chipping used on any particular road should be selected on the basis of anticipated embedment into the existing wearing course under the influence of predicted commercial vehicle traffic load. When specifying a nominal size of aggregate, consideration is given to road hardness, traffic intensity, weight and sometimes speed; often the shape of the chipping is ignored provided a maximum flakiness index is not exceeded.

The ALD (average least dimension) of the chipping is a much better measure of shape than the flakiness index alone. It takes account of both the flakiness index and the grading of the chippings and allows the engineer to predict texture depth, space available for the binder, and the rate of chipping application necessary for a particular shape of chip. The mechanical interlock between the chippings, which is so important for surface treatments, can also be assessed.

The colour of the aggregate is also important not only for defining traffic lanes but also for night driving. The reflectance of a chipping can show great advantage in urban areas where lighting is so important for safety.

18.4.2 Skid resistance

The skid resistance of a road is determined by the microtexture of the aggregate and the macrotexture of the road surface itself (see figure 18.9).

The microtexture, or degree of polish of the exposed aggregate, is primarily dependent on the petrographic characteristics of the aggregate and the intensity of traffic. The resistance to polishing of the aggregate is measured by the PSV test[16]. This test subjects the aggregate to simulated trafficking in an accelerated polishing machine. A high PSV indicates good resistance to polishing. The majority of road aggregates have a good microtexture prior to trafficking; consequently most road surfaces have a high skid resistance when new. Under traffic the exposed aggregate surfaces rapidly become polished and within a year the skid resistance falls to an equilibrium level.

There are many sources of high PSV aggregate in the U.K., mostly gritstones. Unfortunately, high PSV tends to be associated with poor abrasion resistance or poor resistance to weathering and as a result only a few of the high PSV aggregates are suitable for surface dressing. Thus, blanket specifications imposed by some local authorities for all sites, are a waste of valuable resources, and more thought should be given to the type of site to enable more abundant lower PSV aggregates to be used effectively. Catt of Warwickshire county council has developed a rational approach[17] to the selection of suitable aggregate types. The method takes into account all the factors relating to skid resistance and can demonstrate the suitability of aggregates which have previously been considered unsuitable.

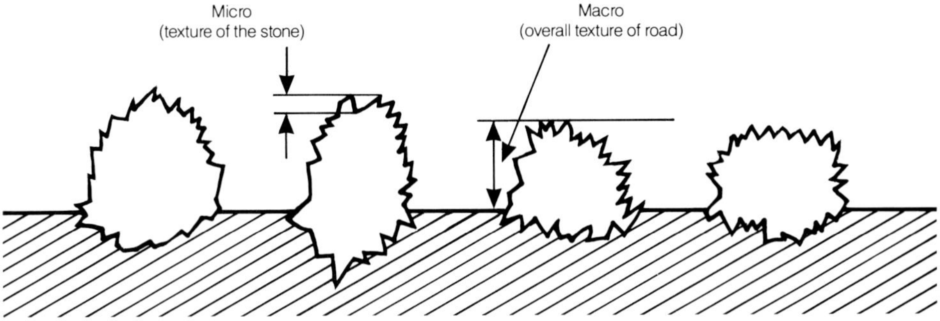

Figure 18.9 – Microtexture and macrotexture of chippings

The macrotexture is determined by the nominal size of aggregate used. However, resistance to abrasion is also important. Aggregates with poor abrasion resistance, (determined by the aggregate abrasion value (AAV) test[16]) will quickly be worn away, with a consequent loss of macrotexture. Maximum AAVs for various traffic categories are specified in Department of Transport (DTp) Technical Memorandum H16/76[18]. The macrotexture of the road surface is measured using the sand-patch test[19]. In this test a known volume of sand is spread onto the road surface to form a circular patch so that the 'valleys' are filled to the level of the'peaks' (see figure 18.10). The 'texture depth' of the road is determined by dividing the volume of the sand by the cross sectional area of the patch, viz:

$$Td = \frac{40\,v}{\pi \times d^2}$$

where Td = texture depth, mm
v = volume of sand, ml
d = diameter of the patch, cm

British Standard (BS) 598: Part 3: 1985[19] recommends that 50 ml of sand is used; for this fixed volume of sand the equation becomes:

$$Td = \frac{636.6}{d^2}$$

As stated earlier, at low vehicle speeds (up to about 50 km/h) the influence of the microtexture is predominant. As the vehicle speed increases, the macrotexture becomes increasingly important. The rate at which skid resistance falls off with increasing vehicle speed is influenced by the macrotexture. The TRRL[20] has shown that there is a correlation between macrotexture, expressed in terms of average texture depth, and skid resistance, measured by percentage decrease in braking force coefficient (BFC) between 50 km/h and 130 km/h, viz:

Texture depth, mm	*Percentage reduction in BFC,* %
0.5	30
1.5	10
2.0	negligible

Mature surface dressings in good condition typically have texture depths between 1.5 mm and 2.0 mm. Thus the high-speed skid resistance of the surface dressing is maintained throughout the life of the dressing.

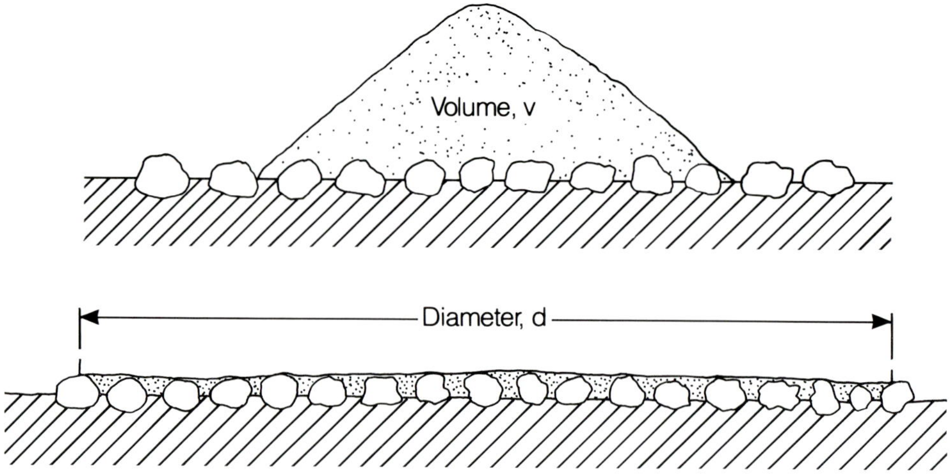

Figure 18.10 – The 'Sand Patch' test(19)

Photograph 18.5 – The texture of a good surface dressing after 5 years service.

It has been shown[21] that there is a strong correlation between skidding resistance (measured by sideway force coefficient, SFC), PSV and intensity of traffic. Defining the relationship between these three variables has provided the rational basis for the selection of suitable aggregates for the road surface. In 1972 the TRRL[4] introduced the concept of 'risk rating' which is a measure of the relative skidding accident potential at an individual site. Within each category there is a range of target SFC values, the target value for each site depending upon its risk rating. The SFC values are not the lowest-at-any-time values used in earlier standards but the lowest acceptable 'mean summer' values.

This concept was incorporated into Technical Memorandum H16/76[18] which specifies the aggregate properties required for surfacings on new trunk roads. The PSV levels specified in the memorandum will give the SFC values corresponding to the lowest risk rating in each site category in LR 510.

18.4.3 Binder/chipping adhesion

Good adhesion of the binder to the aggregate is essential if premature stripping failures are to be avoided. The risk of this type of failure can be significantly reduced if *clean* dry chippings are used. Consideration should be given to the use of double washed chippings to reduce the possibility of failure occurring due to dusty chippings. This is particularly important for emulsions, as the breaking mechanisms of the binder are influenced by the polarity and water absorption of the aggregate used. Alternatively, if a cutback bitumen is used, the problem of dusty chippings can also be overcome by lightly pre-coating the chipping with a 50 pen bitumen. The presence of the coating on the chipping aids the subsequent bonding of the chipping to the hot binder on the road, thus significantly reducing the possibility of a stripping failure. It is important that the chippings are only lightly coated, excess bitumen results in chippings which stick and will not flow. Partial coating and good flow characteristics are far more important than total coating which is often accompanied by poor handling characteristics. Pre-coated chippings must not be used with emulsions.

BS 63[22] lays down precise limits for the amount of fine dust that can be present on chippings supplied for surface dressing; however, dusty chippings are still the cause of a number of failures each year.

There are various theories regarding the mechanism of adhesion and the relative importance of mechanical forces, interfacial energy, surface texture of the aggregate and chemical reactions which may occur. It has, however, been shown conclusively that wetting agents considerably improve binder/aggregate adhesion. A number of proprietary wetting agents is available, which, although acting in a variety of ways, fall into one of two classes: passive or active wetting agents.

Passive wetting agents improve the adhesion characteristics of the binder and provide an improved long-term resistance to subsequent detachment of the binder film in the presence of water.

Active wetting agents modify the surface tension at the binder/aggregate interface and allow preferential wetting of chippings by the binder in the presence of water. However, their effectiveness deteriorates after a few hours at elevated temperature and they are, therefore, generally mixed with the binder immediately before spraying.

Wetting of clean dry aggregate by a binder, at a suitably low viscosity, provides an adhesive bond which generally exceeds the cohesive strength of the binder film. Wetting of the aggregate surface is believed to be principally by capillary action into cracks and pores and along surface microscratches. This is opposed by viscous forces, the time, t, taking to wet a given surface whose dimensions constitute a constant, Z, being given by:

$$t = \frac{ZV}{S \cos A}$$

where V = the viscosity of the binder
S = the surface tension of the binder
A = the contact angle on the solid surface.

Thus, if A is less than 90°, the time to wet the surface will be finite and spontaneous spreading can occur. If A is 90° the time will be infinite and there will be no spreading.

A study of binder/aggregate adhesion(23) showed that the contact angle of binder on an aggregate reduced with increasing viscosity. Thus, there is a critical temperature below which rapid wetting will not occur. It is for this reason that at low ambient temperature (less than 10°C) there is a significantly greater risk of premature adhesion failure. Consequently, chipping application must take place quickly so that the binder temperature is above the critical value when contact with the chipping occurs. Heated chippings will reduce the viscosity of the binder locally and enable rapid wetting to take place.

If chippings are damp, little wetting will occur (other than when emulsions are used) until they are reasonably dry, by which time the binder temperature is well below the critical adhesion temperature. Therefore other mechanisms of adhesion must be utilised. Rolling partly overcomes viscous resistance to spreading of the binder into the microtexture of the aggregate. Thus, under these conditions more emphasis must be placed on additional rolling and aftercare to promote wetting and adhesion before the dressing is opened to fast moving traffic.

18.5 Surface dressing equipment

Improved design of surface dressings and advances in binder technology and aggregate quality have been major contributory factors towards improving the performance of surface dressings. Advances in the design of application machinery in harmony with the above has enabled the whole process to be co-ordinated with respect to speed of operation, ability to comply with specification and continued improvement to standards of safety.

18.5.1 Sprayers

The uniformity of transverse distribution has dramatically improved over the last ten years; however, further improvements can be made, especially at the ends of the bar and at the centre of an expanding bar, where respectively heavy and light application problems can occur. The current distribution requirement for ±15% variation is insufficient for major roads, especially when heavy application appears in the wheel-tracks; however, most contractors provide bars with ±10% or even better. The use of hydraulically expanding spray bars enables the joint to be maintained at the centre of the road while varying the width at the road edge, thus eliminating unsightly and sometimes dangerous joints from the wheel-tracks. Although most conventional binders are still sprayed from distribution bars fitted with 'swirl' jets, the higher initial viscosities of the 'thermosetting' binders necessitate the use of 'slot' jets.

Grade of cutback binder, sec	Spraying temperature, °C	
	Swirl jets	Slotted jets
50 100 200	145 ± 5 155 ± 5 165 ± 5	140 ± 10 145 ± 10 155 ± 10

Table 18.5 – Spraying temperatures for swirl and slot jets

Advanced designs of slot jets are now available which give similar standards of transverse distribution to those obtained with swirl jets, and it is likely that they will be increasingly used in the U.K. market for conventional binders. By using this type of jet it is possible to reduce the quantity of diluent in a binder with subsequent savings in cost and reduced application temperatures (see table 18.5), resulting in a reduction of fire risk.

A programmable bitumen distributor has been developed which is an extremely sophisticated spray machine using two bars to achieve differential rates of spread across the width of the machine. Thus application rates can be changed for areas of patching, wheel-tracks, fatty areas and hungry areas. The rate of spread is automatically reduced by up to 30% in the wheel-tracks and can be increased between wheel-tracks or in shaded areas. Both longitudinal and transverse binder distributions can be pre-programmed. Vehicle speed, binder temperature, spray bar width and application rates are controlled by a micro-processor.

Photograph 18.6 – A modern surface dressing sprayer *(Photograph reproduced by kind permission of Colas Roads Limited)*

18.5.2 Chipping spreaders

Self-propelled chipping spreaders (see photograph 18.7) are now in regular use and these machines not only allow work to proceed more rapidly, but also spread the chippings more evenly and accurately at the required rate per square metre. They also have the advantage that the chippings are applied much closer to the road surface. As a result, the likelihood of chippings bouncing off the road surface and exposing binder or bouncing to another part of the dressing with the binder uppermost is reduced. These chippers are also intrinsically safer because they are driven forward rather than reversing as is the case with lorry mounted spreaders.

The latest development in self-propelled chipping spreaders include four-wheel drive for pulling heavy tippers up steep inclines without a judder; a mechanism for breaking lumps of lightly coated chippings within the hopper; methods for spraying additive onto the binder film; and incorporation of a pneumatic-tyred roller to ensure that the chippings are pressed into the binder film at the earliest possible moment while it is still hot.

18.5.3 Rollers

Rolling techniques have been substantially improved by increasing use of pneumatic-tyred rollers (see photograph 18.8) which secure the initial bond of the chipping to the binder film without crushing the carefully selected aggregate to some other less desirable series of dimensions, with all the attendant problems of fatting-up later in the life of the surface dressing. The use of vibrating rubber drum-type rollers (see photograph 18.9) improves the initial mosaic formation, although it must be recognised that any rolling technique is largely a preliminary process undertaken before the main stabilisation of the dressing by subsequent slow speed trafficking.

18.5.4 Sweepers

High-speed roads require thorough sweeping to remove all excess chippings from surface dressing after traffic control and before allowing unrestricted traffic onto the dressing. To achieve this, a full-width suction sweeper should be used with a power broom collecting chippings thrown off the dressing. However, 14 mm chippings pose a problem as current full-width machines are unable to collect this large stone. Therefore, conventional single suction-with-sweep machines are used when the dressing is stable and will withstand disruption by the brush. Properly organised and implemented sweeping will do much to alleviate the damage caused by surface dressings in their early life by chipping loss.

18.6 Surface dressing techniques[24]

18.6.1 Pad-coats

On excessively hard, porous or hungry surfaces, it is advisable to use a prior application of surface dressing, employing a small size chipping to act as a pad coat, and artificially create a softer surface for the main chippings to embed, prior to the main surface dressing application. The pad-coat is usually a surface dressing binder plus 6 mm chippings or a slurry seal. The pad-coat is then opened to traffic and may be left from a few months to a few years before the main dressing is applied. Clearly there is a financial benefit to be gained by trafficking the pad-coat for a number of

Photograph 18.7 – Self-propelled chipping spreader *(Photograph reproduced by kind permission of Colas Roads Limited)*

Photograph 18.8 – Pneumatic-tyred roller

years. However, the longer the pad-coat is used for traffic, the greater the possibility of a consistent surface being lost, which will affect the final large stone dressing. It is therefore recommended that pad-coats should be covered within one year. If the pad-coat is to be dressed the same season, it is advisable to use a bitumen emulsion in preference to a cutback to avoid fatting-up problems due to entrapped kerosine.

The pad-coat can then be surfaced dressed using either a conventional single surface dressing or a double surface dressing technique. Either system will embed into the pad-coat and derive immediate mechanical strength in so doing, lessening the demands on the binder.

18.6.2 Double surface dressing techniques

Traditionally surface dressing has been carried out by a single application of binder followed by chippings. Single dressing is still satisfactory for most traffic categories; however, for high-speed and heavily trafficked roads, double chipping applications have been successfully used, with small chippings employed as a secondary aggregate to lock the larger chippings in place.

There are two principal types of double surface dressing systems: the single spray double chip or racked-in method and the double spray double chip method. A schematic representation showing the sequence of spray and chip for these two techniques is shown in figure 18.11.

18.6.2.1 The single spray double chip or racked-in method[25]

This type of dressing uses a single application of binder. The 14 mm or 10 mm principal chippings are spread at approximately 80% to 90% cover which leaves a 'window' of binder between the chippings. This window is filled with small chippings to achieve high mechanical interlock of the principal chippings. The rate of spread of binder on this type of surface dressing is usually slightly higher than would be required on a single size dressing.

The initial surface texture and mechanical strength of the dressing are high. The configuration of the principal chippings is different from that achieved with a single dressing. The mosaic associated with the latter cannot be formed because the principal chippings are locked in place. In time, small chippings not in contact with binder are lost to the system without damaging vehicles or windscreens, resulting in increasing surface texture despite the coincident embedment of the larger chipping.

On a racked-in dressing vehicle tyre contact is principally on the large aggregate with little contact on the small aggregate. Thus the microtexture of the small aggregate is less critical and therefore consideration should be given to the use of cheaper lower PSV aggregates such as limestone for the smaller chippings.

The advantages of the racked-in system are:

- virtual elimination of flying 14 mm chippings;
- early stability of the dressing through good mechanical interlock;
- good adhesion of larger size chippings;
- a rugous surface texture with an initial texture depth of over 3.0 mm.

In 1987 there was a marked increase in the use of racked-in surface dressing[26] on main and heavily trafficked roads. The use of this technique has dramatically reduced the incidence of broken car windscreens.

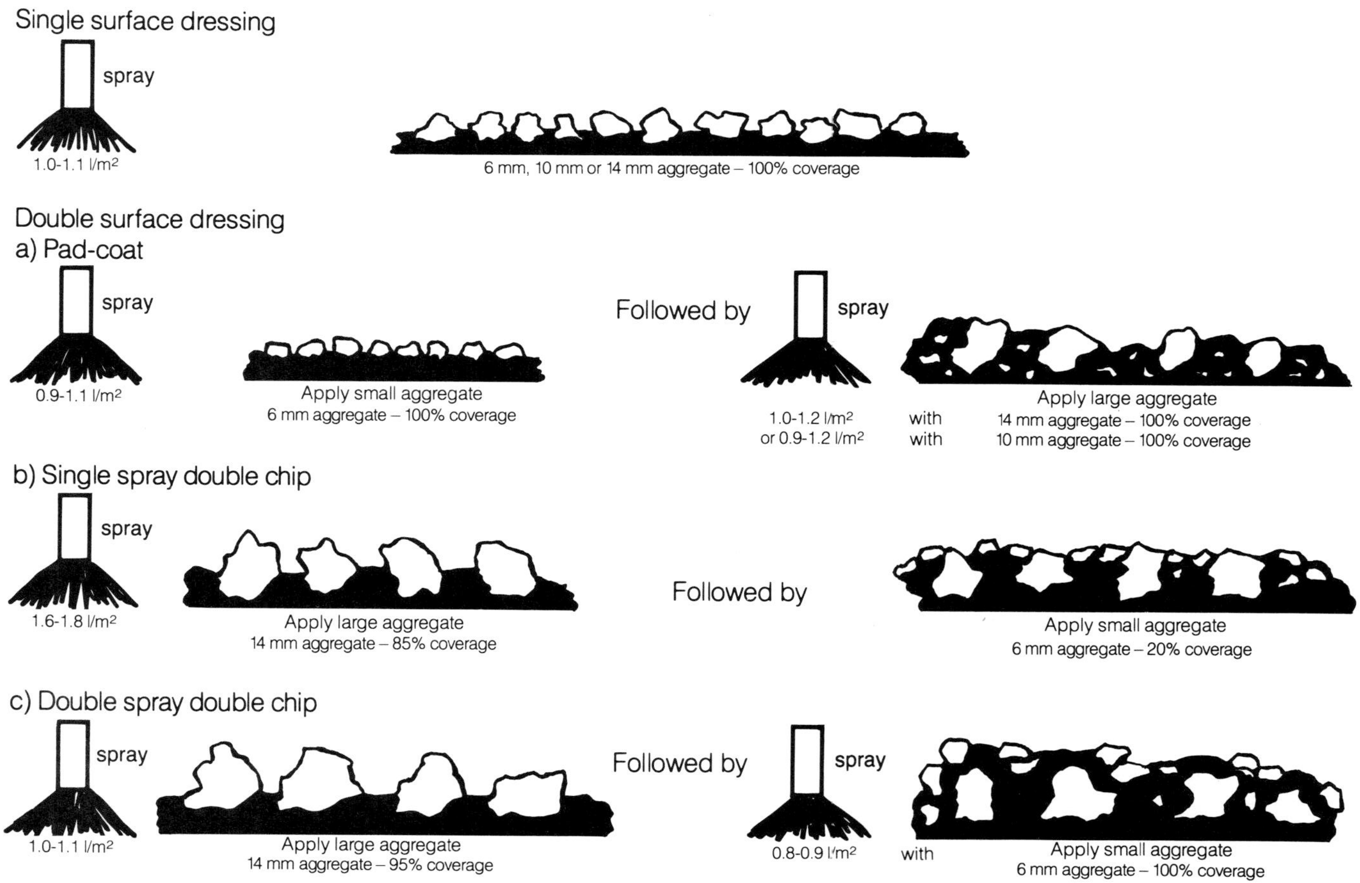

Figure 18.11 – Different surface dressing systems with typical application rates[24]

Photograph 18.9 – Rubber-shod steel-wheel roller

18.6.2.2 The double spray double chip method

This dressing uses two applications of binder and chippings. A typical sequence would be to apply the binder plus 14 mm clean chippings laid at a 95% coverage rate. This dressing is lightly rolled and followed immediately by a second application of binder and clean 6 mm chippings.

This technique provides an excellent dressing with a texture depth over 3.0 mm. This type of dressing is most suitable on roads which can be closed and lanes coned-off, as the initial stability can be rather low.

The advantages of double surface dressing systems may be summarised as follows:

- their initial strength is greater, with less reliance upon the strength of the binder than single surface dressings;
- because they are physically locked in, larger chippings can be selected than otherwise would be the case for a given road hardness, thus prolonging the surface texture of the new dressing and reducing the chance of failure resulting from the onset of cold weather before chipping embedment has taken place;
- the number of non-adhering large chippings lying on the road surface is significantly reduced because they are locked in place by the smaller chippings, reducing the chance of windscreen breakage;
- the initial and long-term surface texture of the single spray double chip system tends to be higher than for the single surface dressing system.

18.7 Traffic control and aftercare[2,27,28]

Traffic control and aftercare is vital to the successful performance of the surface dressing and cannot be over-emphasised. Ideally, when traffic is first allowed onto the new dressing, it should be behind a control vehicle, which should travel at a speed of about 15 mph. If there are gaps in the traffic, it may be necessary to introduce additional control vehicles. The objective should be to ensure that vehicles do not travel at more than 15 mph on the new dressing, and that sharp braking and acceleration of these vehicles does not occur. The length of time involved will depend very much on the type of binder used and the prevailing weather conditions. It is likely to be longest where emulsions are being used and the weather conditions are humid. The use of wet or dusty chippings or the early onset of rain delays adhesion and necessitates a longer period of traffic control. Traffic passing *slowly* over a dressing immediately after completion creates a strong interlocking mosaic of chippings, a result only otherwise obtained by long periods of rolling.

18.8 Types of surface dressing failures[29]

Apart from damage due to accidental spillage of fuel or oil, the faults that develop in a surface dressing are latent from the time it is applied to the road. The majority of surface dressing failures fall into one of the following five categories:

- loss of chippings immediately after laying or during the first few days;
- loss of chippings during the first winter;
- loss of chippings in later years;
- bleeding during the first hot summer;
- fatting-up in subsequent years.

The type and rate of application of the binder and size of chipping has, of course, a large influence on performance; incorrect application rates can result in failures of any of the above types.

Very early loss of chippings may be due to slow break in the case of emulsions, or due to poor wetting in the case of cutback bitumens. The use of high binder content cationic rapid setting emulsions (more than 70% binder) largely overcomes the former problem and the use of adhesion agents and/or precoated chippings, the latter.

Many surface dressing failures are first seen at the onset of the first prolonged frosts when the binder is very stiff. These can normally be attributed to a combination of inadequate binder application rate, the use of too large a chipping or inadequate embedment of the chippings into the old road due to the surface being too hard and/or there being insufficient time between applying the dressing and the onset of cold weather.

Inadequate binder application and/or the use of excessively large chippings will exacerbate this problem. Loss of chippings in the long term usually results from a combination of low durability binders and, again, poor embedment of chippings.

Bleeding may occur within a few weeks on dressings laid early in the season, or the following summer on dressings laid late the previous year. It results from the use of binders of too low viscosity for the high ambient temperatures prevailing, or from high temperature susceptibility binders.

Fatting-up is complex in nature and may result from any or a combination of the following factors:

- *excessive application rate;*

- *gradual embedment of chippings into the road surface*; this causes the relative rise of binder between the chippings to a point where the chippings disappear beneath the surface; this is largely dependent on the intensity of traffic, particularly on the proportion of heavy commercial vehicles in the total traffic; the composition of the binder also affects the process; high solvent content cutbacks can soften the old road surface and accelerate embedment;
- *crushing of chippings*; most chippings will split up or crush under heavy traffic with the loss from the dressing of many small fragments; the binder-to-chipping ratio therefore tends to increase, thus adding to the fatting-up process; a dilemma here is that many of the best aggregates for skid resistance are often the worst for crushing resistance;
- *absorption of dust by the binder*; bitumen binders of high durability tend to absorb dust that falls upon them; soft binders can absorb large quantities of dust, thereby increasing the effective volume of the binder; this effect, coupled with chipping embedment, leads to the eventual loss of surface texture.

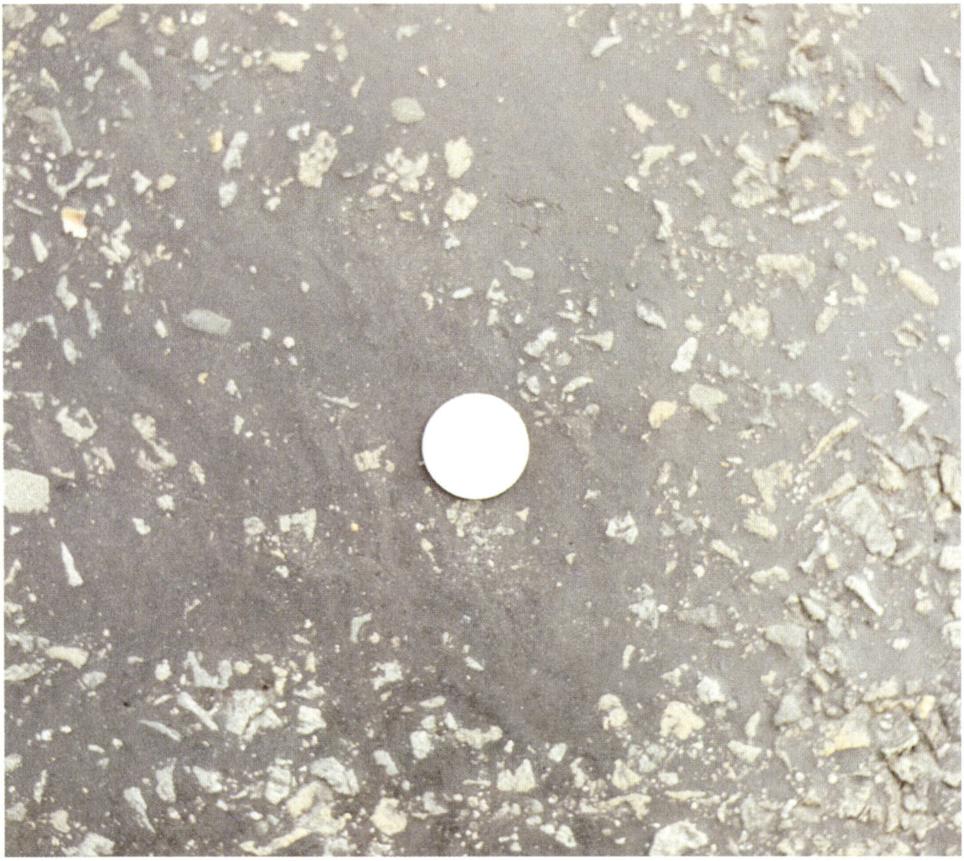

Photograph 18.10 – A fatted-up surface dressing

Photograph 18.11 – Slurry sealing *(Photograph reproduced by kind permission of Colas Roads Limited)*

18.9 Slurry sealing[30]

Slurry sealing was first introduced into the U.K. in the mid fifties to arrest runway deterioration on military airfields. However, it was not until the introduction of quick setting cationic emulsions that its use became widespread on highways.

Two grades of slurry seal are included in BS 434[31,32] and the DTp Specification for Highway Works[33]. One grade is laid at a nominal 1.5 mm thickness and the other is laid at a nominal 3.0 mm thickness.

The main purpose of slurry sealing is to seal the road surface from the ingress of air and water and arrest surface disintegration. In addition the material also fills minor surface depressions providing a more even riding quality and improving the appearance of the road surface. The texture depth achieved with slurry seals is relatively low and therefore the material is not suitable for roads carrying high-speed traffic. Slurry sealing only provides a thin 'veneer' treatment to an existing paved area, thus it does not add any significant strength to the road structure nor will it be durable if laid on an inadequate substrate. The process is therefore usually restricted to lighter trafficked situations such as housing estate roads, footways, light airfield runways, car parks, etc, although specially designed slurry seals providing a higher texture depth have been used on motorway hard shoulders.

The mixed slurry is comprised of aggregate (crushed igneous rock, limestone or slag) an additive and a bitumen emulsion. The additive is usually portland cement or hydrated lime which is added to the mix to control its consistency, setting rate and mix segregation. It normally represents about 2% of the total aggregate. Where a rapid set is required to permit early trafficking or to guard against early damage from wet weather, a rapid setting anionic emulsion, class A4 or cationic emulsion class K3 are used. Where setting time is not critical a slow setting anionic emulsion, class A4 can be used. Depending on the characteristics of the aggregate, between 180 and 250 litres of emulsion per tonne of aggregate are required. The composition of the cured material is about 80% aggregate, 20% bitumen.

Slurry seals are normally applied by specialist continuous-flow machines which meter and deliver aggregate, emulsion, water and additives into a mixer which continuously feeds a spreaderbox towed behind the machine. These machines are capable of laying up to 8,000 m^2 per day. Depending on the composition of the slurry, light rolling with a pneumatic tyred roller may be required. Rolling is applied as soon as the material has set sufficiently to avoid rutting.

18.10 Safety[34,35,36]

The following briefly highlights some of the potential hazards associated with the handling of surface dressing binders. However, it is not intended to be used as a 'Safety Code' and it should not be used as a substitute for health and safety information supplied by individual suppliers or the information contained in the Institute of Petroleum Code of Safe Practice[36].

18.10.1 Handling temperatures

As a result of their viscosity, cutback grades of bitumen have to be heated to enable them to be sprayed through jets. The recommended handling temperatures for cutback grades of bitumen are detailed in table 18.6[34,35]. The temperatures have been calculated on the basis of viscosity measurements and are supported by operating experience. For the normal operations of blending and transferring liquid bitumen, temperatures of 10°C to 50°C above the minimum pumping temperatures are recommended.

Cutback bitumens are commonly handled at temperatures well in excess of their flash point, ie the temperature at which the vapour given off will burn in the presence of air and an ignition source; thus it is essential to exclude sources of ignition in the proximity of cutback handling operations.

The principle to be followed is that, in order to minimise the risks from burns, fumes, flammable atmosphere and fire, every operation should be carried out at as low a temperature as possible, compatible with efficient working, and always below the maxima given in table 18.6. Some bitumen emulsions are applied at ambient temperatures, but the majority of high bitumen content cationic emulsions are applied at temperatures up to 85°C.

18.10.2 Operation of surface dressing sprayers using cutback bitumens

The sprayer should be located on level ground before being heated. The flame tube heater can be used only if a minimum of 150 mm of bitumen covers the heater tubes, and must not be used during product transfer or spraying. The heater should be extinguished and left to cool for at least 30 minutes before spraying. During heating the operator should remain in attendance to ensure that the bitumen is not overheated and there are no operations in the vicinity which might release cutback vapours before the flame tube is cool.

Grade of cutback binder, sec	Temperature, °C		
	Minimum pumping (a)	Spraying (b)	Maximum safe handling (c)
50	65	150	160
100	70	160	170
200	80	170	180

Table 18.6 – Recommended handling temperatures for cutback binders(34,35)

Notes

(a) Based on a viscosity of 2000 cSt (2 Pa.s).

(b) Based on a viscosity of 30 cSt (0.03 Pa.s). Conforms with the maximum spraying temperatures recommended in Road Note 39.

(c) Based on generally satisfactory experience of the storage and handling of cutback grades in contact with air. Subject to the avoidance of sources of ignition in the vicinity of tank vents and open air operations.

To minimise fire risks and avoid excessive fuming, the temperature of the cutback in the sprayer should be kept as near as possible to the optimum for the grade and should never exceed the maxima given in table 18.6. During spraying, clear warning signs against smoking and naked flames should be displayed at appropriate points on site, visible to the general public(36).

18.10.3 Operation of surface dressing sprayers using bitumen emulsions

The same general handling precautions apply as to cutback bitumens. The two major differences are that:

- bitumen emulsions are normally heated to 60°C to 85°C; a maximum temperature of 90°C is recommended because overheating to temperatures approaching 100°C can produce 'boilover' due to the presence of water;
- flammable atmospheres are not generated by bitumen emulsions at normal working temperatures; it is therefore less important that flame tube heaters are thoroughly cooled before product transfer or spraying.

18.10.4 Precautions, protective clothing and hygiene for personnel

When spraying either emulsion or cutbacks all operators should wear protective outer clothing, ie eye protection, clean overalls and impervious shoes and gloves, to protect against splashes and avoid skin contact.

In addition, operators working near the spraybar should wear orinasal fume masks. For work alongside open highways, operators should wear clean high visibility jackets and reflective patches.

Although bitumen emulsions are applied at relatively low temperatures the above protection is essential because emulsifiers are complex chemical compounds and prolonged contact may result in allergic reactions or other skin conditions.

Barrier creams, applied to the skin prior to spraying, assist in subsequent cleaning should accidental contact occur. However, barrier creams are not adequate substitutes for gloves or other impermeable clothing. If any bitumen comes into contact with the skin, operatives should wash thoroughly as soon as practicable and always before going to the toilet, eating or drinking.

18.11 The future for surface dressing

The importance of skid resistance in reducing accident potential was publicly acknowledged in January 1988 by the then Minister for Roads and Traffic, when he announced that the agents who maintain motorways and trunk roads would be continually monitoring levels of skid resistance, to ensure that mean summer sideway force co-efficients on these high-speed roads do not fall below minimum investigatory levels[37,38].

The need to maintain high skid resistance and the intense pressure on local authorities to obtain value for money[39], provide increased technical and economic justification for increasing the use of surface dressings to motorways. However, the advances and innovations in surface dressing technology which have been achieved in the latest 10 years will need to continue if surface dressing on these high profile sites is to become commonplace. The importance of the design of surface dressings, the development of polymer modified binders and new chipping techniques, improved training of operatives and greater attention to aftercare cannot be over-emphasised if the industry is to break down the resistance in some quarters to the use of surface dressing on high-speed roads and make the process more 'socially acceptable' to motorists. Specifying authorities for their part must understand the need for higher initial costs, to implement improved standards of supervision and application techniques and to encourage further innovation of materials and plant, thereby improving long-term performance and reducing full-life costs.

18.12 References

(1) The Road Surface Dressing Association and the County Surveyors Society: *County surface dressing statistics*.

(2) Transport and Road Research Laboratory: *Road Note 39. Recommendations for road surface dressing*. Second edition, 1981.

(3) Ministry of Transport: *Report of the Committee on Highway Maintenance*, London, 1979 (HMSO).

(4) Salt, G F and Szatkowski, W S: *A guide to levels of skidding resistance for roads*. Transport and Road Research Laboratory, Report LR 510, 1973.

(5) Hanson, F M: *Bituminous surface treatments of rural highways*. Proceedings of New Zealand Society of Civil Engineers, vol 21, 1934.

(6) Jackson, G P: *Surface dressing*. Shell International Petroleum Company Limited, London, 1963.

(7) Elborn, M J, Heslop, M F W and Pooley, G R: *Recent developments in surface dressing*. The Highway Engineer, pp 6-19, July 1982.

(8) British Carbonisation Research Association: *The CTRA road-hardness probe*. Carbonisation Research Report 7, June 1974.

(9) Jacobs, F A: *M40 High Wycombe bypass: results of a bituminous surface-texture experiment.* Transport and Road Research Laboratory, Report LR 1065, 1983.

(10) British Standard 3690: Part 1: 1989: *Bitumens for building and civil engineering. Part 1. Specification for bitumens for road purposes.*

(11) Hoad, L and Hoban, T W S: *A cost-performance relationship for surface dressing.* Road Surface Dressing Association, Road Note 39 symposium, March 1982.

(12) Heslop, M F W and Broadbridge, M H: *The trials and tribulations of surface dressing.* The Institution of Highways and Transportation presidential conference, 1986.

(13) Peden, R A: *An evaluation of modern surface dressing.* Paper presented at the national symposium on highway surfacing, The University of Ulster, April 1986.

(14) Hosking, J R and Woodford, G C: *Measurement of skidding resistance. Part 1. Guide to the use of 'SCRIM'.* Transport and Road Research Laboratory, Report LR 737, 1976.

(15) Denning, J H: *Epoxy-resin/calcined bauxite surface dressing on A1, Sandy, Bedfordshire: skid-resistance measurements 1968 to 1977.* Transport and Road Research Laboratory, Report LR 867, 1978.

(16) British Standard 812: Part 3: 1975: *Methods for sampling and testing of mineral aggregates, sands and fillers. Part 3. Mechanical properties.*

(17) Catt, C A: *An alternative view of TRRL's research into skidding resistance.* Asphalt Technology, No 33, July 1983.

(18) Department of Transport: *Specification requirements for aggregate properties and texture depth for bituminous surfacings to new roads.* Technical Memorandum H 16/76.

(19) British Standard 598: Part 3: 1985: *Sampling and examination of bituminous mixtures for roads and other paved areas. Part 3. Methods for design and physical testing.*

(20) *Proceedings of the symposium on skid-resistant roads.* Transport and Road Research Laboratory, Supplementary Report 24 UC, 1972.

(21) Szatkowski, W S and Hosking, J R: *The effect of traffic and aggregate on the resistance to skidding of bituminous surfacings.* Transport and Road Research Laboratory, Report LR 504, 1972.

(22) British Standard 63: Part 2: 1971: *Single-sized roadstone and chippings.*

(23) Coal Tar Research Association: Report 0361, Gomersal, 1966.

(24) Southern, D: *Premium surface dressing systems.* Shell Bitumen Review 60, pp 4-8, March 1983.

(25) Road Surface Dressing Association: *Racked-in surface dressing.* Guidance Note 87-1, March 1987.

(26) Underwood, C: *A review of the 1987 surface dressing season.* Highways and Transportation, vol 35, pp 28-34, March 1988.

(27) Robinson, D A: *Traffic control and 'after-care' in road surface dressing.* Road Tar, vol 28, pp 4-7, 1974.

(28) Refined Bitumen Association: *Surface dressing – a guide to good practice.* Technical Bulletin No 2, November 1978.

(29) Nelson, J R and Hardman, R: *A guide to road surface dressing practice.* Transport and Road Research Laboratory, Supplementary Report 627, 1980.

(30) Road Emulsion Association: *Slurryseal*. Technical data sheet No 9.

(31) British Standard 434: Part 1: 1984: *Bitumen road emulsions (anionic and cationic). Part 1. Specification for bitumen road emulsions.*

(32) British Standard 434: Part 2: 1984: *Bitumen road emulsions (anionic and cationic). Part 2. Code of practice for use of bitumen road emulsions.*

(33) Department of Transport: *Specification for Highway Works*. Part 3, series 900. Sixth edition, August 1986.

(34) Refined Bitumen Association: *Safe handling of petroleum bitumens*. Technical Bulletin No 5, March 1981.

(35) Refined Bitumen Association: *Recommendations to users for the storage and handling of bitumen*. Technical Bulletin No 7, August 1988.

(36) Institute of Petroleum: *Model code of safe practice in the petroleum industry. Part 11. Bitumen safety code*, 1990.

(37) Department of Transport: *Skidding resistance of in-service trunk roads*. Departmental Standard HD 15/87.

(38) Department of Transport: *Skidding resistance of in-service trunk roads*. Departmental Advice Note HA 36/87.

(39) Underwood, C: *Early surface dressing optimises structural maintenance budgets*. Highways, vol 56, No 1942, pp 30-33, October 1988.

Chapter 19 – Miscellaneous uses of bituminous mixes

The vast majority of the 30 million tonnes of bituminous mixes produced annually in the U.K. are used in the construction of roads and airfield runways and taxiways. However, in addition to these well known applications, bituminous mixes are also used for a wide range of specialist applications.

19.1 Hydraulic applications[1,2]

The waterproofing properties of bitumen and bituminous mixes have been recognised from ancient times. The ancient civilisations of Ur, Egypt, Babylon, Assyria and the Indus all used the naturally occurring surface seepages of asphalt for waterproofing and building (see chapter 1, section 1.1). The particularly well-preserved bath at Mohenjo Daro in the Indus basin shows that in this field asphalt can claim a 5000-year life![3]

More recently with the advances in asphalt technology asphalt has been shown to be an effective material for waterproofing dams, reservoirs, canals as well as sea wall defences and coastal groynes. There are now over 200 major structures of this kind worldwide. In the U.K. the following dams have been built using a bituminous mix as part of the construction – Dungonnell, Colliford, Winscar, Marchlyn, Megget, Sulby and most recently at Roadford[4], and the following reservoirs – Shotton, Leamington and Towey.

The two principal properties which make bitumen ideal for hydraulic applications are its impermeability and its chemical inertness. In combination with suitable aggregates in mixes specifically designed for particular applications, bitumen can impart these properties to the structures in which the asphalt is used. In many of these structures it is essential to have impermeability under pressure: asphalt mixes with void contents below 3% can successfully contain water column pressures of up to 200 m.

Reservoir and canal embankments normally have a slope of 1 : 1.75 but often this can be as steep as 1 : 1.25. Thus the mix must have sufficient stability to be laid and compacted on such slopes but without cracking during application and in service. The material must be flexible enough to accommodate differential settlement of the substrate.

One great advantage of asphalt is that it can be laid on the dam or reservoir face continuously, thus eliminating joints. Any joint is a structural weakness and at the high pressures at the base of dams and reservoirs a joint is a positive invitation for the water to find an outlet.

19.2 Railway applications

In the U.S.A., Japan and many European countries, bituminous mixes have been tested and used in the construction of railway track beds for many years. The earliest application dates back to 1894, where an asphalt base approximately 150 mm thick surfaced with a 30 mm wearing course was laid as tramway bed in Visalia, California. In Japan asphalt has been used in railway construction since the late fifties and numerous tests and commercial applications have been carried out in Europe.

In railway construction bituminous mixes are used in one of four ways[5]:

- to stabilise the aggregate ballast;
- an asphalt layer below the aggregate ballast;
- as a direct base for the track sleepers;
- as a direct base for the rails.

The reason for using bituminous layers is to improve the stability of the track, providing an improved running surface for the carriages, and reduce maintenance.

In 1974 an experimental length of track incorporating an asphalt layer was laid in the U.K. at Radcliffe-on-Trent[6]. The experimental track consisted of a 230 mm thick rolled asphalt base, onto which various types of concrete superstructures were laid to support the rails. Since installation the performance of the asphalt track bed has been very good. Despite the good performance, widespread use of asphalt based railway tracks has been limited by the higher initial costs compared to a conventional ballasted track. However, techniques are being developed to simplify anchoring of the rail to the superstructure[5] which makes the process much quicker, less labour-intensive and more cost-effective.

19.3 Coloured surfacings[7]

The appearance of traditional bituminous surfacings, especially when finished with the normal contrasting white lines, is generally very pleasing to the eye. However, there are some locations where a specific colour is desired and there is a number of ways that this can be achieved:

- incorporating coloured pigments into the mix during manufacture;
- application of suitably coloured chippings to the surfacing during laying;
- application of a coloured surface treatment after laying;
- using a conventional bitumen with a coloured aggregate;
- using a suitably coloured aggregate with a translucent binder.

19.3.1 Incorporating coloured pigments

The majority of coloured mixes are manufactured by adding pigment, usually iron oxide, during the mixing process. Appropriately coloured aggregates are used to ensure that, when aggregate is exposed after trafficking, the overall appearance of the material is maintained. The main drawbacks to colouring mixes using conventional bitumens are:

the only acceptable colour that can be achieved is red;

the quantity of iron oxide required to achieve an acceptable red is fairly high, increasing the cost of the mix substantially.

To enable mixes to be pigmented in colours other than red, Shell Research Limited have developed a binder specifically for pigmenting[8]. This is a 'synthetic' binder which, because it contains no asphaltenes and is readily pigmented in virtually any colour[9,10]. It possesses rheological and mechanical properties similar to conventional bitumen, with the exception that high equi-viscous temperatures of this binder are about 15°C lower than the same grade of a conventional bitumen. Coloured mixes manufactured using this binder require 1% to 2% of pigment to achieve a satisfactory colour, whereas only red can be achieved with conventional bitumen, using typically 5% to 6% iron oxide addition.

Photograph 19.1 shows an open graded macadam manufactured using this binder laid in the Liseberg amusement park near Gothenburg, Sweden[9]. To achieve these bright colours relatively large quantities of pigment were used; the red macadam was manufactured using 2% of iron oxide and 1% of a white pigment, and the yellow macadam contains 2% of a yellow pigment and 2% white pigment.

Photograph 19.1 – The Liseberg amusement park surfaced with Shell's pigmentable binder

The Department of Transport recently introduced a novel application for coloured mixes. This is the application of red-pigmented mixes, usually sand carpet, as a protective layer over waterproof membranes laid on bridge-decks. When the wearing course on the bridge-deck needs replacing, contractors planing off the old wearing course will only plane down to the red sand carpet, thereby avoiding damage to the expensive waterproof membrane.

19.3.2 Application of coloured chippings

Hot rolled asphalt and fine-graded macadam (previously termed fine cold asphalt) can have decorative coloured chippings rolled into the surface during compaction. Hot rolled asphalt is suitable for most traffic conditions but fine-graded macadam is only appropriate for lightly trafficked areas, private drives and pedestrian areas.

To provide a decorative finish to both of these types of surfacing, pigmented bitumen-coated or clean resin-coated chippings can be applied during laying. The bitumen or resin coating promotes adhesion to the surfacing and it is not recommended that uncoated chippings are used. However, in the case of fine-graded macadam laid on areas subjected to little traffic, a light application of uncoated chippings, eg white spar, produces a very attractive finish if they are uniformly applied and well embedded.

Decorative chippings cannot be successfully rolled into the surface of high stone content mixes such as dense macadams or high stone content rolled asphalts.

19.3.3 Application of a coloured surface treatment

There are three surface treatments which can be applied to a bituminous wearing course to provide a decorative finish:

pigmented slurry seals;
surface dressing;
application of a coloured paint.

Pigmented slurry seals are available as proprietary products in a range of colours. However, as they are applied in a very thin seal, ca 3 mm, they are only suitable for pedestrian and lightly trafficked areas.

Surface dressings are suitable for most categories of road application[11] but are less suitable for pedestrian situations. The final colour of the surface dressing will be that of the aggregate used.

Proprietary coloured paints are available for over-painting conventional black wearing courses. These are only suitable for pedestrian and games areas, eg tennis courts.

19.3.4 Coloured aggregate bound by a conventional bitumen

When a conventional bitumen is used, the depth of colour achieved is dependent upon: the colour of the aggregate itself, the thickness of the binder film on the aggregate and the rate at which the binder exposed on the road surface is eroded by traffic. In medium and heavily trafficked situations the natural aggregate colour will start showing through fairly quickly, but in lightly trafficked situations, where coloured surfacings are usually specified, the appearance of the aggregate colour may take considerable time.

19.3.5 Coloured aggregate bound by a translucent binder

Several proprietary macadams, in which the binder is a clear resin rather than a bitumen, are available. By selecting appropriately coloured aggregates a range of coloured surfacings can be manufactured. The major advantage of this type of system is that the colour is obtained immediately.

19.4 Bituminous materials for recreational areas[12]

Bituminous mixes are used extensively for recreational areas, eg tennis, netball and volley ball courts, five-a-side football pitches, etc. They are used in two ways:

as the final surfacing over a prepared base;
as a firm base beneath a synthetic surfacing.

By choosing an appropriate mix, a pleasing and durable finish requiring little maintenance can be achieved. Bituminous mixes are generally black or dark grey, providing a good contrast for line paints. If a coloured surfacing is required this can be achieved by using a special pigmentable binder (see section 19.3.1) or by over-painting the surface with a special proprietary paint.

Synthetic sports surfaces are very popular for athletic tracks, football pitches and other specialised applications. Bituminous mixes have been found to be ideal for forming the base for these applications.

Photograph 19.2 – Surfacing of the Lucas-Girling test track *(Photograph reproduced by kind permission of Tilcon Limited)*

19.5 Motor racing tracks[4]

The vast majority of international motor racing circuits are surfaced with bituminous mixes, such as asphalts, macadams or asphaltic concrete.

In view of the very high acceleration, braking and sideways shear stresses imposed by racing cars, the surfacings need to be very durable and skid-resistant. The standards specified by the regulatory bodies for motor racing can, in the main, be met by conventional bituminous mixes, although some proprietary macadams have been found to be ideal for this type of application.

19.6 Vehicle testing circuits

Many vehicle testing circuits have steep banked curves and special techniques have been developed, by experienced contractors, to ensure that the finished surfacing is fully compacted, meets the specified surface tolerances and has an even running surface. Photograph 19.2 shows a surfacing being laid on the Lucas-Girling test track; considerable skill and expertise is required to lay and compact materials satisfactorily under these conditions.

19.7 Asphalt kerbs

Extruded asphalt kerbs are very simple to apply. They are produced by extruding a rolled asphalt mix through specialised equipment directly onto a tack-coated basecourse or wearing course. They are not suitable for the edge of carriageways carrying heavy traffic as they may be damaged if hit by vehicles, particularly commercial vehicles. However, they do provide a convenient means of producing a raised edge on other paved areas, eg perimeter kerbing to drives or car parks. British Standard 5931 : 1980[13] contains guidance on the use of asphalt for producing kerbs *in situ*.

19.8 References

(1) van Asbeck, Baron W F: *Bitumen in hydraulic engineering.* Volume 1. Shell International Petroleum Company Limited, London, 1959.

(2) van Asbeck, Baron W F: *Bitumen in hydraulic engineering.* Volume 2. Elsevier Publishing Company, 1964.

(3) Menzies, I: *Waterproofing with asphalt.* Water and Waste Treatment, vol 31, pp 20-44, November 1988.

(4) British Aggregate Construction Materials Industries: *Miscellaneous uses of asphalts and macadams.* Information sheet No 9, 1987.

(5) Beecken, G: *Railway. Asphaltic base course for railway track construction.* Shell Bitumen Review 61, pp 10-21, June 1986.

(6) Cope, D L: *Construction and experience of asphalt based track on British railways.* Proceedings of the second Eurobitume symposium, Cannes, pp 241-244, October, 1981. Reprinted in a special publication of Asphalt Technology, January 1982.

(7) British Aggregate Construction Materials Industries: *Decorative and coloured finishes for bituminous surfacings.* Information sheet No 4, 1989.

(8) Schellekens, J C A and Korenstra, J: *Shell MEXPHALTE C brings colour to asphalt pavements.* Shell Bitumen Review 62, pp 22-25, June 1987.

(9) Gustafssen, P: *Shell MEXPHALTE C. Liseberg amusement park in Gothenberg, Sweden opened for the 1987 season in Shell colours.* Shell Bitumen Review 63, pp 12-14, April 1988.

(10) Le Coroller, A and Herment, R: *Shell MEXPHALTE C and Colas Colclair. A range of products for all road surfacing processes.* Shell Bitumen Review 64, pp 6-10, September 1989.

(11) Transport and Road Research Laboratory: *Recommendations for road surface dressing.* Second edition. Road Note 39, 1981.

(12) British Aggregate Construction Materials Industries: *Use of asphalts and macadams in the construction of games and sport areas.* Information sheet No 7, 1989.

(13) British Standard 5931 : 1980: *Code of practice for machine laid in situ edge details for paved areas.*

Chapter 20 – Miscellaneous information

20.1 Physical constants of bitumens[1]

20.1.1 Specific gravity

The specific gravity of a bitumen is primarily dependent on the grade of the bitumen and temperature. Typical specific gravities for a range of grades of bitumen are given in table 20.1 and the effect of temperature on specific gravity is detailed in table 20.2. The number of litres per tonne of bitumen at various temperatures and specific gravities is detailed in table 20.3.

20.1.2 Coefficient of cubical expansion

The coefficient of cubical expansion of bitumen is effectively independent of the grade and is virtually constant at 0.00061/°C over the temperature range 15°C to 200°C.

Grade of bitumen	Typical specific gravity at 25°C
Penetration grades	
300	1.010 – 1.020
200	1.015 – 1.025
100	1.020 – 1.030
70	1.020 – 1.030
50	1.025 – 1.035
40 HD	1.025 – 1.035
35	1.025 – 1.035
25	1.030 – 1.040
Oxidised grades	
75/30	1.015 – 1.030
85/25	1.020 – 1.035
85/40	1.010 – 1.025
95/25	1.015 – 1.030
105/35	1.000 – 1.015
115/15	1.020 – 1.035
Hard grades	
H 80/90	1.045 – 1.055
H 110/120	1.055 – 1.065
Cutback grades	
50 secs	0.992 – 1.002
100 secs	0.995 – 1.005
200 secs	0.997 – 1.007

Table 20.1 – Typical specific gravities at 25°C

Temperature, °C	Specific gravity at 25°C					
	1.00	1.01	1.02	1.03	1.04	1.05
15.5	1.006	1.016	1.026	1.036	1.046	1.056
25	1.000	1.010	1.020	1.030	1.040	1.050
45	0.988	0.998	1.008	1.018	1.028	1.038
60	0.979	0.989	0.999	1.009	1.019	1.029
90	0.961	0.971	0.981	0.991	1.001	1.011
100	0.955	0.965	0.975	0.985	0.995	1.005
110	0.949	0.959	0.969	0.979	0.989	0.999
120	0.943	0.953	0.963	0.973	0.983	0.993
130	0.937	0.947	0.957	0.967	0.977	0.987
140	0.931	0.941	0.951	0.961	0.971	0.981
150	0.925	0.935	0.945	0.955	0.965	0.975
160	0.919	0.929	0.939	0.949	0.959	0.969
170	0.913	0.923	0.933	0.943	0.953	0.963
180	0.907	0.917	0.927	0.937	0.947	0.957
190	0.901	0.911	0.921	0.931	0.941	0.951
200	0.895	0.905	0.915	0.925	0.935	0.945

Table 20.2 – The specific gravity of bitumen at various temperatures

Temperature, °C	Specific gravity at 25°C					
	1.00	1.01	1.02	1.03	1.04	1.05
25	995	984	973	963	953	943
45	1010	999	988	978	968	958
60	1020	1009	998	988	978	968
90	1041	1030	1019	1009	999	989
100	1047	1036	1026	1015	1005	995
110	1054	1043	1032	1022	1011	1001
120	1060	1049	1038	1028	1017	1007
130	1067	1056	1045	1034	1024	1013
140	1074	1063	1052	1041	1030	1019
150	1081	1070	1058	1047	1036	1026
160	1088	1076	1065	1054	1043	1032
170	1095	1083	1072	1060	1049	1038
180	1103	1091	1079	1067	1056	1045
190	1110	1098	1086	1074	1063	1052
200	1117	1105	1093	1082	1070	1058

Table 20.3 – The number of litres per tonne of bitumen at various temperatures and specific gravities

20.1.3 Electrical properties

20.1.3.1 Electrical conductivity

Bitumen has a low electrical conductivity and is therefore an ideal insulating material. Harder grades of bitumen have slightly lower electrical conductivity than soft grades. However, these differences are insignificant. The conductivity of all grades increases with increasing temperature, viz:

Temperature, °C	*Conductivity*, S/cm
30	10^{-14}
50	10^{-13}
80	10^{-12}

The influence of fillers on electrical conductivity is negligible, unless conductive fillers such as graphite, coke or metal powders are used in significant quantities.

20.1.3.2 Dielectric strength

The dielectric strength is measured in kV/mm, and depends upon the conditions of measurement and the shape of the electrodes. Hard grades of bitumen tend to have a higher dielectric strength than soft grades; the strength of all grades decreases with increasing temperature, viz:

Temperature, °C	*Dielectric strength*, kV/mm (flat electrodes)
20	20 – 30
50	10
60	5

20.1.3.3 Dielectric constant (permittivity)

The dielectric constant of bitumen is about 2.7 at 25°C rising to about 3.0 at 100°C. The dielectric losses in bitumen increase with increasing temperature but decrease with increasing frequency.

The Transport and Road Research Laboratory have shown that the ability of bitumen to weather under the combined effects of oxygen, rain and oil deposition from traffic and ultra-violet light in sunlight is related to the dielectric constant or permittivity of the bitumen (see chapter 7, section 7.2.3).

20.1.4 Thermal properties

20.1.4.1 Specific heat

The specific heat of bitumen is dependent on both the grade of bitumen and on temperature. Values of specific heat vary from 0.4 cal/g/°C to 0.43 cal/g/°C at 0°C. The specific heat is increased by 0.0004 to 0.0006 cal/g/°C for every 1°C increase in temperature.

20.1.4.2 Thermal conductivity

Bitumen is a good thermal insulating material and typically has a thermal conductivity of 0.13 to 0.15 kcal/(m.h.°C).

20.2 Conversion factors for viscosities

The viscosity of bitumen is usually measured by either capillary (kinematic viscosity) or cup viscometers (see chapter 5, section 5.3). Kinematic viscosity is determined by timing the flow of bitumen through a glass capillary viscometer at a given temperature. The product of flow time and a calibration factor gives the kinematic viscosity in centistokes (cSt) or mm^2/s. In cup viscometers the

Known viscosity	To obtain unknown viscosity multiply by:						
	Kinematic, mm²/s	Redwood I, sec	Redwood II, sec	Saybolt Universal, sec	Saybolt Furol, sec	Engler, °E	Standard Tar Viscometer, sec
Kinematic, mm²/s	–	4.05	0.405	4.58	0.458	0.132	0.0025
Redwood I, sec	0.247	–	0.1	1.13	0.113	0.0326	–
Redwood II, sec	2.47	10	–	11.3	1.13	0.326	0.0062
Saybolt Universal, sec	0.218	0.885	0.0885	–	0.1	0.0287	–
Saybolt Furol, sec	2.18	8.85	0.885	10	–	0.287	0.0054
Engler, °E	7.58	30.7	3.07	34.81	3.48	–	–
Standard Tar Viscometer (10 mm cup), sec	400	–	162	–	183	528	–

Table 20.4 – Conversion factors for viscosities

time is recorded in seconds for a standard volume of bitumen to flow out through the orifice in the bottom of the cup. Table 20.4 gives a number of factors for converting the results of different types of cup viscometer, or for converting from cup viscosity to kinematic viscosity, or vice-versa.

20.3 Blending charts and formulae

Bitumen may be blended with a wide variety of crude oil based fractions for different applications. Volatile light fractions, eg white spirit, are used where rapid drying is required. Less volatile fractions, such as cutback kerosine, are used for the manufacture of cutback bitumens. Figures 20.1 and 20.2 show the effect of adding cutback kerosine and diesel to 100 pen, 200 pen and 300 pen bitumen respectively. Similarly figure 20.3 shows the effect on penetration when diesel is added to 50 pen bitumen.

Bitumens are miscible with each other in all proportions. The penetration and softening point of a blend of two bitumens can be estimated using the blending charts shown in figures 20.4 and 20.5. These charts should be used only for blends of bitumen of the same type, ie the same penetration index. They should not be used for blends of oxidised with penetration bitumen.

As an alternative to these charts the following formula can be used to estimate the penetration and softening point of a blend:

$$\log P = \frac{A \log P_b + B \log P_b}{100}$$

where P = penetration of the final blend

P_a = penetration of component a

P_b = penetration of component b

A = percentage of component a in the blend

B = percentage of component b in the blend

and

$$S = \frac{AS_a + BS_b}{100}$$

where S = softening point of the final blend

S_a = softening point of component a

S_b = softening point of component b

A = percentage of component a in the blend

B = percentage of component b in the blend.

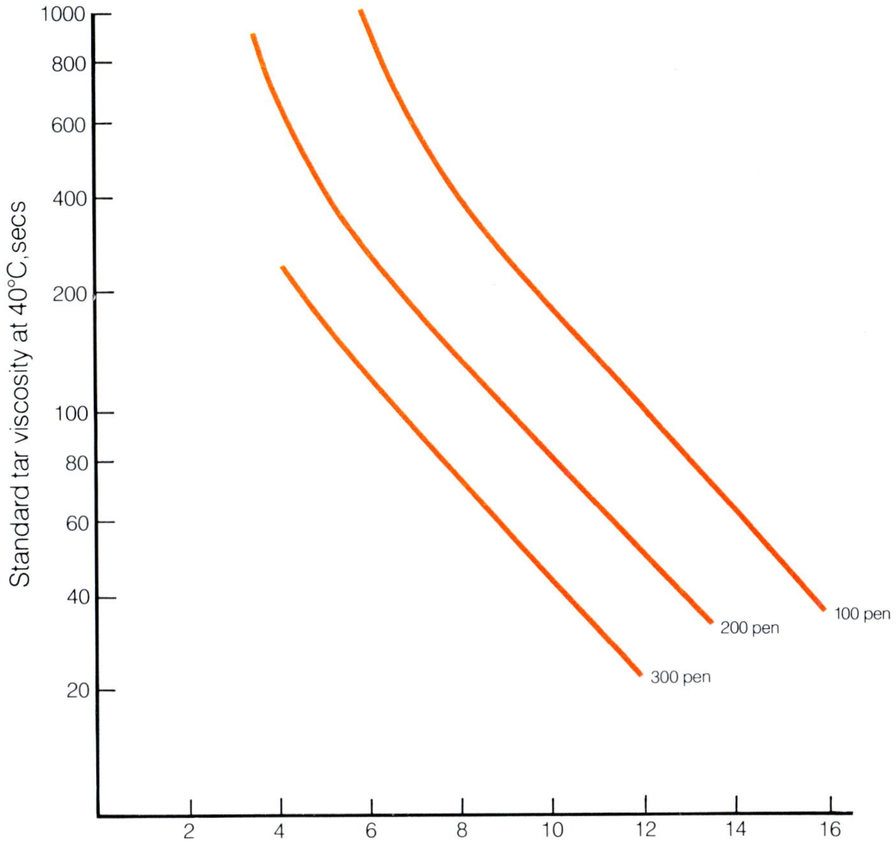

Figure 20.1 – Typical blending curves for 100 pen, 200 pen and 300 pen bitumen and cutback kerosine

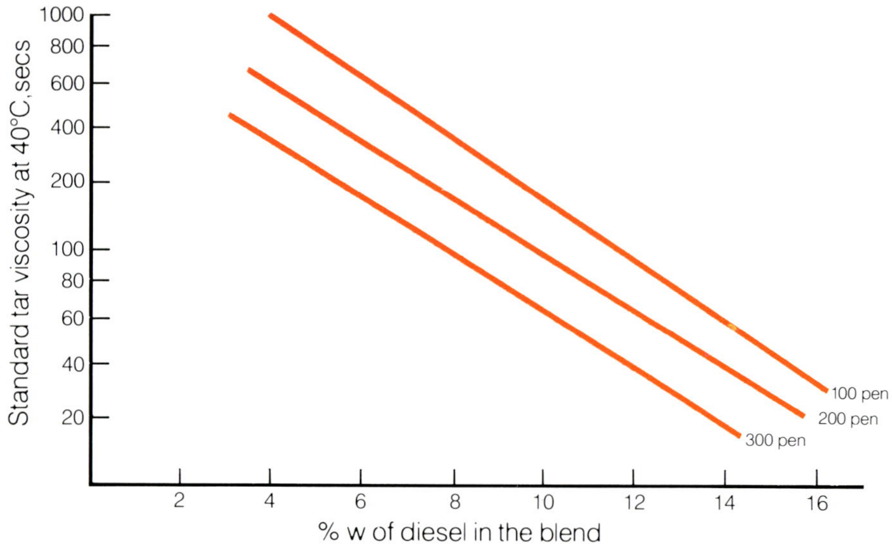

Figure 20.2 – Typical blending curves for 100 pen, 200 pen and 300 pen bitumen and diesel

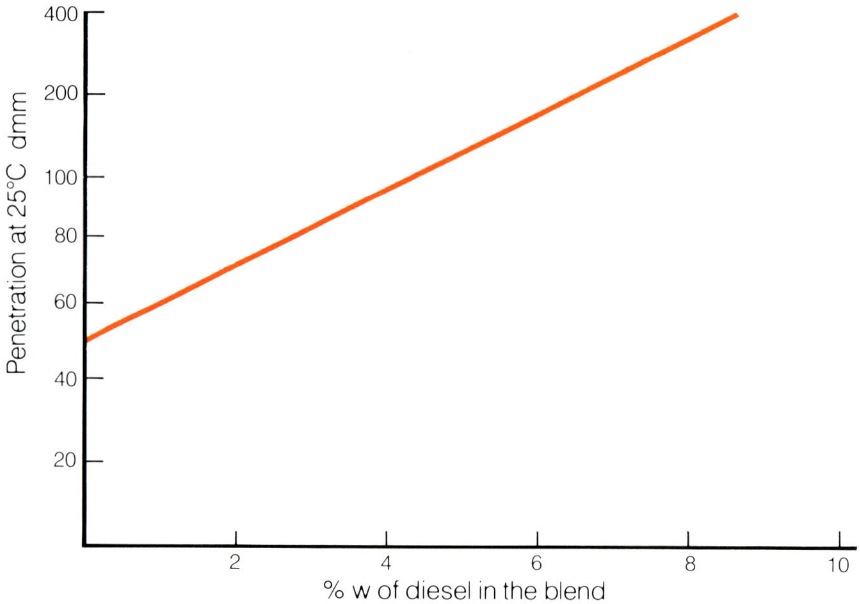

Figure 20.3 – Typical blending curve for 50 pen bitumen and diesel

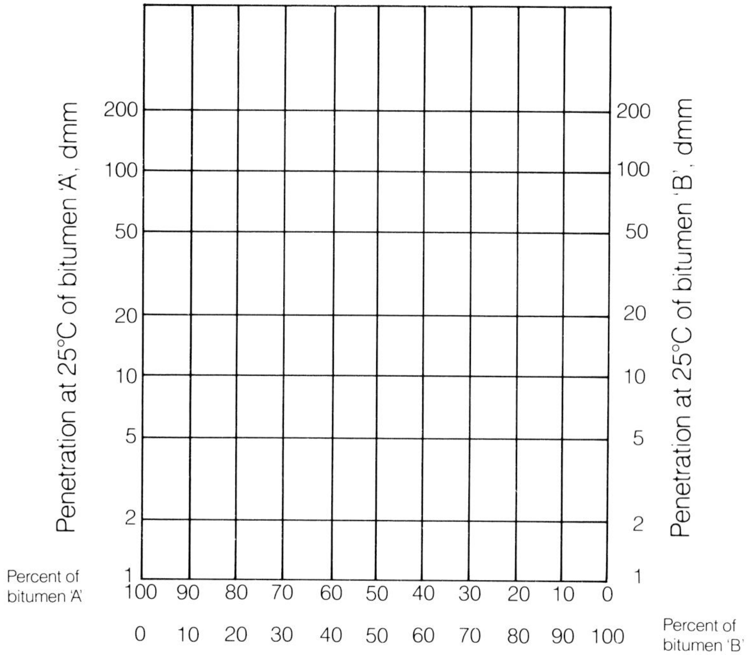

Figure 20.4 – Chart for estimating the penetration of a blend of two bitumens

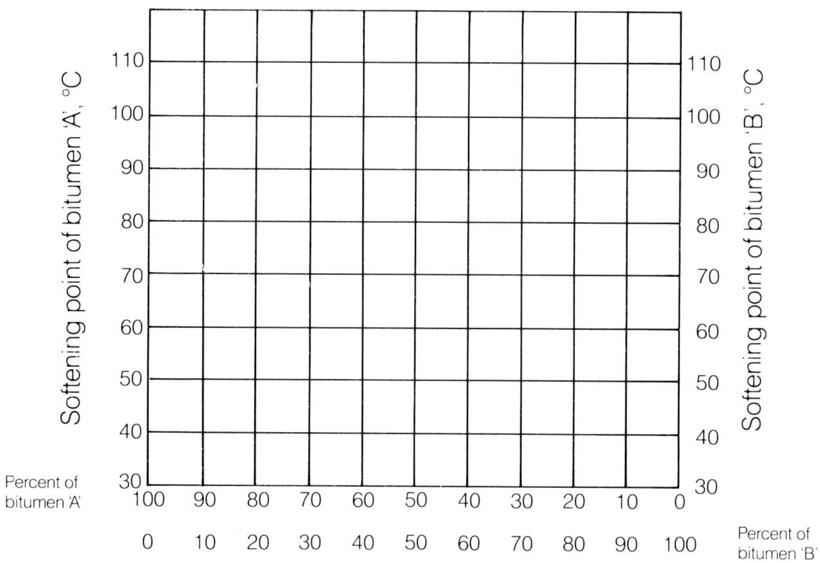

Figure 20.5 – Chart for estimating the softening point of a blend of two bitumens

20.4 Calculation of bitumen film thickness in a bituminous mix

It is possible to estimate the surface area of aggregate by coating it with an oil and measuring the quantity of oil required for complete coating. Alternatively, the surface area of an aggregate can be calculated by assuming a specific aggregate particle shape. Hveem calculated surface area factors, assuming a spherical particle shape and a specific gravity of 2.65. Typical surface area factors are as follows:

Sieve size, mm	*Surface area factor, m^2/kg*
0.075	0.3277
0.150	0.1229
0.300	0.0614
0.600	0.0287
1.18	0.0164
2.36	0.0082
> 4.75	0.0041

The surface area of the aggregate is calculated by multiplying the total percent passing each sieve size by the appropriate surface area factor, and adding the products together. All the factors must be used and if different sieves are used different factors are necessary. The theoretical bitumen film thickness is then calculated as follows:

$$T = \frac{b}{100\text{-}b} \times \frac{1}{SG_b} \times \frac{1}{SAF}$$

where T = bitumen film thickness, mm

SG_b = specific gravity of bitumen

SAF = surface area factor

b = bitumen content, %

20.5 The future

"To stand still is to retreat". This applies to the technology of bitumen as much as to the world of commerce and industry. Laboratories in many parts of the world, with Shell traditionally in the forefront, are working on new answers to old problems as this book is being written; and new ideas, new discoveries, new techniques will undoubtedly form the subject of new chapters in future editions of this book.

20.6 Reference

(1) Pfeiffer, J Ph: *The properties of asphaltic bitumen*. Elsevier Publishing Company, 1950.

Index

A

ALD, *see* Average Least Dimension
Adhesion
– failures 136
– in surface dressing 303, 304
– of bitumen 39, 135-148
– testing 143-145
Ageing of bitumen 132, 133
Analytical pavement design 274-277
Antiquity (earliest uses of bitumen) 13
Aromatics 90
Asphalt, hot rolled, *see* Hot rolled asphalt
Asphalt, mastic, *see* Mastic asphalt
Asphalt Institute mix design method 247-251
Asphalt kerbs 323
Asphaltenes 90
Asphaltic concrete 178
Average Least Dimension 287-290, 300
Axle loads 267, 268

B

BISAR and BISAR-PC 277
BISTRO 277
BS 594 : 1973 mortar design method 252
BS 598 : 1985 design method 252-254
Basecourse 266
Batch mixing plant
– directly heated 193, 194
– indirectly heated 189-193
Bitumen blending charts 328, 329
Bitumen film thickness in bituminous mix 332
Bitumen macadam, *see* Coated macadam
Bitumen properties
– influence during construction 100-104
– influence on mix performance 104-108
Bitumen Test Data Chart 68-74
Bituminous mixes
– general 169-187
– manufacture 189-196
– testing 196-199
Bituminous substitution 278
Blowing of bitumen 22-25

C

California Bearing Ratio (CBR) 269
CARIPHALTE DA and DM 160, 165, 167
Chemcrete 153-154
Chemical composition of bitumen 91
Chipping spreaders (for surface dressing) 306
Coated chippings on hot rolled asphalt 122, 173
Coated macadam 175-177
Coating tests 145
Cohesion 39
Coloured surfacings 320-322
Compaction
– of bituminous materials 104, 207-215
– specification 213-215
Constitution of bitumen 89, 90
Consumption of bitumen in UK 15
Continuous mixing plants 194-196
Cracking 109-113
Crude oil
– fractional distillation 21, 22
– types of 18, 19
Cutback bitumens
– general 36
– in surface dressing 293-296

D

DoE (Airfields Branch) mix design method 249, 250
Deformation, *see* Permanent deformation of bituminous materials
Design and specification of bituminous mixes 245-264
Design of bituminous pavements 265-280

Disbonding 140-142
Double surface dressing 308-311
Drum mixer 194
Ductility 39, 41
Durability of bitumen 40, 123-134
Dynamic stiffness of bituminous materials 235-237
Dynamic stiffness tests 226

E

Electrical properties of bitumen 327
Empirical and semi-empirical pavement design methods 269-273
Emulsifiers 45, 46
Emulsions
– adhesivity 50
– breaking rate 54
– general 45-59
– in road mixes 57
– in surface dressing 292, 293
– manufacture 46
– miscellaneous uses 57, 58
– particle size distribution 55, 56
– stability 47
– storage stability 54, 55
– viscosity 51-54
Engineering properties of bitumen 75-87
Environmental aspects, *see* Health, safety and environmental aspects
Epoxy-resin thermoset binders for surface dressing 299, 300
EROPHALT 300

F

Fatigue
– characteristics of bituminous mixes 241-243
– strength of bitumen 86, 87
– test 226-229
Fatting-up 311, 312
Fine-graded macadam (formerly fine cold asphalt) 178
Floating screed paver 205-207
Foundation 266
Fraass breaking point test 64
Fretting 112, 114-118
Fiction course, *see* Pervious macadam
Full-depth asphalt, *see* Bituminous substitution

G

Gussasphalt 171

H

HDM, *see* Heavy duty macadam
HRA, *see* Hot rolled asphalt
Hard bitumens 36
Hardening
– evaporative 40, 123
– exudative 42, 124
– general 123-131
– oxidative 40, 123
– physical 123
Health, safety and environmental aspects of bitumen 28-33, 315, 316
Heavy duty macadam 177, 277, 278
Hot rolled asphalt 171-174
Hydraulic applications of bituminous mixes 319

I

Immersion tests
– chemical 144
– dynamic 143
– mechanical 144
– static 143
– trafficking 144, 145

L

LR 1132 271-273
Lake asphalt 16
Laying bituminous materials 104, 205-207
Low-temperature performance of bituminous mixes 109-118

M
Macadam
– coated, *see* Coated macadam
– heavy duty, *see* Heavy duty macadam
– pervious, *see* Pervious macadam
Macrotexture 300-303
Manufacture of bitumen 21
Manufacture of bituminous mixes, *see* Bituminous mixes, manufacture
Marshall properties (alternative approach to specifying) 260-262
Marshall test 230-232
Mastic asphalt 171
Mechanical properties
– of bitumen 65-87
– of bituminous materials 235-244
Mechanical testing
– of bitumen 61-64
– of bituminous mixes 223-234
MEXPHALTE C 324
Microtexture 300-303
Mix design, *see* Design and specification of bituminous mixes
Mixing plants 189-196
Modified bitumens 149-168
Motor racing tracks 323

O
Organo-manganese compounds in bituminous mixes 152-154
Oxidised bitumens 35, 36

P
PSV 300, 303
Pad-coats 306, 307
Pavement design, *see* Design of bituminous pavements
Penetration grade bitumens 35
Penetration Index 74-77
Penetration test 60-62
Percentrage refusal density 213-215
Permanent deformation of bituminous materials 105, 108, 237-239
Permittivity 119, 327
Pervious macadam 178-184
Pigmentable binders 324
Polymer modified binders for surface dressing 297, 298
Polymers in bitumen
– thermoplastic rubbers 155-165
– thermosetting 165-167

Q
QUALAGON 38-40, 42
Quality 36-42

R
Railway applications of bituminous mixes 319, 320
Recipe specifications 246
Recreational areas 322
Recycling bituminous materials 201-203
Repeated load indirect tensile test 224
Repeated load triaxial test 223, 224
Resins 90
Rheology of bitumen 92, 93
Road Note 29 267, 269-271
Road Note 39 283-287
Roadbase 266
Rock asphalt 17
Rollers
– deadweight 209, 210
– for surface dressing 306
– pneumatic tyred 306, 307
– vibratory 210-212, 307
Rolling thin film oven test 40, 41, 133
Rubbers in bitumen 152

S
Safety, *see* Health, safety and environmental aspects
Sampling 31, 196
Sand patch test 301, 302
Saturates 90

SHELL EPOXY ASPHALT 166, 167
Shell Pavement Design Manual 274
SHELLGRIP 299, 300
SHELPHALT 36, 293
Skid resistance 300
Sliding plate rheometer 82
Sliding plate viscometer, *see* Viscometer, sliding plate
Slurry seal 313, 314
Softening point test 62, 63
Specific gravity of bitumen 325-327
Sprayers (for surface dressing) 304, 314, 315
SPRAYGRIP 299
Stiffness
– dynamic, *see* Dynamic stiffness of bitumen
– of bitumen 75-86
Storage and handling of bitumen 25-27
Structural layers in a bituminous pavement 265
Structure of bitumen 91-94
Sulphur in bitumen 151, 152
Surface dressing
– failures 311, 312
– general 281-318
– safety 314-316
Surface texture
– general 217-219
– in surface dressing 290-292
– influence of bitumen 118-119
Sweepers (for surface dressing) 306

T

Tar 17
Temperature susceptibility of bitumen 74
Tensile strength of bitumen 87
Texture depth
– in surface dressing 292, 301-303
– in wearing courses 217-219
Thermal properties of bitumen 74, 327
Thin film oven test 133
Transport of bituminous materials 205
Trinidad lake 15, 16
True boiling point gas liquid chromatography 40, 199, 200

U

Unconfined static uniaxial creep compression test 224

V

Vehicle testing circuits 323
Viscometer
– capillary 68
– cup 68
– kinematic 65
– sliding plate 67
– standard tar 66
– U-tube reverse flow 67
Viscosity
– conversion factors 327-328
– of bitumen 65-74

W

Wearing course 266
Wheel-tracking tests 229
Workability of bituminous materials 215-217